ACTIVITY NETWORKS

ACTIVITY NETWORKS:

Project Planning and Control by Network Models

SALAH E. ELMAGHRABY

North Carolina State University
Raleigh, North Carolina

A Wiley-Interscience Publication
JOHN WILEY & SONS
New York • London • Sydney • Toronto

Library of Congress Cataloging in Publication Data

Elmaghraby, Salah Eldin, 1927–
Activity networks.

"A Wiley-Interscience publication."
Bibliography: p.
Includes index.
1. Network analysis (Planning) I. Title.

T57.85.E42 658.4'032 77-9501
ISBN 0-471-23861-9

Printed in the United States of America

10 9 8 7 6 5 4 3 2 1

To Dean Andrew Schultz, Jr.
of Cornell University

PREFACE

The theory of project planning and control has undergone serious scrutiny and considerable expansion since the early models of CPM (critical path method) and PERT (program evaluation and review technique), which were advanced in 1959 (see Kelley and Walker (1959) and Malcolm et al (1959). Furthermore, the scope of interest and the power of analysis of such theory have expanded well beyond the originally envisaged temporal considerations to include many more topics, such as the study of optimal project "compression," the optimal planning and acquisition of resources, and optimal scheduling under constrained availability of scarce resources, and the analysis and synthesis of projects under conditions of uncertainty in their structure. More significantly, the conceptual problems posed and the approaches devised for their resolution transcend the area of project planning and control, even though they had their genesis in, and were motivated by, problems originating in it. Throughout this book the reader will discover ideas that are equally pertinent to other fields of study, ranging from transportation to communication and from population studies to waiting-line analysis, in particular, problems that are meaningfully modeled as directed networks of the genre treated here.

An author who undertakes the writing of a book usually has several objectives in mind. Mine are three: First, to provide a unified and up-to-date treatment of the relevant "technology" in the field of project planning and control through network models (which I have termed "activity networks," ANs); second, to discuss in more depth than is usually available in books on the subject the theoretical foundations of the OR models offered, their underlying assumptions as well as their shortcomings; and third, to pose some of the more pressing problems that are still outstanding and are in need of considerable research effort.

The first objective is addressed to the practitioner, be he a manager, a consultant, or an engineer. The second objective is relative to the pedagogy of the subject that takes place in the classroom. The third objective is addressed to the researcher.

The majority of past books written in this area seem to have addressed themselves almost exclusively to the practitioner, ignoring or almost ignoring the last two categories of readers. This posture implicitly assumes that the theory is well established so that all that remains to be done is to "bring it down to the level of elementary exposition" for the layman to understand, very much like, say, semiconductor theory or laser technology. These assumptions take a great deal for granted. In reality, the recent developments in the learned journals and research monographs are not widely known. The state of the art is such that what is needed is not merely to explain to the managers in the most elementary terms what is already established, but rather to alert the managers to the hidden imperfections of the tools they are handed so that they can use them with discretion and imagination, at the same time as the theory is being rigorously established!

Moreover, the expansion of the horizons of modern ANs leaves expositions that are content with a recitation of the (now "classical") CPM–PERT calculations uninterestingly sterile. It reminds one of George Bernard Shaw's remark, when advised of a novel which ended with the hero and heroine getting married, that "but it is just then when the real story begins." The "real story" of ANs actually starts after the few temporal considerations of CPM–PERT have been answered.

At this juncture it is perhaps appropriate to insert a brief historical note on the development of the area to its present state.

The study of interacting activities for the purpose of effecting managerial controls is not of recent vintage. Undoubtedly, since time immemorial the problem of control of large projects must have faced every project leader who devoted his talents and efforts to the accomplishment of what we call a "project" in the modern parlance. But the formalization of this practice into the structure of network models had to wait until the midfifties of this twentieth century and the work of Kelley on CPM and Malcolm, Roseboom, Clark, and Fazar on PERT. Since the appearance of these two pioneering reports and the glamorous successes claimed for their early applications, the important role played by projects[†] in the national economy seems to have been discovered. Consultants sprang up everywhere to advise management on the new "technology"; the United States government (particularly, the Department of Defense) adopted the models as a standard requirement in tendering bids on government contracts; considerable interest was exhibited by the operations research community

[†]The word "project" connotes different things to different people. It is used here in the sense of medium- or large-scale undertakings that are one of a kind in nature (more or less), requiring special talents and consuming resources over their durations. If the identity of an object is better understood by the mention of its antithesis, then I would contrast the concept of project with assembly-line production in which the same product is repeated for a long period of time.

Webster's Third New International Dictionary defines the word "project" as: (1) a specific plan or design; (2) a planned undertaking, and (3) a vast enterprise, usually sponsored and financed by the government. The dictionary also mentions that, in dialect, the word "project" is used to connote: to go about idly with no particular purpose; to fool around!!

in the proposed models; and literally hundreds of reports, books, pamphlets, and other publications appeared. The outcome of all this activity, which spanned some 16 years, is the emergence of a body of knowledge that refines the original concepts and expands them to a considerable degree.

This body of knowledge forms the subject matter of this book.

It seems redundant to state that a great deal of the "planning and control" of projects takes place long *before* the project comes into being and the formalism treated in this book is applied. Actually, we seem to need constant reminding of this fact. Indeed, the future development of a project may have already been preordained by the prior activities of contract bidding; financial commitments; subcontract negotiations; labor, engineering, and other contractual agreements; the structure of the organization; and many other conditions that have been agreed on before, or that are in effect at the time of "existence" of the project as such.

Regrettably, this manuscript had to be content with a discussion of a segment, and admittedly a small segment, of the total picture. The reason is not my lack of awareness of the true nature of the planning and control function, but rather that the enormity of the subject is well beyond one treatise.

The audience to which this book is addressed includes those who are professionally concerned with a fundamental understanding of ANs. They include students at the advanced undergraduate and graduate levels in the curricula of operations research, management science, quantitative methods in business administration, and industrial engineering; operations research personnel in government and industry; management consultants; and managers of technical projects and/or technical establishments who are themselves technically trained.

As to be expected in a treatise of this nature, the level of presentation is necessarily uneven: the treatment of a problem whose tools of analysis have been well developed (e.g., linear programming models) is very different from the treatment of a problem area that is still in need of research. In the former case I am interested in familiarizing the reader with the most sophisticated approaches, and I may call on some advanced concepts. By contrast, in the latter case there is no established theory; therefore, I am interested in bringing the problem to the attention of researchers. This can be accomplished only through a presentation of the "scenario" of the situation, which is usually mundane and written in simple prose.

Since there seems to be no escape from this diversity in the level of presentation, it appears that the ideal course of action is to have the majority of the book available to the reader with a minimal background, with a small portion specialized and addressed to a select few. These specialized sections are identified with an asterisk at their beginning. Thus a reader with one introductory course in OR, one course in linear programming, and the usual college undergraduate background in probability, statistics, and calculus should have access to 85% of the book. The remaining 15% may require more advanced knowledge in methods of OR (in particular, nonlinear programming and stochastic processes).

For the serious reader the problems at the end of each chapter should be considered as an integral part of the presentation. Frequently, a concept that was not properly developed in the text because of the constraints on the size of the manuscript is relegated to a problem. In some instances a problem may summarize a very recent research which, because of the time lag inherent in preparing a manuscript such as this, could not be incorporated in the text.

This book is an outgrowth of my lecture notes on the subject over the past 8 years. The material is approximately 20% in excess of what may be comfortably covered in one semester of 45 contact hours. This leaves the instructor the freedom to highlight some topics at the expense of others and still have enough material to cover in a semester. I have done so myself over the years in which the manuscript has taken shape. I find this strategy also helpful in keeping the instructor's interest alive in the subject over a number of years.

The subject matter of this book may be divided into two broad categories: the deterministic case (giving rise to deterministic activity networks —DANs) and the probabilistic case (giving rise to probabilistic activity networks—PANs). The former have their genesis in the CPM, the latter in the PERT Model.

Chapter 1 is devoted to the introduction of notation and terminology and to an exposition of the two "elementary" models of CPM and PERT. The chapter terminates with a discussion of the problems of managerial cost accounting in project-type activities—a subject that is of central and vital importance to managers but is rarely mentioned in books on the subject (except for books that are explicitly devoted to cost considerations, such as Novick [1965]).

DANs are more intensively studied in Chapters 2 and 3. In Chapter 2 the problem of the optimal reduction in project duration (the so-called "project compression") is considered. This is a problem of *synthesis* (as opposed to the problems of *analysis* treated in Chapter 1), in which the following question is asked: if it is desired to reduce the total duration of the project from its "normal" duration to a "specified" duration, what is the most economical reduction in individual activity durations to achieve the desired objective? The various assumptions on the cost-duration function of individual activities (such as linearity, convexity, and concavity) give rise to the different specialized approaches described in the chapter.

Chapter 3 carries the discussion further into the important topics of planning under scarce resources. This is the fundamental managerial problem relative to the resources' acquisition and utilization over the life of the project. Again, a variety of approaches are proposed under different sets of conditions.

The thread of discussion of PANs, which are introduced in Section 3 of Chapter 1, is picked up again in Chapter 4, which presents an in-depth analysis of the assumptions and derivations of the PERT model. Here the reader will encounter the startling facts about the implications of these assumptions and simplifications on the results obtained, such as the probability of completing a set of tasks by a specified time—which is one of the early, and most important, claims of PERT.

Chapter 5 extends probabilistic models beyond the "classical" PERT model into GAN (generalized activity networks), GERT (graphical evaluation and review techniques), and its various simulation languages, generically denoted by GERTS. These recent developments are capable of modeling decision situations that involve nondeterministic branching—a basic assumption in all the models treated thus far. The utility of these models transcends the field of ANs into the general domain of stochastic systems, wherever their origin may be.

Viewed in its totality, the discussion of each topic proceeds from the simple to the complex—from the deterministic to the probabilistic, from analysis to synthesis, and from CPM to PERT to GANs. This mode of presentation has the appeal of following the historical development of the subject. Also, it has the appeal of stimulating the reader to ask: what would be the logical step, or steps, to follow?

To assist the reader in his literature search, the references are organized in three ways:

(i) At the end of each chapter is a list of the references that are most intimately related to the subject matter of the chapter.

(ii) Segregated at the end of the book are two categories of references that do not fit neatly in any chapter. These are (a) books written on the subject of ANs, and (b) general treatises on the underlying methodologies (such as probability theory, statistical inference, and network flow theory).

(iii) Finally, at the end of the book there is an alphabetical listing of all the references.

For the sake of economy, the complete reference is given only once in its alphabetical listing. Other listings, either at the end of each chapter or in category (ii) mentioned above, give only the author(s) and the year of the reference.

It is my pleasure to acknowledge the help I have received from friends and institutions over the years during which this manuscript took shape. I am indebted to Professors Willy Herroelen of the Katholieke Universiteit,

Leuven, Belgium, and Joseph J. Moder of the University of Miami, Florida, for having read the complete manuscript and given me their valuable criticisms. I am also indebted to the two reviewers (whose identity is unknown to me) who evaluated the original version of the manuscript. Their comments were instrumental in prompting me to effect the changes that were made in subsequent versions. To my fellow professor Henry L. W. Nuttle and my students A. A. Elimam and Subhash Sarin go my thanks for many helpful comments and suggestions. I am also grateful to Ms. Lilian Hamilton for her patience and care in typing the rather awesome-looking expressions strewn throughout this manuscript. Finally, I wish to acknowledge the support I received over the years in research related to the subject matter of this book from the Army Research Office at Research Triangle Park, N.C., the Office of Naval Research in Arlington, Va., and the National Science Foundation in Washington, D.C.

SALAH E. ELMAGHRABY

Raleigh, North Carolina
March 1977

CONTENTS

ABBREVIATIONS

A-on-A	Activity-on-arc mode
A-on-N	Activity-on-node mode
$\mathcal{A}(i)$	Set of nodes connecting *from* i (i.e., occur *a*fter i)
ALB	Assembly-line balancing
AN	Activity network
BB	Branch and bound
$\mathcal{B}(i)$	Set of nodes connecting *to* i (i.e., occur immediately *b*efore i)
CA	Critical activity
CDF	Cumulative distribution function
CP	Critical path
CPM	Critical path method
CS	Setup cost
CSN	Currently scanned node
CV	Variable cost
CVT	Control variate
DAN	Deterministic activity network
DB	Decision box
DF	Distribution function or Density function
DP	Dynamic program or dynamic programming
Eq.(.)	Equation number (.)
ESS	Earliest start schedule
EST	Earliest start time
Excl.-or	Exclusive-or
FT	Function type
G/A	General and administrative (costs)
GAN	Generalized activity network
GERT	Graphical evaluation and review technique
GERTS	GERT-simulation language
IBM	International Business Machine Co.
iff	If and only if
ILP	Integer linear program or integer linear programming

Incl.-or	Inclusive-or
ITP	Interval transition probability
$\mathcal{L}$.b.	Lower bound
LOB	Line-of-balance
LP	Linear program or linear programming
$\mathcal{L}$.s.	Left side
LSS	Latest start schedule
$\mathcal{L}_z$	The z-transform
PAN	Probabilistic activity network
PDF	Probability distribution function
PERT	Program evaluation and review technique
$\mathcal{P}_j$	Set of activities preceding node j (in A-on-A mode)
PPBS	Planning, programming, and budgeting system
Pr(.)	Probability of event (.)
PS	Parameter set
RV	Random variable
r.s.	Right side
SFG	Signal flow graph
SMP	Semi-Markov process
s.t.	So that or such that
u.b.	Upper bound
UIF(.)	Undeleted immediate followers of (.)
Var(.)	The variance of (.)

SYMBOLS

$\triangleq$	Defined as
$\forall$	For all
$\{ \}$	A set contained within the braces
$\prec$	Precede
$\equiv$	Identity
$\bigtimes$	Cartesian product
$\times$	Ordinary multiplication
$\Rightarrow$	Implies or yields
$\leftrightarrow$	Corresponds to
$\mathcal{E}$	Mathematical expectation

ACTIVITY NETWORKS

1

THE BASIC MODELS: STRUCTURE AND TERMINOLOGY

In this treatise we view a project as a collection of *activities* and *events*. An *activity* is any undertaking that consumes time and resources (although we sometimes find it expedient to speak of *dummy* or *pseudo activities*, which consume neither). Examples of activities are transport of an item from one location to another, design of a stabilization system, construction of the foundation of a building, and so on. An *event* is a well-defined occurrence in time; examples are shipment received, design completed, trial run successful, and so on.

Perhaps of paramount importance in the models of concern to us here is the concept of *precedence* that occurs naturally because of technological and other considerations. It is the representation of this aspect through the use of network models that distinguishes *activity networks* (ANs) from previously established models for the scheduling of activities of the Gantt chart and other varieties.[†]

A more precise definition of ANs is as follows. Let a denote an activity. Since a occurs over time it has a duration, denoted by $y_a \geqslant 0$ (with $y_a = 0$ only for dummy activities). Historically speaking, the model for the critical path method (CPM) assumes that the durations $\{y_a\}$ are deterministically known, while the model for program evaluation and review technique (PERT) assumes that some of, or all, the durations are only known in a probabilistic sense. We see later in the present work that this characterization has been expanded into several directions and in fundamentally novel approaches, to the point where we find it more convenient and precise to abandon the CPM–PERT categorization, while retaining the dichotomy between deterministic and probablistic models, which we refer to as *deterministic activity networks* (DANs) and *probablistic activity networks* (PANs).

Furthermore, each activity consumes certain *resources*: (a) financial resources, (b) labor, engineering, and managerial skills, (c) machinery and equipment, and (d) natural resources (energy, land, water, materials, etc.). Given such entities as activities, events, durations, and precedence relations, (we relegate the treatment of resources to Chapter 3), our discussion in this chapter proceeds in the following fashion. In the present section we describe the manner in which an AN model is constructed and in Section 2, the temporal considerations relevant to the scheduling of the activities. Here, in our discussion of PANs, the treatment is limited to the approach originally suggested by the originators of the PERT model. This view is modified in Chapters 4 and 5, but for the moment it serves two important functions. First, it acquaints the reader with the "classical" model, which is the one usually treated in other texts on the subject. Second, it serves as the basis for our critical analysis of the PERT model (in Chapter 4), and for launching the more elaborate models of PANs (e.g., GAN and GERT, presented in Chapter 5). Finally, because of its central importance, we present in Section 3 a discussion of the problem of financial and accounting controls in project management. The present chapter concludes with an appendix giving the theoretical background of the network model adopted in ANs.

[†]Prior to the advent of network diagraming the most commonly used visual aid to the planning and scheduling of activities was the Gantt chart and its varieties, see the work by Clark (1954).

§ 1. REPRESENTATION

We have stated that the modeling of *precedence* is perhaps the most distinguishing aspect of the modeling of projects via ANs. We preface this chapter with two remarks on the concept of precedence†:

1. Precedence is basically a binary relationship: an activity a precedes another activity b, which we write as $a \prec b$, meaning that *a must be completed before b is started*; or a set of activities A precedes another set B, also written $A \prec B$, in which case the internal order of activities in either set is immaterial, but *no activity in B may be started before all activities in A have been completed.* Of course, we may transform this binary relation between sets into a binary relation between elements of the sets in the following manner. Let l_A denote the activity accomplished *last* in the set A, and let f_B denote the activity ranked *first* among the activities in the set B. Then the condition $A \prec B$ is equivalent to $l_A \prec f_B$.
2. Precedence is a transitive relation; in other words, if $a \prec b$ and $b \prec c$, then $a \prec c$. This property plays an important role in the structure of the network.

We proceed next to elaborate on such structure.

There are two possible modes of representation of projects with the elements mentioned above: (a) the *activity-on-arc* representation and (b) the *activity-on-node* representation.

§ 1.1 Mode 1: The Activity-on-arc Representation (A-on-A)

An activity is represented by an arc, and an event by a node. Each branch‡ bears an *arrow* leading from one end node of the branch to the other. Thus, the generic element of representation is a *directed* branch as shown in Fig. 1.1. The event (node) at the tail of the arrow represents the start of the activity, and the event (node) at the head of the arrow represents the termination of the activity. The fact that activity a precedes some other activity b (or b succeeds a) is represented by having the terminal node of the branch representing a coincide with the start node of the branch

†The concept of precedence, so dominant in ANs, has been adapted for the management of production. The most outstanding example is the approach called *line of balance* (LOB); see Richardson, Butler, and Coup (1972), Turban (1968), and Wilson (1964), and Problem 27 at the end of the present chapter.

‡We reserve the term "branch" to designate the arc *with* its two end nodes. The "arc" refers to the branch *without* its end nodes.

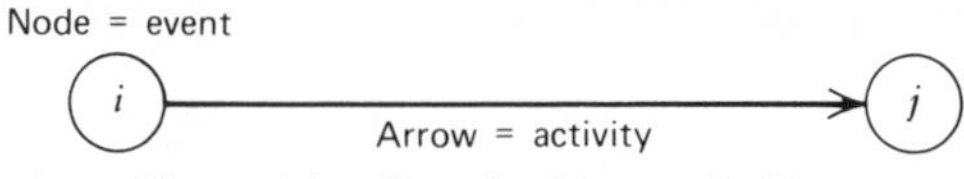

Figure 1.1. Generic element of ANs.

representing b (or of another branch representing an activity that succeeds a but precedes b).

The result of such representation is a directed graph (or *digraph*), the so-called activity network (AN) (see Fig. 1.2) in which, for example, $a \prec \{c,d\}$; $\{d,i\} \prec l$; $\{f,g\} \prec \{i,j,k\}$, and so on. Such AN has the following characteristics:

1. The nodes are numbered such that an arrow always leads from a small number to a larger one. This is always possible in projects which contain no feedback activities (i.e., activities that return the status of the project, or portions thereof, to a preceding event). We assume this to be the case and remove such assumption only in Chapter 5.

   In fact, throughout Chapters 1–4 we insist on such a numbering scheme and forfeit it only when it is practically inconvenient, such as in the case when new activities are continuously introduced into the project and it is not feasible to renumber the old activities. In such a case, special precautions have to be taken to avoid circularity (i.e., the presence of cycles)—see the discussion on page 15.

   An immediate consequence of such a numbering scheme is that the *adjacency matrix* representation of the network is always upper triangular with zero diagonal. The adjacency matrix is an $n \times n$ matrix, where n is the number of nodes in the network and has an entry $+1$ for element (ij) if an arrow leads *from* node i *to* node j, and 0 otherwise. (The adjacency matrix is thus a *Boolean* matrix.) Such matrix representation is a convenient and concise representation of the structure of the network, especially for machine computation. The adjacency matrix of the network of Fig. 1.2 is shown in Fig. 1.3.

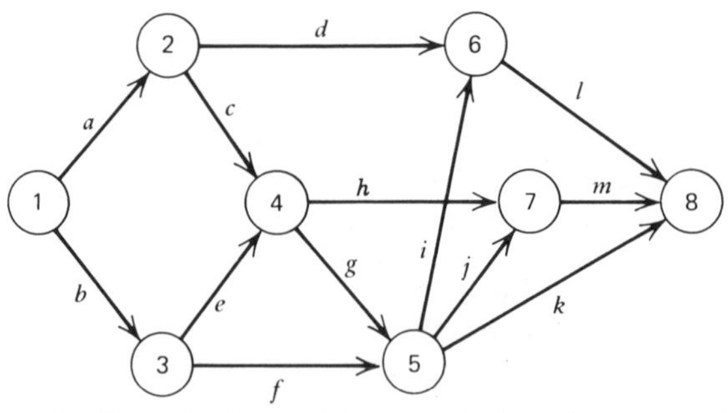

Figure 1.2. Example of an activity network (A-on-A representation).

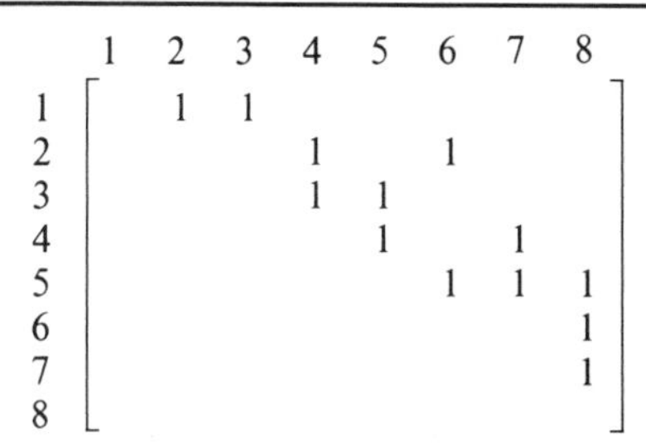

	1	2	3	4	5	6	7	8
1		1	1					
2				1		1		
3				1	1			
4					1		1	
5						1	1	1
6								1
7								1
8								

Figure 1.3. Adjacency matrix of network of Fig. 1.2.

2. Each node (event) must have at least one arc leading into it and one arc going out of it, with the exception of the origin (node *1*) and the terminal (node n). The former has only arcs leading out of it and the latter arcs leading into it.

   For machine computation an activity is designated by its end nodes. Consequently, any two nodes may be connected by at most one arc. This constraint can be satisfied in one of two ways: (a) if several activities must take place simultaneously (i.e., in parallel) between two nodes, perhaps they can be combined in one global activity or (b) use *dummy activities* (of zero duration and zero consumption of resources) and *dummy events* as shown in Fig. 1.4. Incidentally, this is the first (of three) instances in which resort to dummy activities and events is necessitated in this mode of representation. Note that the adjacency matrix of every digraph is a square upper triangular Boolean matrix (entries 0 and 1) with zero diagonal and, conversely, that any such

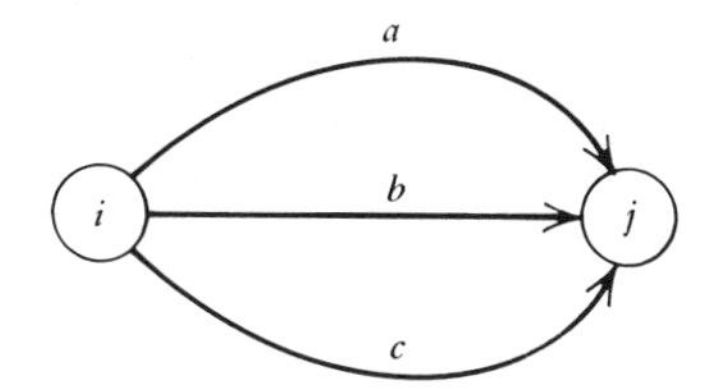

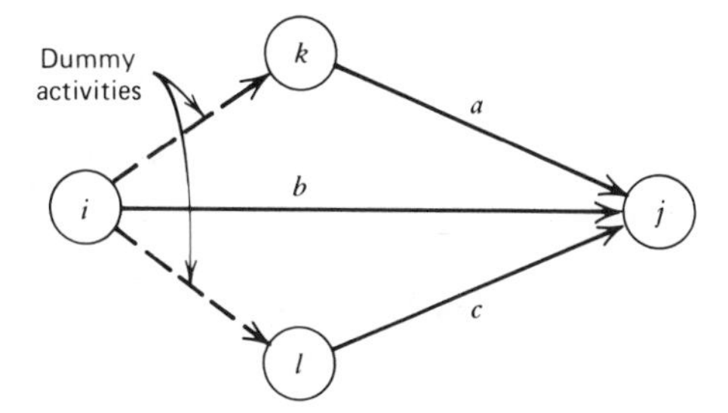

Figure 1.4. Representation to preserve uniqueness of activity designation.

matrix satisfying the condition that any two columns (or any two rows) be either identical or orthogonal, is the adjacency matrix of some digraph in the A-on-N representation described below. The justification for this condition is given in Appendix A.

3. By construction, the network contains no cycles, since otherwise by tracing around the cycle we would conclude, by the transitivity property of precedence, that a task must precede itself, which is an impossibility. In other words, in any forward path (i.e., in the direction of the arrows) from origin to terminal each node on the path appears once and only once.

   We now impose the constraint that each network must possess one and only one origin and one and only one terminal. This condition can be always satisfied by adding one dummy event, node *1*, which connects by dummy activities to all nodes of the network with no predecessors, and a dummy event, node n, which connects by dummy activities from all nodes of the network with no successors. (This constitutes the second use of dummy activities and dummy events.)

   As a consequence of such construction we can state that each node (event) of the network is connected from the origin by at least one path and is connected to the terminal by at least one path.

4. Since the precedence relationship is transitive, each node (event) along a path must have all preceding activities completed (and all preceding

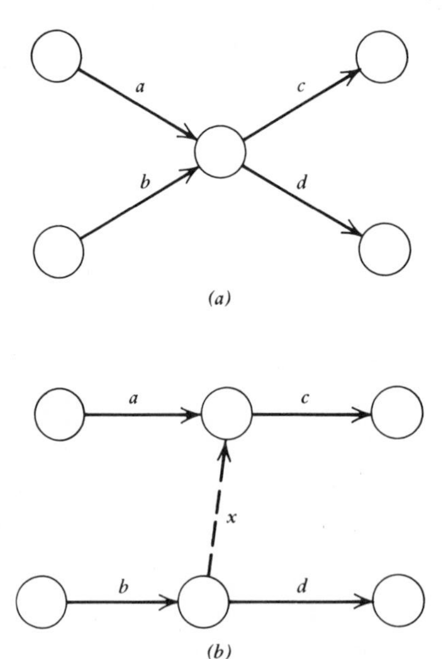

Figure 1.5. Representation to preserve the logical relations: $a, b \prec c; b \prec d$.

events realized) before it can be realized. In fact, we take such a condition as the definition of "*the realization of a node.*"

Because of this strict logical interpretation of the arrows, great care must be taken in drawing the network lest a precedence relationship is either added to, or omitted from, the original specifications of the project. For instance, consider four activities a, b, c, and d related by precedence as

$$a \prec c, b \prec c, b \prec d$$

On first blush one might construct the network of Fig. 1.5a. However, close scrutiny reveals that such a representation *also implies that* $a \prec d$, which was not specified. The correct representation is shown in Fig. 1.5b, in which x is a dummy activity of zero duration. (This exemplifies the third use of dummy activities.)

§ 1.2 Mode 2: The Activity-on-node Representation (A-on-N)

An activity is represented by a node and the precedence relationship between two activities is represented by a directed arc in which the direction of the arrow specifies the precedence. In this mode of representation one may consider an "event" to be represented by the arc since, in some sense, "movement" from one node to another can take place only after the completion of the former task. Such state of completion defines an event.

The (A-on-N) representation of the (miniscule) project of Fig. 1.2 is shown in Fig. 1.6, and the corresponding adjacency matrix is shown in Fig.

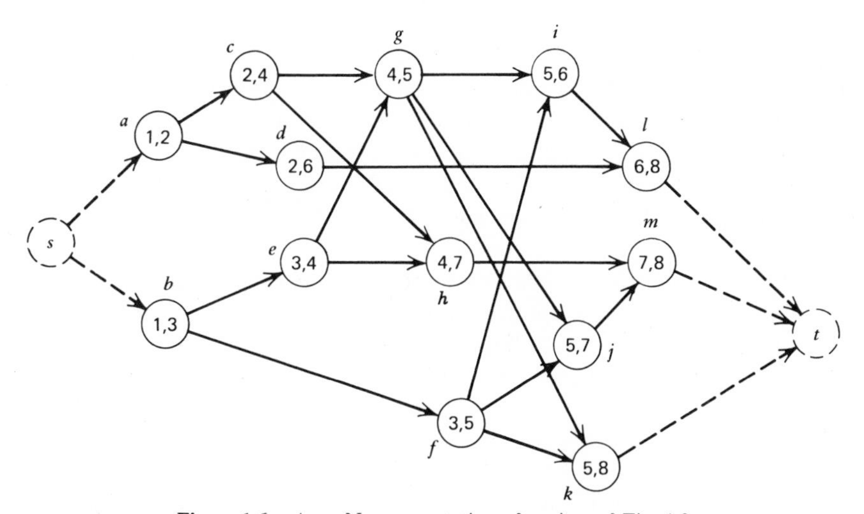

Figure 1.6. A-on-N representation of project of Fig. 1.2.

1.7. Notice that if the project has A activities, the adjacency matrix in this mode of representation is an $A \times A$ matrix with entry $+1$ if activity u precedes activity v, and 0 otherwise (hence, with the proper numbering of the activities, the matrix is upper triangular).

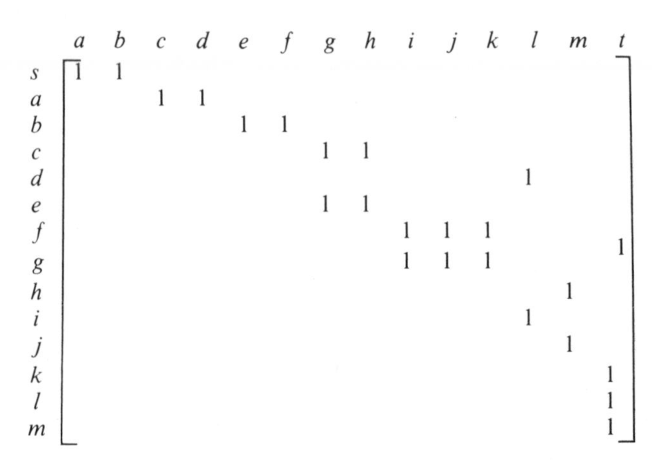

	a	b	c	d	e	f	g	h	i	j	k	l	m	t
s	1	1												
a			1	1										
b					1	1								
c							1	1						
d												1		
e							1	1						
f									1	1	1			1
g									1	1	1			
h													1	
i												1		
j													1	
k														1
l														1
m														1

Figure 1.7. Adjacency Matrix of Network of Fig. 1.6

A remarkable feature of this mode of representation is that dummy activities are not needed except in one instance only: to comply with the requirement that the network possess only one start node and one terminal node. For instance, in Fig. 1.6 we added the nodes s and t and the dummy events (s,a), (s,b), (l,t), (m,t) and (k,t) to achieve this desired objective.

It could not have escaped the reader that the two modes of representation discussed above are, in some sense, "duals" of each other. In fact, there is a strong relationship between the two representations, which warrants a closer look at the two modes from an abstract point of view as two arrow diagrams. This is accomplished in Appendix A.

§ 1.3 Some Remarks on Representation

Thus far we have discussed two modes of graphical representation of ANs, namely, the A-on-A mode and the A-on-N mode. Curiously enough, these two modes are referred to in the construction trade as *arrow diagrams* and *precedence diagrams*, respectively; two terms that are confusing as well as nondescriptive, since *both* diagrams contain arrows and *both* represent precedence! But such are the wiles and pitfalls of "trade" languages.

Apart from such confusing vernacular, it is important to note that adaptations of the two original network diagraming modes proliferate in industry. These variations are mostly multicolored *bar charts*, or *node-and-arrow* diagrams, with elaborate visual aids designed to assist the management in "seeing the whole picture" in as short time as possible. Several consulting firms offer specialized (and proprietory) computer codes that generate reports and diagrams that are tailor-made to a particular audience and to specific applications. At the time of this writing it seems that a good segment of the management consulting industry is devoted to the refinement of the visual representation of a project and the reporting of its progress in comprehensible fashion. These efforts are, indeed, welcome since they facilitate the acceptance of the theory of ANs by the practitioners—a most essential step if that theory is not to remain sterile. We cannot emphasize enough this important task, and we urge the reader to devote a great deal of attention to the presentation of results of his analysis in a manner most appropriate to his audience, since it may spell the difference between success and failure in the practical implementation of his expert knowledge.

To illustrate this point and perhaps help drive it home, consider the following situation, which emanated in a practical environment in the construction industry [Ponce-Campos and Kedia (1976)].

It has always been recognized in the analysis of ANs that two or more activities may "overlap" in time, in the sense that they are performed concurrently. Now suppose that a new element is introduced into the picture as follows: we wish to restrict the domain of overlapping between two activities u and v specifying one or more of the following "lags": (a) start-to-start lag (SS lag), (b) finish-to-finish lag (FF lag), (c) start-to-finish lag (SF lag), or finish-to-start lag (FS lag), and (d) start-to-start and finish-to-finish lags (SS/FF lags). The first restriction, the SS lag, specifies the shortest time that should elapse between the start of u and the start of v; the two activities may overlap any time thereafter. This situation is diagramed in the "bar chart" of Fig. 1.8*a*.† It may arise naturally in a construction project in which activity u represents, say, the pouring of concrete for a foundation, activity v represents the erection of various structures, and the SS lag represents the minimal time that must elapse for the first segment of the foundation to dry. The second restriction, the FF lag, specifies the shortest time that should elapse between the finish of u and the finish of v; the activities may overlap sometime therebefore. This situation is diagramed in the "bar chart" of Fig. 1.8*b*. The last three

†In Fig. 1.8*a* to *e* we indicate on each diagram the *specified* lag, such as the SS lag in *a* and the FF lag in *b*, and so on, as well as the *actual* lag, L, such as L_{ss} for the SS lag constraint and L_{FF} for the FF lag constraint. Some additional information is also exhibited on these figures, such as the time of overlap, the start and/or the finish time of activities u and v, and the reflection of the lag constraint on the earliest start and/or finish time of activity v.

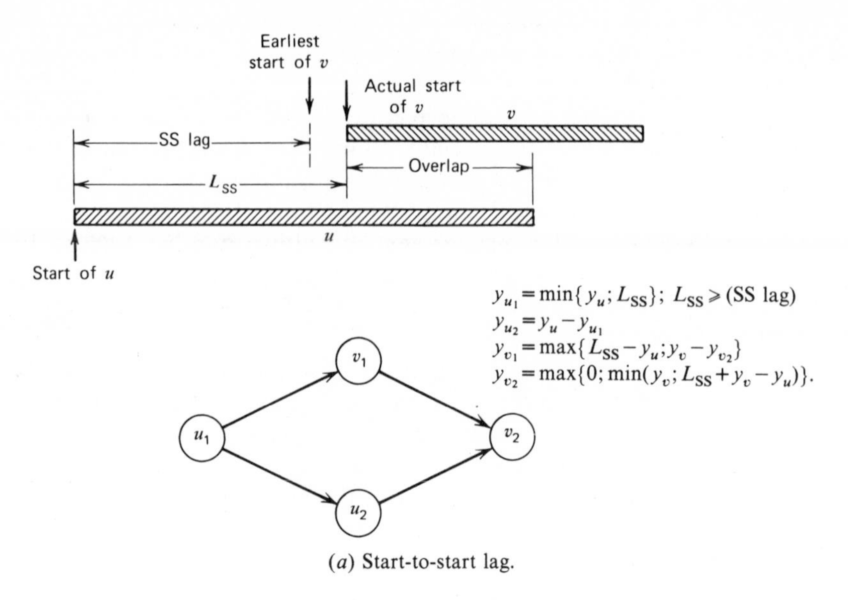

(*a*) Start-to-start lag.

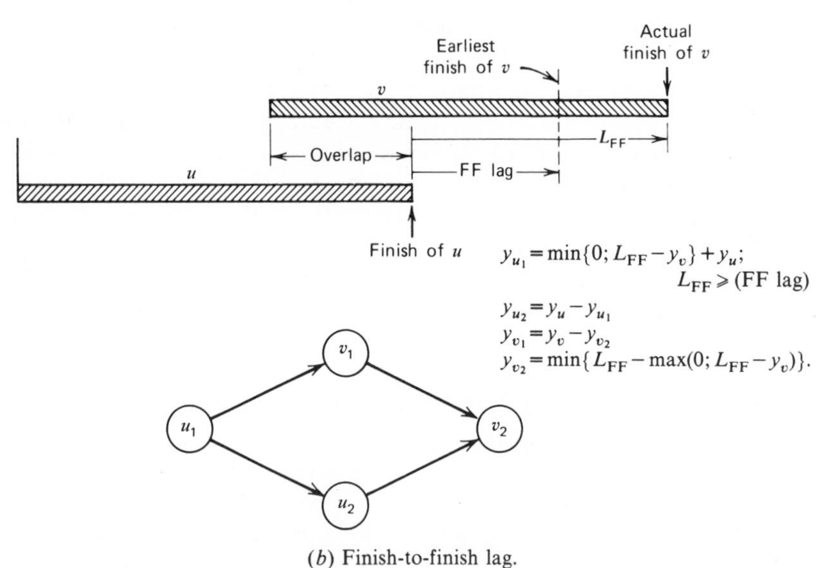

(*b*) Finish-to-finish lag.

Figure 1.8. Lag restrictions.

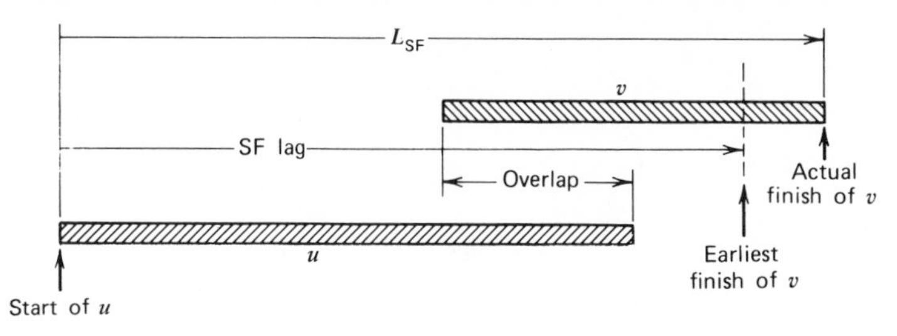

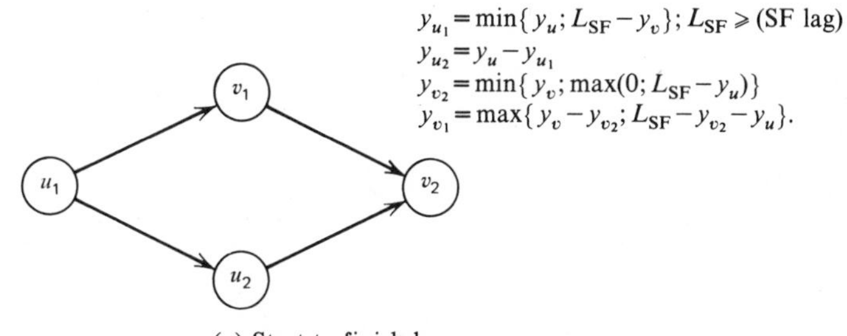

(c) Start-to-finish lag.

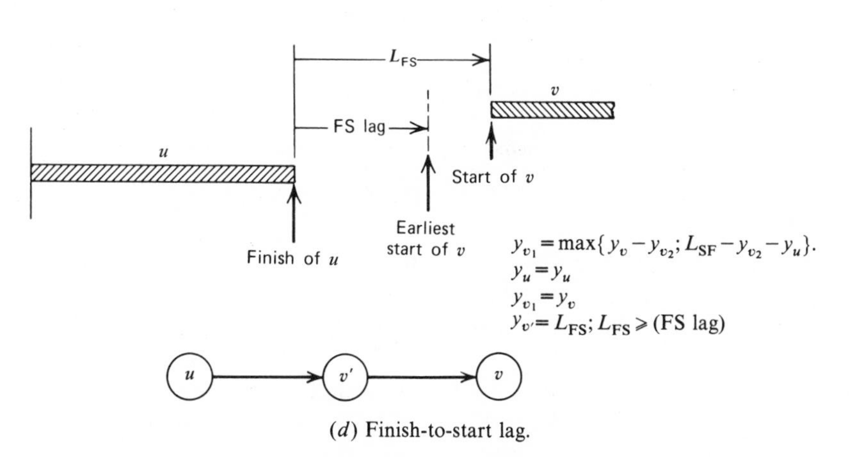

(d) Finish-to-start lag.

Figure 1.8. (Continued)

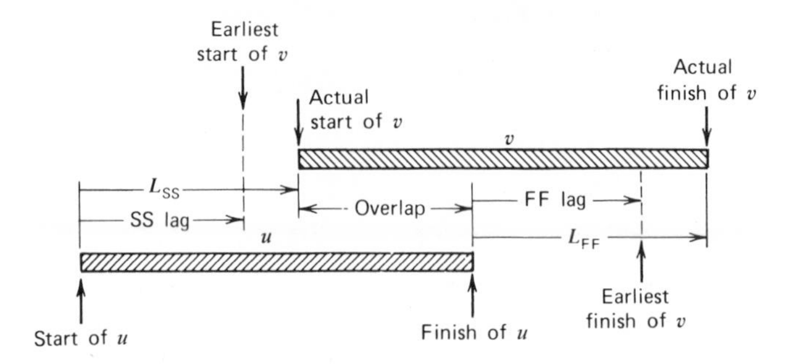

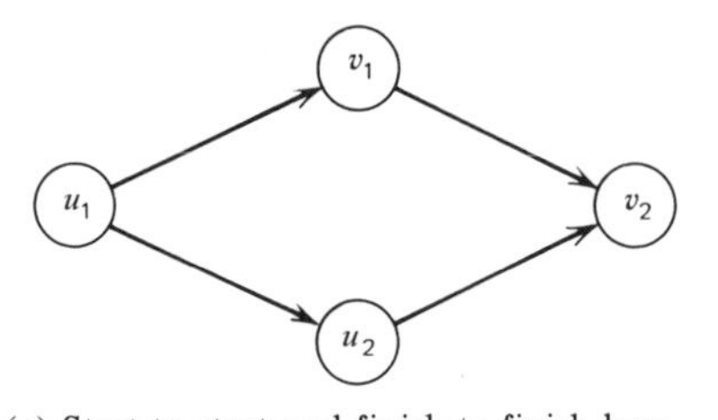

(*e*) Start-to-start and finish-to-finish lags.

$y_{u_1} = \max\{\min(L_{SS}, y_u); \min(y_u + L_{FF} - y_v, y_u)\}$

$L_{SS} \geqslant (\text{SS lag}); L_{FF} \geqslant (\text{FF lag})$

$y_{u_2} = y_u - y_{u_1}$

$y_{v_1} = \max\{L_{SS} + y_v; L_{FF} + y_u\} - y_{v_2}$

$y_{v_2} = \max\{\min(y_v, L_{FF}); \min(y_v, L_{SS} + y_v - y_u)\}.$

Figure 1.8. (Continued)

restrictions, namely, the SF, FS, and the SS/FF lags, are also diagramed in Figs. 1.8*c–e*, respectively; their interpretations and the description of circumstances in which they arise in practice are left as an exercise to the reader.

The problem of representation is simply that the two standard modes of representation, namely, A-on-A and A-on-N, are obviously quite capable of accommodating the additional restrictions, but the price paid for conforming to a standard mode would be a large expansion in the number of activities, a loss of the identity of the individual activities, and a vastly complicated AN; all three features are thoroughly distasteful to management. Can a diagrammatic representation be devised that exhibits the added restrictions without incurring any of the penalties mentioned above? Note that the problem relates to the *visual representation* of the information, not to its analysis.

Before responding to this question we wish to substantiate the statement made above, namely, that the standard modes of representation are, indeed, capable of accommodating these added restrictions. We adopt the A-on-N mode of representation and exhibit in Fig. 1.8 the network representation alongside the "bar chart" for each condition. We illustrate the analysis in reference to condition *c*—the SF lag.

It is evident that one must distinguish four subactivities, namely, the two parts of u and v that overlap and the part of each activity that does not overlap. Therefore, it is natural to define four subactivities: u_1, u_2, v_1, and v_2, with u_2 and v_1 as the (possibly) overlapping subactivities. The precedence is as shown in Fig. 1.8*c*, where u_1 must precede u_2 and v_1 must precede v_2. Furthermore, since v_1 and u_2 are the only two subactivities that may run concurrently, we must have u_1 preceding v_1 and u_2 preceding v_2. Now consider the durations of the various subactivities and recall that the expressions must be derived for arbitrary values of y_u, y_v, and SF lag. Slight reflection reveals that there are only four possibilities shown in Fig. 1.9; in cases (*a*)–(*c*) the two activities may overlap (recall that the SF lag restricts the earliest completion of v relative to the start of u), while in case (*d*) it is impossible for the two activities to overlap. Furthermore, it is better to interpret case (*a*) as an FS lag condition with the roles of activities u and v exchanged in order to avoid negative durations in the subsequent expressions. We assume that to be the case and ignore case (*a*) in our further analysis of the SF lag.

The expressions for the durations of the various subactivities are given in Fig. 1.9 (as well as in Fig. 1.8*c*), together with the segments on the "bar charts" representing the various subactivities. The reader should satisfy himself that the A-on-N representation with the specified durations does, indeed, respect the restrictions imposed by the SF lag.

In Problem 20 at the end of the chapter we pose the question of diagraming the other restrictions and verifying the accuracy of their representations.

Having satisfied ourselves that the A-on-N mode of representation is capable of accommodating these various lag restrictions, we return to the question of alternative representation that avoids the disadvantages mentioned on p. 12. A possible model may be constructed by retaining the simplicity and elegance of the basic A-on-N mode of representation but adding the requisite symbolism that indicates the existence of the various lags. Such symbolism is offered in Fig. 1.10. Others are possible, and the choice is subject to taste as well as the degree of sophistication in the art of graphics. But whichever symbolism is chosen, the number of activities has not been increased, the durations of the activities are still retained at their original values, and the resultant AN (or bar chart) is no more complicated. *For graphical representation*, and for that purpose only, a new "language" has just been invented to preserve the original simplicity. Of

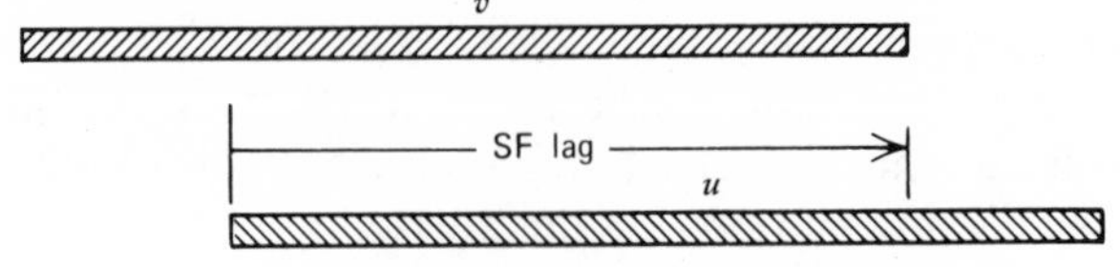

(a) SF lag $< y_u \Rightarrow$ condition better interpreted as FS lag with u and v exchanged

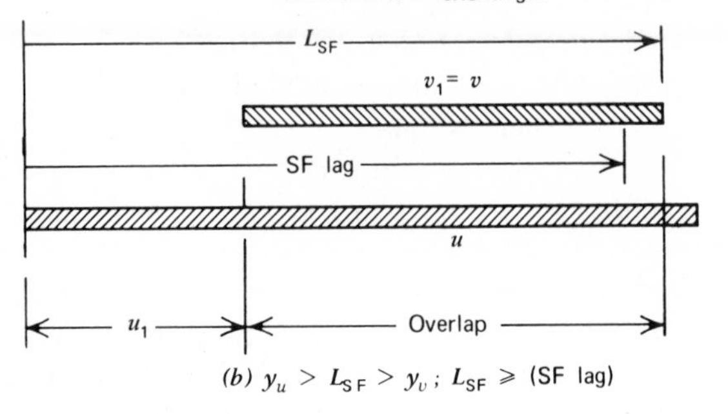

(b) $y_u > L_{SF} > y_v$; $L_{SF} \geqslant$ (SF lag)

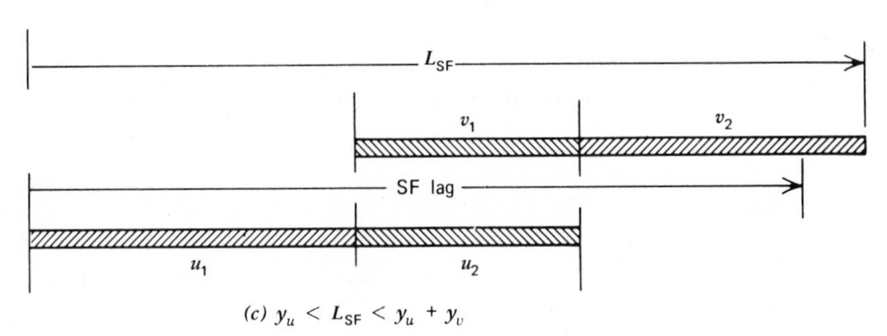

(c) $y_u < L_{SF} < y_u + y_v$

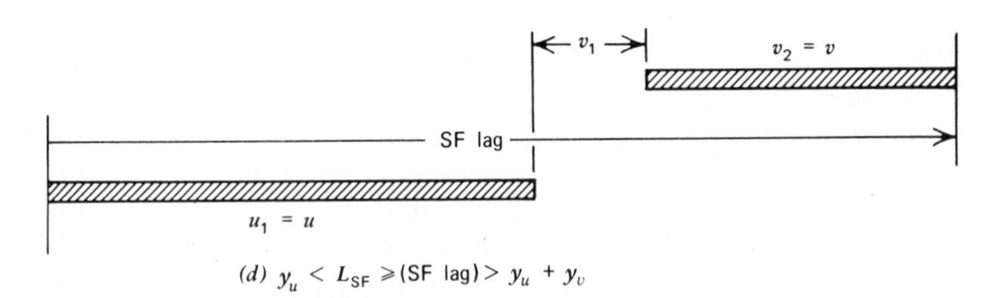

(d) $y_u < L_{SF} \geqslant$ (SF lag) $> y_u + y_v$

Figure 1.9. The case of SF-lag restriction:
$y_{u_1} = \min\{y_u; L_{SF} - y_v\}; L_{SF} \geqslant \text{(SF lag)}$
$y_{u_2} = y_u - y_{u_1}$
$y_{v_2} = \min\{y_v; \max(0; L_{SF} - y_u)\}$
$y_{v_1} = \max\{y_v - y_{v_2}; L_{SF} - y_{v_2} - y_u\}$.

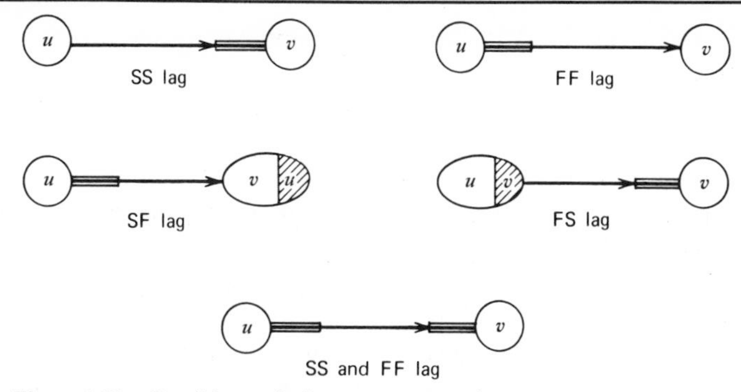

Figure 1.10. Possible symbolism to represent lags on A-on-N networks.

course, for *analysis* purposes, there is no escape from the complexity imposed by the added restrictions. But such analysis is the domain of the experts, for whom graphical representation occupies a lesser priority.

§ 1.4 On the Checking of Consistency in Adjacency Matrices*

An important practical problem in the development of large-scale activity networks is to ascertain the consistency of the precedence relationship throughout the (probably very large) network. Basically, one must ascertain that the precedence relationships specified by the adjacency matrix do not lead to the conclusion, due to the transitive character of the precedence relationship, that an activity a precedes itself!

An immediate and most obvious approach is to *insist* on the numbering scheme suggested in point (1) (p. 4), namely, that the nodes (events) be numbered such that an arrow leads from a smaller numbered node to a larger one. Then, the adjacency matrix must be upper triangular with zeros along the diagonal. It is then a trivial task to check the adjacency matrix for entries on or below the diagonal.

In case such numbering is not practically feasible the problem takes on serious proportions, because the inconsistency rarely appears explicitly in the primary set of relations ($a \prec b$ and $b \prec a$); it is too easily detected. The contradiction generally occurs as an implication (due to transitivity), usually of high order.

Obviously, clarification of all of the implications will reveal any inconsistencies. This can be achieved by "squaring" the adjacency matrix (i.e., multiplying it by itself) at most $n-1$ times. If the matrix is consistent, it is

*Starred sections may be omitted on first reading.

nilpotent† of index $n-1$ or less, since, in the absence of loops, the maximum spacing between any two nodes is $n-1$. A positive entry after $n-1$ such self-multiplications indicates the presence of a loop, that is, an inconsistency.

This approach, while mathematically valid, is impractical, especially for large matrices ($n \geqslant 500$). A much more elementary procedure, due to Marimont (1959), runs as follows.

Consider the adjacency matrix of A-on-A representation; it must have at least one column of zeros (corresponding to a source node) and at least one row of zeros (corresponding to a terminal node). Delete every row or column of zeros and the same numbered line (i.e., column or row); that is, if row (column) i is deleted because it is all zero, also delete column (row) i. This step will yield a new submatrix with new sources and terminals, namely, those nodes previously linked to the (now deleted) previous sources and terminals. This process is repeated until either (a) every line in the matrix has been deleted, indicating a consistent set, or (b) a submatrix with no zero lines remains, indicating an inconsistent set. Condition (b) is called a *residual* submatrix. As an example, consider the adjacency matrix shown in Fig. 1.11*a* and assume that it is inconvenient to construct the corresponding graph. Is this matrix consistent?

It is easy to see that columns 1 and 3 and rows 8, 11, and 12 are zero lines (indicated by an asterisk). Deleting all lines (i.e., all rows and columns) numbered 1, 3, 8, 11, and 12, we obtain the residual matrix of Fig. 1.11*b*. Again, column 2 and rows 4 and 10 are zero lines and can be deleted, to yield the residual submatrix of Fig. 1.11*c*. There are no zero lines, and we conclude that the original adjacency matrix is inconsistent.

It is left as an exercise to the reader to draw the graph corresponding to the adjacency matrix of Fig. 1.11*a* and verify that the successive deletions of zero lines do, indeed, leave intact the cycles in the original graph.

The procedure outlined above is easily programed for a digital computer. If the adjacency matrix contains n rows and columns, approximately $2n^2$ additions and $n^2/2$ subtractions will be necessary for a consistent matrix and somewhat less for an inconsistent one. This is to be compared with the n^3 multiplications and n^3 additions required in squaring an $n \times n$ matrix, repeated $n-1$ times! The method is as follows:

1. Compute all row and columns sums (real arithmetic sums, not Boolean sums) — $2n^2$ additions

†A square matrix M is termed *nilpotent* if $M^n = 0$ for some positive integer n, and of index k if $M^k = 0$ but $M^{k-1} \neq 0$.

	1*	2	3*	4	5	6	7	8	9	10	11	12
1		1										
2				1						1		
3						1						
4											1	
5						1						1
6				1					1			
7					1							
8*												
9							1					
10								1			1	
11*												
12*												

(*a*) Original adjacency matrix

	2*	4	5	6	7	9	10
2		1					1
4*							
5				1			
6		1				1	
7			1				
9					1		
10*							

(*b*) Residual submatrix: First iteration

	5	6	7	9
5		1		
6				1
7	1			
9			1	

(*c*) Residual submatrix: Second iteration.

Figure 1.11. Testing for consistency

2. For every zero row sum, subtract the corresponding column from the column of row sums and delete both row and column	Average of $n/2$ subtractions in row
3. Same as 2, interchanging the role of rows and columns	Same
4. Repeat 2 and 3 until either all lines are deleted or a residual submatrix of no zero lines remains	At most n repetitions

A major objection to the method outlined above for detecting inconsistency is that the exact nature of the closed cycle or cycles is not immediately obvious from the residual submatrix, while that task was relatively easier in the "squaring" procedure due to the appearance of a 1 along the diagonal whenever a cycle is formed. This issue is still open to investigation.

§2. TEMPORAL CONSIDERATIONS

By *temporal considerations* we refer to the multitude of questions of analysis that are concerned with the timing of activities and events. In other words, given the structure of the **AN** and the duration of the various activities, we seek answers to questions such as: (a) how long the project (or portions thereof) will take, (b) how early a particular activity may be started, and (c) how far can an activity be delayed with no delay in the total duration of the project. As we see in the text that follows, the answers to these and other questions involve some novel and rather important concepts (from a managerial point of view) such as the "critical path," the "event slack" and the "activity float" in DANs and their counterpart notions in PANs, such as the probability distribution function (PDF) of the duration of the project and "optimistic" and "pessimistic" estimates of durations.

The reader will notice that we limit ourselves in the present chapter to questions of *analysis* (as opposed to *synthesis*) and that we retain the sharp distinction between deterministic and probabilistic analyses. Questions of *synthesis*—such as the determination in an optimal manner (where optimality is defined relative to some criterion) of the duration of the activities or the very structure of the network, are relegated to subsequent Chapters 2–5 (see Chapter 2 for the determination of activity durations and 5, for structural considerations).

As always, analysis is easier than synthesis. To determine how long the project will take (given the structure of the network and the durations of the activities) is much easier than to estimate the duration of each activity for the total duration of the project not to exceed a specified value.

§2.1 Deterministic Arc Durations

The logical structure of an AN compels us to conclude that a node (event) cannot be realized unless *all* nodes (events) and arcs (activities) of the subnetwork preceding that node have been realized—see remark (iv) on p. 6. (We are adopting the A-on-A mode of representation in this discussion.) Let $t_i(E)$ denote the *earliest* possible time of realization of node i. We assume always that $t_1(E)=0$, that is, the *start* node is always taken at the time origin. Suppose that more than one forward path leads from *1* to i. Denote the paths by $\pi_1, \pi_2, \ldots$. The duration of path π_k, denoted by $T(\pi_k)$, is given by $T(\pi_k)=\Sigma_{a\in\pi_k} y_a$. Then it is obvious that

$$t_i(E)=\max_k \{T(\pi_k)\}, \quad t_1(E)\equiv 0; \qquad i=2,3,\ldots,n \tag{1.1}$$

Equation (1.1) can be considered as the *definition* of $t_i(E)$, for certainly it is not an operational formula for the numerical evaluation of $t_i(E)$ in large networks due to the very large number of such paths. To achieve such a computing formula, it is better to proceed iteratively. Let $\mathcal{A}(j)$ denote the set of nodes that connect *from* j (i.e., occur after j) and let $\mathcal{B}(j)$ denote the set of nodes that connect *to* j (i.e., occur before j). Then

$$t_j(E) = \max_{i \in \mathcal{B}(j)} \{t_i(E) + y_{ij}\}; \quad t_1(E) \equiv 0; \quad j = 2, 3, \ldots, n \tag{1.2}$$

In particular, $t_n(E)$ is the earliest realization of the project given the set of arc durations defined on the AN.

As the reader may suspect by now, certain activities can be delayed with no effect on the completion time of the last node n (i.e., on the total duration of the project). Such activities are usually referred to as possessing *slack*. In the determination of such slacks it is safer and more accurate to work with the nodes of the network rather than the arcs, as becomes apparent presently. The latest time an event must be accomplished in order to avoid a delay in node n is denoted by $t_i(L)$. We specify that either $t_n(L) = t_n(E)$ and then $t_1(L) = t_1(E) \equiv 0$ or $t_n(L) = \tau > t_n(E)$, where τ denotes a specified value, and then $t_1(L) > 0$. Note that a value $\tau < t_n(E)$, as specified by, say, contractual constraints, is *infeasible* (in the sense of being unrealizeable) *except* when some or all activities in the project can be shortened. Normally, such "crash" programing is obtained at a premium cost. It must then be determined which activities are to be "crashed", and by how much, in order to complete the project at time τ as specified, to incur minimum cost. This is the "time-cost trade-off problem" treated in Chapter 2.

Assuming that $\tau \geqslant t_n(E)$ the set of values $\{t_i(L)\}$ can be easily determined as follows. By similar reasoning to above, and *moving backward* from node n to node 1, we evaluate $t_i(L)$ either from the expression

$$t_i(L) = \min_k \{\tau - T(\pi_k)\}, \quad i = n-1, n-2, \ldots, 1 \tag{1.3}$$

where π_k is now a path from n to i when all the arrows of the network are reversed; or recursively from

$$t_i(L) = \min_{j \in \mathcal{A}(i)} \{t_j(L) - y_{ij}\}; \quad t_n(L) = \tau; \qquad i = n-1, n-2, \ldots, 1 \tag{1.4}$$

when the arrows are retained in their original orientation. The difference

$$t_i(L) - t_i(E) = s_i \geqslant 0, \quad i \in N \tag{1.5}$$

is the *slack time* of event i. It represents the possible delay in the realization of i that causes no delay in $t_n(L)$.

Starting from the origin *1* and moving *forward* in a stepwise fashion, following Eq. (1.2), we evaluate $t_i(E)$ for $i=2,3,\ldots,n$. This specifies the earliest completion time of the terminal, $t_n(E)$. Now assuming $\tau=t_n(L)=t_n(E)$, we can move *backward* toward the origin *1* [see Eq. (1.4)], and evaluate $t_i(L)$ for $i=n-1,\ n-2,\ldots,1$.

Activity Float (Slack)

The node slack is given by Eq. (1.5); we now turn to a discussion of *activity slack* (or activity *float*, as it is often called), a subject to which some attention has been given in the literature.

Realizing that each branch (ij) has two end points, i and j, and that each node possesses two time values; $t_i(L)$ and $t_i(E)$, with $t_i(L)\geqslant t_i(E)$; all i; we immediately deduce four different relations:

$$S_{ij}^1=t_j(L)-t_i(E)-y_{ij} \tag{1.6a}$$

$$S_{ij}^2=t_j(L)-t_i(L)-y_{ij} \tag{1.6b}$$

$$S_{ij}^3=t_j(E)-t_i(E)-y_{ij} \tag{1.6c}$$

$$S_{ij}^4=\max(0;t_j(E)-t_i(L)-y_{ij}) \tag{1.6d}$$

Equation (1.6d) requires some explanation. Recall that, by construction,

$$t_j(L)\geqslant t_i(L)+y_{ij} \quad \text{and} \quad t_j(E)\geqslant t_i(E)+y_{ij}.$$

Hence, S_{ij}^1, S_{ij}^2, and S_{ij}^3 are assured to be $\geqslant 0$. However, the same cannot be stated concerning $t_j(E)-t_i(L)-y_{ij}$ since there is no guarantee that even $t_j(E)-t_i(L)$ is $\geqslant 0$. Consequently, we must take $S_{ij}^4=0$ if $t_j(E)-t_i(L)-y_{ij}$ is <0, although a negative value of this last expression may be used to highlight a potential conflict in that particular activity. Furthermore, it is obvious that S_{ij}^1 is the *largest* and that S_{ij}^4 is the *smallest* float; that is,

$$S_{ij}^1\geqslant\max\{S_{ij}^2,S_{ij}^3,S_{ij}^4\}; \quad \text{and} \quad S_{ij}^4\leqslant\min\{S_{ij}^1,S_{ij}^2,S_{ij}^3\}.$$

The relative ranking of S_{ij}^2 and S_{ij}^3 is indeterminate, depending on the network.

The four activity floats given in Eq. (1.6) have been given names by Battersby (1967) and Thomas (1969) as follows:

$$\begin{aligned} &\text{Total float} &&= S_{ij}^1 \\ &\text{Safety float} &&= S_{ij}^2 \\ &\text{Free float} &&= S_{ij}^3 \\ &\text{Interference float} &&= S_{ij}^4. \end{aligned}$$

However, at least in this context, names are unimportant; the essential consideration is the relative position of the preceding as well as the succeeding activities. Thus S_{ij}^1 assumes that all activities prior to activity (ij) are accomplished as early as possible while all activities succeeding it are started as late as possible. Therefore, S_{ij}^1 measures the maximum possible "freedom" in activity (ij). At the other extreme, S_{ij}^4 assumes that the preceding activities are accomplished as late as possible, while the succeeding activities are started as early as possible. This situation will necessarily "squeeze" activity (ij) in the smallest domain, perhaps even yielding a negative value to the difference $t_j(E)-t_i(L)-y_{ij}$. The other values S_{ij}^2 and S_{ij}^3 lie between these two extremes, as indicated above. We leave their interpretation as an exercise to the reader.

We thus conclude that while there is one unambiguous node slack, there are four activity floats. Still, we can assert that in the forward and backward movements leading to Eqs. (1.2) and (1.4) there must be at least one path whose slack (i.e., the slack of every node, and all four types of float of every arc along that path) is equal to zero. This is called the *critical path* (CP); denoted by π_C, for the simple reason that an increase in any arc duration along this path would certainly cause an increase in $t_n(E)$, the total duration of the project.

It is evident that if management wishes to reduce the time of completion of a project it should concentrate its effort on activities that lie along the critical path(s). Since other activities do not affect $t_n(E)$, expenditure of effort on them would be a waste. (Of course, this is true until significant changes have been made in the activities along the CP that cause $t_n(E)$ to be reduced to a point where a new critical path(s) is created that includes these activities.) This brings to light the fallacy of "across-the-board" speedup of activities.

Figure 1.12 summarizes the calculations of Eqs. (1.2), (1.4), (1.5), and (1.6) for a simple network. Notice that three activities possess positive float *even though all nodes of the network have zero slack*! (All four types of activity float are equal in this example.) This is a subtle point that escapes

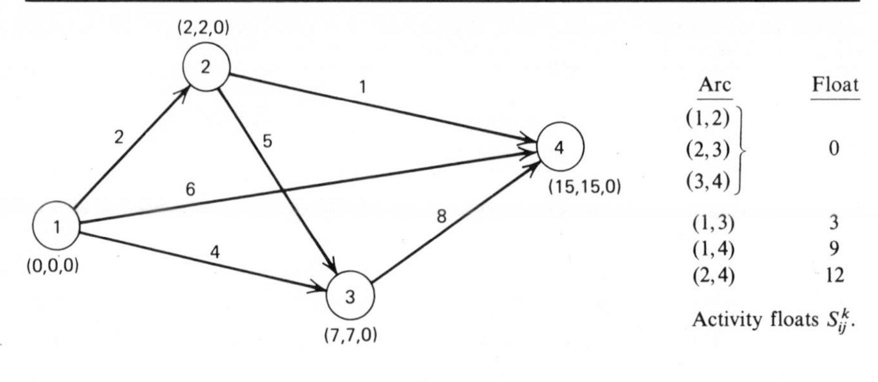

Arc	Float
(1,2)	
(2,3)	0
(3,4)	
(1,3)	3
(1,4)	9
(2,4)	12

Activity floats S_{ij}^k.

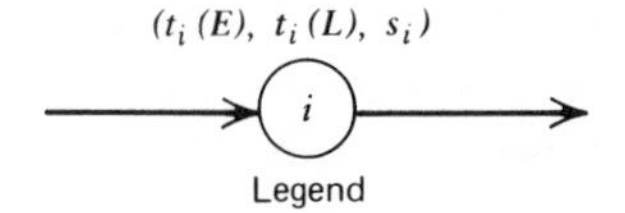

Figure 1.12. Node slacks and activity floats.

many analysts in this field, and the reader is cautioned about statements concerning activity float since many of them are valid only under certain conditions.[†]

As a final footnote to the above analysis the reader should recognize that although the present and most other investigators have emphasized the CP, one may wish to consider extensions of the concept to the second, third,..., CP; where the kth CP is defined as the path(s) from node 1 to node n of the (ordered) kth longest duration.

The advantages of determining the k longest paths are as follows. First, such information provides a means of assessing the sensitivity of the chosen CPs duration to possibly suboptimal decisions (e.g. concerning its "compression" or "prolongation"; see Chapter 2). Second, one may be interested in a class of solutions and not just in a single solution, because it may bring to the forefront some concepts that a single CP is incapable of accomplishing (e.g., the concept of "critical activities"; see Problem 11 of the present chapter, and Section 4.1 of Chapter 4). Third, the k longest paths provide a measure of the robustness of the underlying model when the data are approximate (which is usually the case).

The determination of the k longest paths has been extensively studied in the more recent literature [Minieka (1975), Shier (1974a and b)]. In

[†]The most notorious examples of such misuse of activity floats manifest themselves in heuristic procedures for ranking the activities at any point of time (see, e.g., the discussion of Section 2 Chapter 3).

Problem 5 we ask the reader to devise his own procedure for such determination and answer some relevant questions.

*A Flow-network Interpretation for the Determination of the CP**

Activity networks of concern to us here possess no "flow" of any uniform commodity through them. In fact, the network is a representation of a heterogeneous number of activities or tasks that combine to realize certain events. However, for the purpose of determining the CP it is possible to view the network as a flownetwork in which a unit flow enters at the origin *1* and exits at the terminal n. The duration of each activity, y_{ij}, can then be interpreted as the time of "transportation" of the commodity from node i to node j or, better still, as the "utility" of such transportation. Naturally, all intermediate nodes act as "transhipment" centers. Under such interpretation, determining the CP, which is the longest path, is equivalent to determining the path of *maximum utility* from *1* to n.

The advantages of such interpretation are three: first, a linear programing (LP) formulation—and in particular a dual algorithm—is immediately available; second, the whole body of literature on network flows and LP, which is both fascinating and important, can be utilized to answer any other questions (e.g., sensitivity analysis to activity durations) deemed of interest in the study of such networks; and third, this (rather elementary) model serves as an introduction to the more complex and more meaningful model of "chance constrained programing" discussed on p. 28 in regard to PANs.

The following LP formulation is facilitated if the reader refers to Fig. 1.13. Let $x_{ij} \geqslant 0$ denote the quantity of the commodity flowing in arc (ij) in the direction $i \rightarrow j$. Then the (primal) LP is (recall that y_{ij} is a constant)

$$\text{maximize} \quad \sum_{(ij)\in A} y_{ij} x_{ij};$$

s.t.

$$\begin{aligned} x_{ij} &\geqslant 0 \\ \sum_{i\in\mathcal{A}(1)} x_{1i} &= 1 \\ -\sum_{i\in\mathcal{B}(j)} x_{ij} + \sum_{k\in\mathcal{A}(j)} x_{jk} &= 0, \quad j=2,3,\ldots,n-1 \\ -\sum_{i\in\mathcal{B}(n)} x_{in} &= -1 \end{aligned} \tag{1.7}$$

*Starred sections may be omitted on first reading.

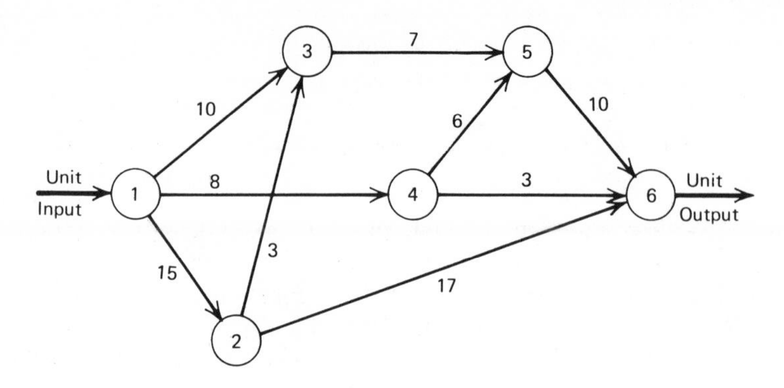

(*a*) Original activity network.

(*b*) Primal LP.

$$\begin{array}{lr}
\text{Max.}\ 15x_{12}+10x_{13}+8x_{14}+3x_{23}+17x_{26}+7x_{35}+6x_{45}+3x_{46}+10x_{56} = & z \\
\text{s.t.} & \\
x_{12}+x_{13}+x_{14} = 1 & w_1 \\
-x_{12}+x_{23}+x_{26} = 0 & w_2 \\
-x_{13}-x_{23}+x_{35} = 0 & w_3 \\
-x_{14}+x_{45}+x_{46} = 0 & w_4 \\
-x_{35}-x_{45}+x_{56} = 0 & w_5 \\
-x_{26}-x_{46}-x_{56} = -1 & w_6
\end{array}$$

(Right-hand column: Dual variables)

(*c*) Dual LP.

$$\begin{array}{l}
\text{Min. } g = w_1 - w_6 \\
\text{s.t.} \\
w_1 - w_2 \geqslant 15 \\
w_1 - w_3 \geqslant 10 \\
w_1 - w_4 \geqslant 8 \\
w_2 - w_3 \geqslant 3 \\
w_2 - w_6 \geqslant 17 \\
w_3 - w_5 \geqslant 7 \\
w_4 - w_5 \geqslant 6 \\
w_4 - w_6 \geqslant 3 \\
w_5 - w_6 \geqslant 10
\end{array}$$

(*d*) Dual optimal solutions.

$$w_1^* = 0, w_2^* = -15,\ w_3^* = -10,\ w_4^* = -8,\ w_5^* = -25,\ w_6^* = -35$$

Figure 1.13. Example of LP model for determination of CP.

(e) Primal optimal solution.

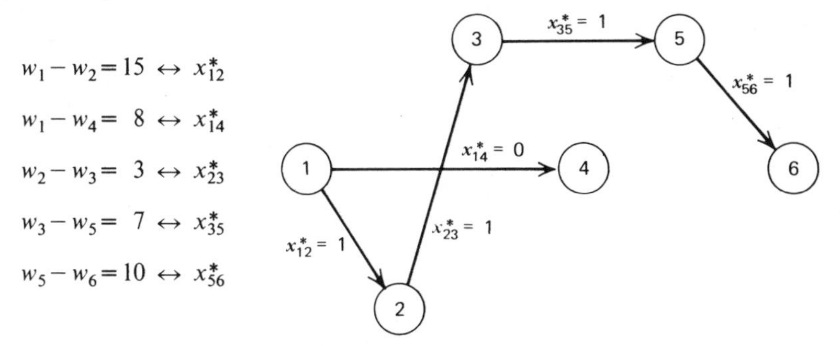

Figure 1.13 (Cont'd) Example of LP Model for The Determination of CP.

These equations represent the "flow conservation constraints" in which we adopted the convention that flow into a node is negative flow and flow out of a node is positive flow.

The solution of this LP is readily obtained through the dual formulation. Let $\{w_i\}$, $i = 1, \ldots, n$ be the dual variables (there are exactly n such variables, sometimes referred to as the "node numbers" for obvious reasons) and note that the coefficient of x_{ij} is a column vector with only two entries: $+1$ and -1. Then the dual LP is

$$\text{minimize} \quad w_1 - w_n$$

s.t.

$$w_i - w_j \geqslant y_{ij}, \qquad \text{for each } (ij) \in A; \tag{1.8}$$

in which the w_i's are not restricted in sign.

Since each inequality in the dual formulation contains two and only two variables, the minimization can be achieved by inspection in one scanning of the A inequalities. In particular, if w_1 (or w_n) is put equal to any arbitrary value (usually 0), the remainder set of constraints is triangular in the variables $\{w_i\}$, and the optimal solution is easily calculated.

The optimal solution to the primal LP is derived from the optimal solution of the dual LP by invoking the complementary slackness theorem of LP. In particular, each inequality constraint in the dual LP satisfied as strict inequality must have its corresponding x_{ij} varible$=0$. Moreover, the theorem uniquely determines the optimal basic solutions of the primal LP, $\{x_{ij}^*\}$. This is accomplished by first noting that any basic feasible solution to the (primal) LP of Eqs. (1.7) defines a spanning tree[†] of the network and

[†]That is, a connected graph over all the nodes of the network that contains no loops; [cf. remark (3) on p. 6].

then considering the subset of primal variables $\{x_{ij}\}$ that correspond to dual constraints satisfied as *equalities*; those that define a path from the origin, node *1*, to the terminal, node n, are given the value 1; the rest are given the value 0. The path defined by the variables $\{x_{ij}=1\}$ is the CP that may be accorded the following interpretation: the unit input flow is "transhipped" as an entity along the CP. The fact that $x_{ij}=0$ or 1 is a direct consequence of the "unimodular property"† of the matrix of coefficients of the LP of Eqs. (1.7).

Each optimal solution to the LP defines a unique CP, while alternate CPs are discovered through the enumeration of the alternate optima of the LP. Note that the number of primal variables $\{x_{ij}\}$ in the LP of Eqs. (1.7) is equal to the number of *arcs* in the network, while the number of dual variables $\{w_i\}$ is equal to the number of *nodes*. The reader should satisfy himself, using the numerical example of Fig. 1.13a, that except for an additive constant the node number w_i is identical to $t_i(E)$, the earliest completion time of node i.

The steps of this procedure are illustrated in Fig. 1.13. The simple AN exhibited in Fig. 1.13a gives rise to the primal LP of Fig. 1.13b which, in turn, results in the dual LP of Fig. 1.13c. The optimal dual solution is exhibited in Fig. 1.13*d*, and the corresponding optimal primal variables and solution are exhibited in Fig. 1.13*e*. It is easily verified that the w values are obtained in "one scanning" of the dual LP and that the primal solution is obtained directly from the dual constraints that are satisfied as equalities.

§2.2 Probabilistic Arc Durations

There are several good reasons for considering the arc durations to be random variables (RVs) rather than fixed constants. We expound on these reasons in Chapter 4, so it suffices here to briefly state some of them. The activities may be research and development activities whose durations are in fact unknown entities at the start of the project, or they may depend on the availability of certain resources or the existence of certain conditions (e.g., suitable weather) that are themselves chance occurrences. Or the process of estimation of the activity durations may be subject to random error (e.g., when it is based on work sampling), and so on.

In any event, and whatever the reason may be, when the durations of some of the arcs (activities) are RVs, denoted by Y_a for arc a, the original approach proposed by the originators of PERT [Malcolm et al. (1959)] ran

†A matrix is unimodular if the determinant of every submatrix in it is equal to -1, 0 or $+1$. It is enjoyed by few matrices, among which are the *transportation* and *transhipment* matrices of LP.

as follows. Let $\bar{y}_a = \mathcal{E}(Y_a)$ be the expected value of the RV Y_a. Substitute everywhere the (so-called certainty equivalent) expected value and determine the CP following the standard procedure of DANs. Let π_C denote the CP(s). Assume that:

1. The activities are independent.
2. The CP contains a "sufficiently large" number of activities so that we can invoke the Central Limit Theorem.
3. We may ignore all activities not on the CP(s).

The length of the CP, $T(\pi_C)$, is identical with the time of realization of node n, denoted by T_n. Furthermore, the length of the CP is itself a RV, being the sum of random variables,

$$\mathrm{T}_n = \sum_{a \in \pi_C} Y_a = T(\pi_C)$$

The above first two assumptions lead to the conclusion that T_n is approximately Normally distributed with mean

$$g_n \cong \sum_{a \in \pi_C} \mathcal{E}(Y_a) = \sum_{a \in \pi_C} \bar{y}_a;$$

and variance (1.9)

$$\sigma_n^2 = \sum_{a \in \pi_C} \sigma_a^2;$$

where σ_a^2 is the variance of the RV Y_a.

The new concept introduced in the analysis of networks with probabilistic time estimates is that of probability statements (i.e., confidence statements) concerning the duration of the project. In particular, if $\Phi(x)$ denotes the Standard Normal (probability) distribution function (PDF), that is,

$$\Phi(x) = \int_{-\infty}^{x} (2\pi)^{-1/2} \exp\left(\frac{-r^2}{2}\right) dr$$

then the probability that event n will occur on or before a specified time $t_n(s)$ is given by

$$Pr\{\mathrm{T}_n \leqslant t_n(s)\} = \Phi\left\{\frac{t_n(s) - g_n}{\sigma_n}\right\}$$

where g_n and σ_n are given above, and Pr denotes the "probability of."

For the practical application of these ideas, the originators of PERT suggested that instead of asking for the PDF of Y_u, for all u, or asking for "educated guesses" about the mean $\bar{y}_u$ and the variance σ_u^2, for all u,† it is easier and possibly more meaningful to make the following assumptions and simplifications in addition to the three assumptions made above:

4. The PDF of Y_u, denoted by $F_u(y)$, can be approximated by the beta‡ distribution; that is,

$$dF_u(y) = K(y-a)^{\alpha}(b-y)^{\beta}; \qquad a \leqslant y \leqslant b$$

   where a and b are "location" parameters and α and β are "shape parameters," and
5. The mean of the beta DF can be approximated by $y_u \cong (a_u + 4m_u + b_u)/6$ and the variance, by $\sigma_u^2 \cong (b_u - a_u)^2/36$; where m_u is the mode of the DF of Y_u (the so-called most likely duration).§

Acceptance of these two additional assumptions and simplifications leads one to conclude that for the application of PERT one needs to obtain the best estimates of the three parameters a, m, and b, which are labeled, respectively, the "optimistic," "most likely," and "pessimistic" estimates of the duration of the activity. These parameters determine the estimates $\bar{y}_u$ and σ_u^2 which, in turn, are used in the determination of the CP and the evaluation of the probability of project duration.

The original PERT model depended rather heavily on the above mentioned assumptions and simplifications, as well as on the assumption that the three parameters, a, m, and b can be extracted from the knowledgeable people for each activity in the network. In Chapter 4 we see that the assumptions and most of the conclusions based on them are open to serious questions that cast a grave shadow of doubt on the validity of the results obtained.

*A Chance-constrained Formulation for Event Realization**

In DANs we achieved an LP formulation for the determination of the CP. Now, by interpreting the activity durations as random variables, such as in the PERT model, a simple extension of the LP formulation is achieved

†The subscript of Y is arbitrary. Thus far we have designated an activity by $a \in A$. We now alternatively use the subscripts u as well as ij (thus write Y_u or Y_{ij}) to designate an activity. The change in the subscript is necessary to avoid confusion with other symbols used in the text, such as the symbol a for the "optimistic" estimate of the activity duration.

‡A detailed discussion of the pertinent statistical properties of the beta distribution is given in Appendix C to Chapter 4.

§A possible rationale for the variance expression may be that the estimate b_u is not exceeded more than 0.1% of the time, hence it is about 3σ away from the mean value (assuming the Normal distribution). Similarly, a_u is not bettered more than 0.1% of the time; hence it is about -3σ away from the mean. Thus $b_u - a_u = 6\sigma$, which leads to the given expression.

*Starred sections may be omitted on first reading.

through chance-constrained programming that is meaningful as well as solvable. For instance, suppose management is considering a contract for a certain project. The precedence among the activities is known (i.e., the structure of the network is known), but their durations are not known except in probability. Before contracting for a target completion date—with possible resulting delay penalties—management would like to control, with preassigned confidence level, the realization times of crucial phases of the project. It is then meaningful to inquire about the completion times to be quoted in the contract that are guaranteed to be realized with the given level of confidence.

The LP model of Eq. (1.7) called for maximizing

$$\sum_{u \in A} y_u x_u$$

s.t.

$$x_u \geqslant 0$$

$$\sum_u \delta_{iu} x_u = b_i; \qquad i = 1, 2, \ldots, n$$

where $\delta_{iu} = +1$ if the arrowtail of arc u is at node i, $\delta_{iu} = -1$ if the arrowhead of arc u is at node i, and $\delta_{iu} = 0$ otherwise; $b_i = +1$ for $i = 1$, $b_i = -1$ for $i = n$, and $b_i = 0$ for $i \neq 1$ or n. Recall that y_u designated the duration of arc u. This LP has the dual LP

$$\text{minimize} \qquad w_1 - w_n$$

s.t.

$$\sum_i w_i \delta_{ij} \geqslant y_u, \qquad \forall u \in A$$

The w_i's are the *node numbers* designating the earliest realization time of node i.

In the case of probabilistic arc durations the dual problem can be stated as:

$$\text{minimize} \qquad w_1 - w_n$$

s.t.

$$Pr\left\{ \sum_i w_i \delta_{iu} \geqslant y_u \right\} \geqslant q_u, \qquad \forall u \in A \tag{1.10}$$

where q_u is a positive number less than 1, which designates the desired confidence level in the node numbers w_i and w_j (the end nodes of activity

u). The ℓ.s. of the inequality measures the probability that the difference between the node numbers $w_j - w_i$ for each $i \in \mathfrak{B}(j)$ and for all $j \in N$ is equal to or larger than its respective duration y_{ij}. The constraint itself requires that this probability is at least equal to $q_{ij}(\equiv q_u)$.

The interpretation of the constraints of (1.10) is both illuminating and important. Essentially, we are requiring that the node numbers, $\{w_i\}$, will *simultaneously* "bracket" the activity duration with the specified probabilities, $\{q_u\}$. This is no trivial requirement, especially since the q_u values are usually in the high 90s percentile. For instance, consider the network of Fig. 1.13; we may require that

$$Pr\{w_2 - w_1 \geqslant y_{1,2}\} \geqslant q_{1,2} = 0.95$$

$$Pr\{w_3 - w_1 \geqslant y_{1,3}\} \geqslant q_{1,3} = 0.98$$

$$Pr\{w_5 - w_3 \geqslant y_{3,5}\} \geqslant q_{3,5} = 0.97$$

and so forth.

Since Y_u is assumed a RV of known PDF, $Pr\{y_u \leqslant \Sigma_i w_i \delta_{iu}\} = F_u(\Sigma_i w_i \delta_{iu})$, where $F_u(\cdot)$ is the PDF of the duration of arc u. Hence the constraints can be stated as

$$F_u\left(\sum_i w_i \delta_{ij}\right) \geqslant q_u, \qquad \forall u \in A$$

By the monotonicity and right continuity of the PDF, F_u, we can express the constraints in terms of the inverse PDF. Hence the equivalent (deterministic) LP is

$$\text{minimize} \qquad w_1 - w_n$$

s.t.

$$\sum_i w_i \delta_{iu} \geqslant F_u^{-1}(q_u), \qquad \forall u \in A \tag{1.11}$$

For any given set of PDFs and confidence levels $\{q_u\}$, this LP can be solved easily, yielding the desired node numbers $\{w_i\}$. These node numbers are times of realization of the nodes, similar in vein to the node numbers of Eq. (1.8), except for the following important distinction: the probability that the node numbers give rise to values $\Sigma_i w_i \delta_{iu}$ that are exceeded by the actual duration of the activity, y_u, is less than $1 - q_u$, for all arcs $u \in A$.

In Problems 14–16 we ask the reader to solve for such node numbers in a small project, and compare the results with those obtained from a straightforward application of the original PERT model.

§3. COST CONSIDERATIONS[†]

While developmental programs (referred to in this text as *projects*) differ significantly from standard repetitive work, the fact still remains that each developmental task must be performed on time (to maintain the delivery schedule) and for the estimated cost (to meet the program budget) and required quality (to accomplish the program mission)—just as with repetitive work.

The previous section addressed itself to the scheduling aspects of projects. In this section we turn our attention to the financial and accounting aspects. In particular, we attempt to determine whether a tool can be provided to management for the control of cost in project-type activities, and what impact such controls would have on the overall management function. Our discussion follows closely the guidelines as spelled out in *PERT-Cost*, a document issued in June 1962 under the aegis of the U.S. Department of Defense and the National Aeronautics and Space Administration [DOD and NASA (1962)]. This is followed by a criticism of the proposed approach from both theoretical and practical points of view.

At all times the reader should maintain a clear distinction between the financial, or economic, aspects of the project and its accounting aspects. The *financial aspects relate to the management of cash flows* in and out of the project, which include: (a) borrowing money (at an interest), (b) the investment of extra funds (at an interest), and (c) the terms of payment of debts and collection of receipts. Indeed, the management of cash flows may spell the difference between the economic success and failure of a project, irrespective of its technical merits. On the other hand, the *accounting aspects relate to the timely and accurate reporting of expenditures and receipts* and to the correct allocation of expenditures to their respective causes as a prelude to decisionmaking.

§3.1 The Approach

In *PERT-Cost* the title is really a misnomer, since it involves no probabilistic considerations whatsoever. Therefore, the reader should interpret the acronym as "an approach for incorporating cost considerations in project control." To a great extent, the ideas presented below closely parallel the traditional analysis of "cost variations" in standard cost accounting for manufacturing.

[†]It behooves the serious student, especially of cost controls, to familiarize himself also with the new governmental procedures for planning called "Planning, Programming and Budgeting System (PPBS)"; see the works by [Lyden and Miller (1967), Novick (1965)].

In this approach the overall project is divided into successively smaller pieces of prime hardware, support equipment, facilities, and services for costing purposes. Estimates are made of manpower, material, and other resources necessary to perform groups of activities referred to as *work packages.* These are then converted into dollars. The result is a *budgeted cost and work value,* represented schematically as in Fig. 1.14. While budgeted cost may be familiar to all, it is possible that "work value" may be less widely understood. The idea here is that as a part of the original planning and contract negotiations, the standard cost of each work package is established. For example, in an aerospace development program each preliminary design, each "black box" development, each system design, procurement, fabrication, and environmental test, each launch- and flight-data analysis task is cost estimated, negotiated with the government, and then made into task, or work package, budgets. It follows that if "now," represented by t_0 in Fig. 1.14, 260 of these discrete work packages have been completed then the value of work performed to date, the "work value" for short, is the originally budgeted cost of those specific 260 tasks. And if the money spent to date is greater than the negotiated cost of these 260 tasks, then the project is, by definition, overrun (in cost). Thus the curve of Fig. 1.14 provides not only the budgeted costs, but also the planned, or budgeted, work value to be simultaneously created.

The ensuing discussion would be further clarified if the reader bears in mind that we are dealing with a three-coordinate system: time, cost, and work value. While time is uniquely measured, both cost and work value have two measures each: budgeted and actual. Both measures on the two coordinates are necessary if meaningful interpretation of status reports is to be achieved. The graphs for actual work and actual cost, as functions of time, are schematically shown in Fig 1.15, together with the budgeted cost and work value curve of Fig. 1.14.

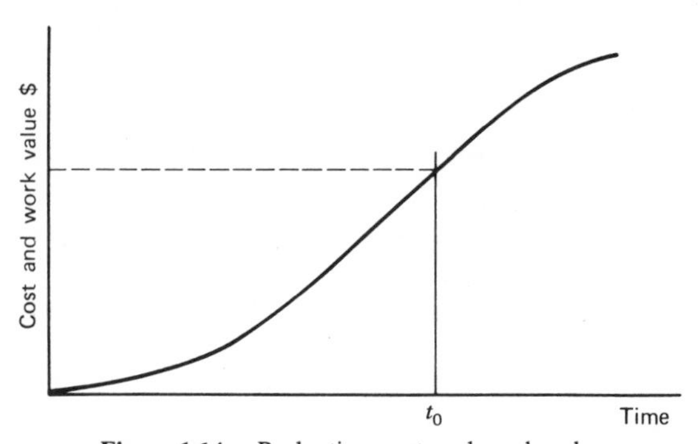

Figure 1.14. Budgeting cost and work value.

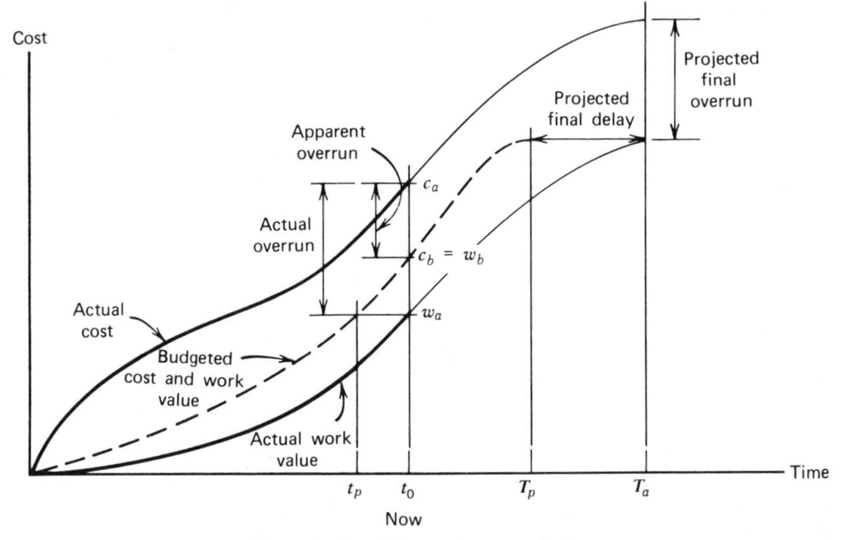

Figure 1.15. Time and cost variations.

Time t_0 represents "now." At t_0, status is measured in terms of four parameters: (a) actual cost incurred, c_a, (b) budgeted cost, c_b, (c) actual work value performed, w_a, and (d) budgeted work value, w_b.† In Fig. 1.15 the apparent cost overrun is $c_a - c_b$. But this is a misleading figure, since while the budgeted outlay of cost at time t_0 is, indeed, c_b, that expense is intended as payment for a certain amount of work value. In particular, the work content w_a was planned to be realized at time t_p. Hence the project has incurred the cost c_a for *less than planned* work content at time t_0, and we may conclude that the actual cost overrun is given by the total difference $c_a - w_a$.

The *delay* in the project (or, if the reader wills, the *time* overrun) is evidently equal to $t_0 - t_p$. Note that the basic dimension of measurement here is the work content.

At time t_0 it is possible to make a projection of future progress based on current status and any additional information. This is shown in Fig. 1.15 by the light solid lines. Notice that there are *two* such lines—one as extrapolation of the actual cost curve, and the other as extrapolation of the actual work content curve. The projected final cost overrun and time delay can then be read off the figure directly.

The above discussion, which was couched in terms of the total project, could be applied with equal validity to either major portions of the project

†Both w_a and w_b are measured in the same monetary units as the cost parameters.

(e.g., major subassemblies), or major organizations (e.g., engineering, manufacturing, advanced systems research). In such application the above analysis would serve to pinpoint sources of difficulty or responsibility for delays. We do not pursue this subject any further here; the interested reader is referred to the lucid article by Paige (1963).

§ 3.2 Critical Evaluation

Despite its apparent logic and simplicity, the approach of *PERT-Cost* is open to serious criticism on both theoretical and practical grounds. Perhaps the major criticism of this approach from a theoretical point of view is the fact that the budgeted cost and work-value graph of Fig. 1.14 is a simplistic view of the situation. In reality, there exists not one but an infinite number of such graphs. This is due to the presence of activities with *float* which, by its very definition, implies the possibility of delaying these activities with no delay in the final completion time of the project! Consequently, there are *two limit curves*, one drawn with all activities completed as early as possible and the other drawn with all activities completed as late as possible. There is an infinite number of curves lying between these two curves depending on the delay in the noncritical activities (see Fig. 1.16). This fact undermines the whole concept of cost

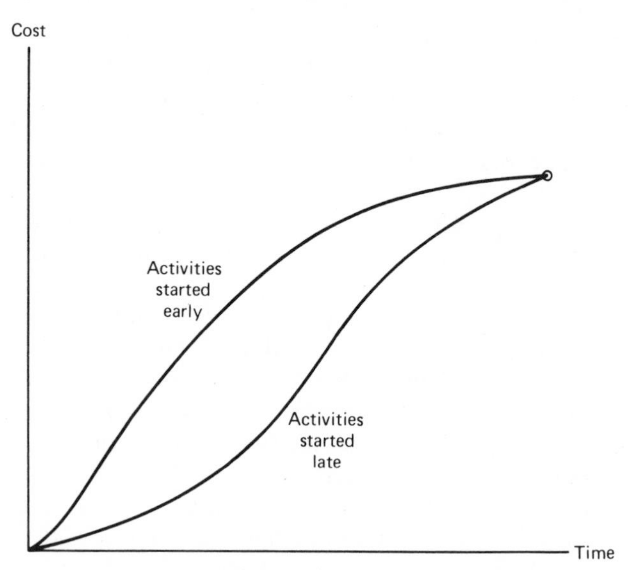

Figure 1.16. The limiting cost curves in DANs.

comparisons described above, since it is not immediately obvious which cost–time curve is the one chosen for budgeting purposes.

If, in addition, one wishes to consider the uncertainty in the activity durations (which, the reader will recall, was the raison d'être of PERT), it becomes evident that the above procedure completely breaks down. For now there is a possibility of random variation in both *duration and cost* (the latter as a function of duration) of the individual activities and total project as well. Hence every point on the cost curve—any cost curve, including the two limiting ones—may lie within an ellipse (or a rectangle) (see Fig. 1.17) whose abscissa (i.e., time) represents the variation in the *time* of realization of a particular node and whose ordinate (i.e., cost) represents the variation of cost as a function of duration. Therefore, it is even less obvious how the budgeting process is to be structured following the *PERT-Cost* dicta and which corrective decisions are to be made (optimally?) should an "overrun" situation be diagnosed! To the best of our knowledge the problem of optimal budgeting in project-type enterprises under conditions of uncertainty in the duration as well as in the cost of the activities has not been treated to date.

Apart from the theoretical considerations stated above, the *PERT-Cost* procedure is open to criticism from a practical point of view. These considerations are based on findings derived from field investigations in several companies during early attempts to implement the system, as reported by Hill (1966). It should be remembered that some of these difficulties are inherent in more traditional cost-accumulation systems, while others are partly peculiar to *PERT-Cost* with its requisites for increased detail for end items. No attempt has been made to arrange the problems in order of probable significance, since such a comparative measure is largely indeterminate at this time.

From a managerial control point of view, the accounting of actual cost is helpful only to the degree that it sheds light on the reasons for its variation

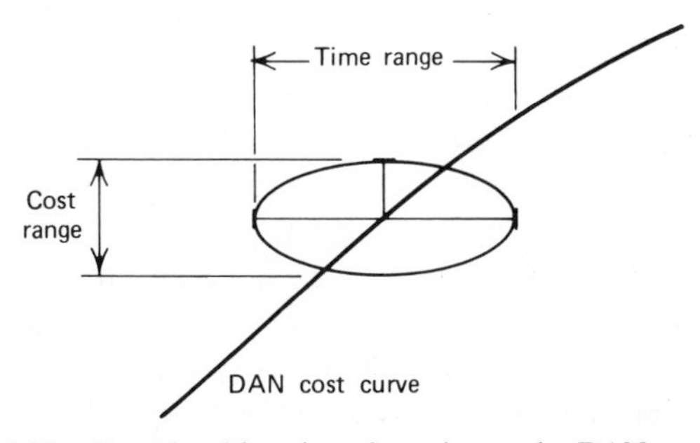

Figure 1.17. Domain of location of a point on the DAN-cost curve.

from the standard costs. But the analysis of such cost variance (=standard −actual) is itself extremely difficult because it has four dimensions: (a) time, (b) work content (=amount of work accomplished to date), (c) cost of resource content in work accomplished (material, labor, power, facilities, etc.), and (d) quality content. We now take a closer look at some of these dimensions.

Direct Labor.

There are two problems relative to the costing of direct labor: (a) level of detail required by *PERT-Cost* (b) costing (i.e., valuation) of the direct labor effort. For there is little doubt that *PERT-Cost* requires the definition of work orders at a lower level than other systems. Unfortunately, the finer the breakdown, the less accurate the charging activity is likely to be. This could lead to the distortion of the *PERT-Cost* accounting procedure, unless extensive policing procedures are introduced, a normally expensive undertaking. On the other hand, the conversion of direct labor hours to dollars may not be easy. This stems from the fact that direct labor rates are normally recorded at a (managerial) level different from the level at which they are budgeted or forecasted. This leads to a variance between the budgeted and the actual values, even under the strictest adherence to plans, and it is not evident how one can determine the false nature of such variance.

Material

For discussion purposes, the term "material" is defined to include such items as raw materials, purchased parts, and electrical and mechanical equipment. In many companies the problem may stem from the lack of identification of material with end products or work packages, since records on material are usually matintained by commodity rather than usage. Supposing that this hurdle is overcome through the establishment of the direct allocation of material to work packages, there are still remaining two factors that detract from the efficacy of the procedure. First, it is a more elaborate and hence more expensive system to administer. Second, it is not clear *when* the allocation of cost should take place—should it be when the material is *committed* or when it is *actually used*? The long time lapse between the planning of a work package and its actual undertaking may play havoc with either approach; costing against commitments may introduce errors when the actual material usage is shifted from one work package to another due to a shift in priorities, and it may distort the picture of cash flow (unless separate records are kept). On the other hand, costing against actual usage may curtail the predictive (and corrective) capability of the cost-control system.

In fact, it is evident that if effective control is the desired objective, the project manager is interested in four critical dates: (a) the date when the purchase is planned, (b) the date the commitment (i.e., the purchase order) is made, (c) the date the material is received, and (d) the date of payment. Each of these four dates and the information associated with it has a particular significance to the manager, and each helps him in a different phase or a particular aspect of the control process. (We leave it as an exercise to the reader to specify in detail the uses of each piece of information.)

To the best of our knowledge, no control mechanism based on such information has been suggested or incorporated in the *PERT-Cost* system.

Work Packages.

While it may be easy for one to glibly state that the total effort of the project should be divided into "reasonable" work packages—where "reasonableness" is defined as being of approximately three months' duration and \$100,000 in value—the practical implementation of such dictum poses several problems. For one, the level of detail in the definition of the work package is as much a function of the individual company responsible for its execution as it is of the project itself. The a priori definition of work packages in reference only to the project may lead to the creation of another subsystem of cost accounting in the company different from that envisaged by *PERT-Cost.* The relation between the two systems may be at best fuzzy if not downright contradictory.

Furthermore, several activities (defined either in the strictest sense of DANs, e.g., flight testing, or in a more liberal sense, e.g., maintainability and quality assurance) do naturally occur over a longer period of time —perhaps as much as ten times longer than the desired three months—and span a large number of "conventional" work packages. It is not immediately obvious how to "package" such activities for purposes of accountability and control. Treating them as overhead, which is oftentimes done, deprives the project manager from the knowledge of the allocation of effort. On the other hand, their division into smaller work packages may distort rather than illuminate the true picture.

Finally, the concept of a work package as a neatly defined entity over time breaks down in the presence of scarce resources shared by several, if not all the packages. Such breakdown is due to the possible need to subdivide a package into smaller entities to respect the resource availabilities. (See Chapter 3 for a more extensive discussion of the impact of scarce resources.) Consequently, the accounting must then be made in terms of the smaller units rather than the original work packages. Since this eventuality cannot be preplanned on the basis of the technical content of the project, it raises the specter of the futility of the effort in defining the work packages in the first place!

Overhead.

This is reported as a single line item in *PERT-Cost*, and overhead activities are usually not included in the network. Yet, burden costs are conventionally about equal to labor expense in many companies.

A first step toward better control of overhead might be an expansion within the *PERT-Cost* reporting framework of the overhead line item into its basic categories (e.g., indirect labor, operating supplies). Because of the differences in definition and allocation of overhead by various contractors, it is all the more necessary to include indirect costs. Consideration should also be given to aggregation of burden costs into fixed and variable categories. Indirect charges are no less contributory to overruns than direct costs. The same comment may apply to the general and administrative costs (G/A) and fee, where the inclusion of one line item only in the report sheds little light on the ingredients of such a charge. Fortunately, in this case it may not be significant, because G/A costs usually are minor compared to other overhead expenses. Fee should clearly be included, at least at the total contract level. Inclusion of fee does add a measure of complexity to the system, particularly when an incentive-type contract using a sliding scale for the allowable fee is incorporated.

§ 3.3 Organizing for Project Control

The reader will note that we have not discussed the important problem of the *administrative organization* of projects within the context of already existing manufacturing enterprises. The latter are normally organized along traditional "functional" lines, such as engineering, drafting, machine shop, foundry, personnel, and quality control. A project usually "cuts across" such functionally organized administration. Hence, the responsibility for the success or failure of a project may be distributed among several (possibly numerous) individuals, any one of whom may have the power to hinder its progress but none the authority to ensure its success! This would be contrary to the elementary principles of good management, which demand that responsibility and authority for an undertaking go hand in hand and be well defined from the outset.

Evidently, designating one man as the "project director" begs the issue since he cannot be held responsible for something over which he has no control (a delay in the machine shop may be well justified by the shop manager, but the project manager will find no solace in the perfectly legitimate reasons that caused *his* project to be delayed). The alternative, of course, is to give the project manager the requisite authority over all functional activities, which is tantamount to making him the plant manager. But a moment's reflection reveals that such is no solution at all because, first, such arrangement merely shifts the burden to the other foot; the problem now is thrown in the lap of the people responsible for nonproject activities! Second, it provides no answer to the case of involvement in several projects simultaneously, which is not an uncommon occurrence; will each project leader have *his* plant to manage?

A possible solution to this dilemma is the so-called matrix organization of administrative authority. Simply stated, it operates as follows: the project managers as well as the functional managers are at the same "level" in the executive hierarchy of the firm. If arranged in a two-way table, or a matrix, each "cell" in the table represents the interface between a particular project and a specific function, such as "Project A and engineering" or "Project D and test facilities". The responsibility for success lies with *both* managers, and failure is interpreted as joint failures —the managers "float or sink together."

Such an administrative structure requires strong leadership from the general manager, who must be assisted by an efficient management-review system (such as described here). This organizational structure was adopted at one time by the Martin-Marietta Company of Orlando, Florida, the prime contractor for several Army Missile Systems. For more details on the operation of the organization structure and the review procedure, see the work by Soistman (1966).

APPENDIX A. Digraphs and Line Digraphs†

A directed graph, or *digraph*, D has $N=N(D)$ as its set of points and $A=A(D)$ as its set of directed arcs. We designate the arc from i to j by (ij) in which case point i is adjacent to point j and j is adjacent from i. Two arcs, u and v, are said to be adjacent if u and v have one node in common. In particular, u is adjacent to v if the terminal point of u is the initial point of v, in which case v is adjacent from u. An example of a digraph is Fig. 1.2.

†This discussion follows closely that by Geller and Harary (1968).

To each digraph D we define the line digraph $L(D)$, which itself is a directed graph, as follows: $N[L(D)]=A(D)$, that is, the points of $L(D)$ are the arcs of D, and for points u, v, $\in N[(L(D))]$ we have $(u,v)\in A[L(D)]$ if arc u is adjacent to arc v in D, that is, if u precedes v in D. A digraph D is a line digraph if it is the line digraph of some digraph E, that is, if $D=L(E)$. An example of a line digraph is Fig. 1.6, which is the line digraph of the digraph of Fig. 1.2.

The following theorem characterizes a line digraph, and is due to Harary and Norman (1961). A collection S_i of subsets of a set S, some of which may be empty, is called a *partition* of S if $\cup S_i = S$ and $S_i \cap S_j = \phi$, the empty set, for $i \neq j$.

Theorem A.1: A digraph D is a line digraph if for some m there exist two partitions $\{A_i\}$ and $\{B_i\}$ of $N(D)$ into m subsets s.t.

$$A(D)=\bigcup_{i=1}^{m} A_i \times B_i$$

The theorem is illustrated in Fig. A.1, which is a redrawing of the digraph of Fig. 1.6 (exclusive of nodes s and t) with the partitions A_i and B_i specified. It is worth remarking that the definition, as well as the characterization, of a digraph do not exclude the presence of cycles in the graph.

A different characterization of line digraphs can be given by the following theorem, which combines the results of Richards (1967) and Geller and

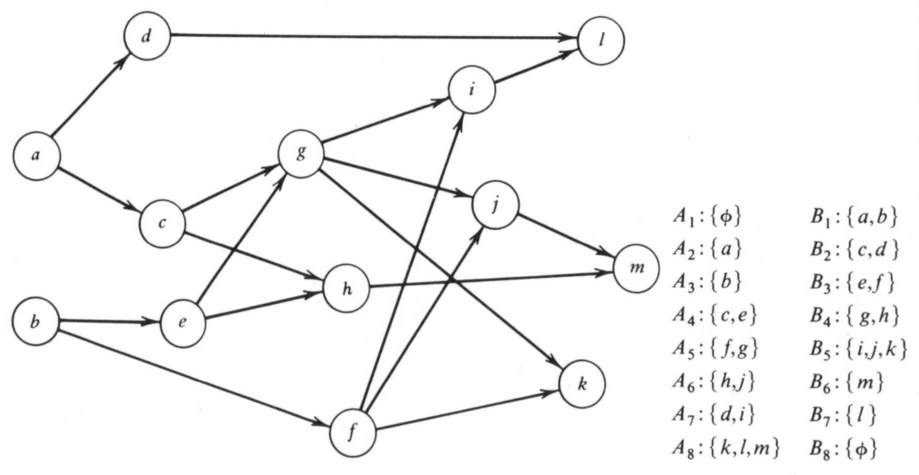

Figure A.1. Example of partitions of Theorem A.1:

Harary (1968). It is the basis for the condition previously made (p. 61) concerning a Boolean matrix (i.e., with entries 0 and 1) possessing a digraph. Furthermore, any digraph possesses a line digraph, by the construction already outlined.

Theorem A.2: A square Boolean matrix M possesses a line digraph iff any two columns of M (or any two rows of M) are either identical or orthogonal. Equivalently, suppose that the matrix M possesses a digraph D (hence, M must have zero diagonal). Then D is a line graph iff for any two points u and v, whenever there is a point adjacent to both u and v, then for all points w, w is adjacent to u if it is adjacent to v.

The second version of the theorem is really a statement concerning the columns of M, since it simply states that columns u and v must be identical if there exists one other node w adjacent to both points u and v. Of course a similar statement can be made about the *rows* of M, and would read as follows:

Corollary. D is a line digraph iff for any two points u and v, whenever there is a point *adjacent from* both u and v, then for all points w, w is adjacent from u if it is adjacent from v.

Theorem A.2 and its corollary are illustrated by the adjacency matrix of Fig. 1.7. An immediate consequence of such characterizations is to exclude from the class of line graphs those which contain the digraph shown in Fig. A.2. It is easy to verify that such a digraph violates the conditions of both Theorems A.1 and A.2. In other words, it is not possible to define partitions (A_i) and (B_i) such that the arcs of the digraph of Fig. A.2 are given by $\cup_i A_i \times B_i$. Furthermore, it is obvious that the columns (or rows) of the corresponding Boolean matrix are neither identical nor orthogonal.

It is illuminating to understand why the digraph of Fig. A.2 is *not* a line digraph. The reason is that we defined a line digraph in terms of a "parent" digraph. It is easy to verify that it is not possible to construct digraph E whose $L(E)=D$ is given by Fig. A.2. In fact, interpreting the points u_1, u_2, v_1, v_2 of the digraph of Fig. A.2 as activities, it is not possible to construct a digraph of activities-on-arcs representing the precedence relationship of Fig. A.2 without adding a dummy activity x, as shown in Fig. A.3a. With this modification, the corresponding line digraph is as shown in Fig. A.3b, which differs from the digraph of Fig. A.2 in the node x. The reader can easily verify that the digraph of A.3b is indeed a line digraph satisfying the conditions of Theorems A.1 and A.2. Thus, the

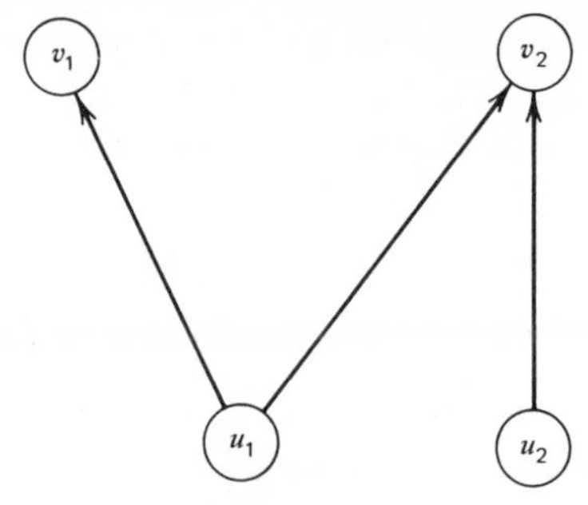

Figure A.2.

modification of a digraph to correspond to the conditions of the theorems is not a difficult task.

In essence, the characterization of digraphs and line digraphs is an abstraction, and generalization, of the necessary and sufficient conditions that guarantee the integrity of the graphic representation of precedence relations. Here, the reader is reminded of the discussion concerning Fig. 1.5*a* and the fact that it is not a true representation of the precedence relations: $a \prec c$, $b \prec c$, and $b \prec d$. The need to introduce the dummy activity x in Fig. 1.5*b* corresponds precisely to the introduction of node x in Fig. A.3*a*.

Necessarily, an important question prior to proceeding with the analysis of activity networks (in either modes of representation) is concerned with the verification that the given arrow diagram is, indeed, an acyclic digraph or a line digraph. Such a question usually resolves itself into the verification of the structure of the corresponding M matrix. In this respect, the reader is referred to the discussion on p. 16.

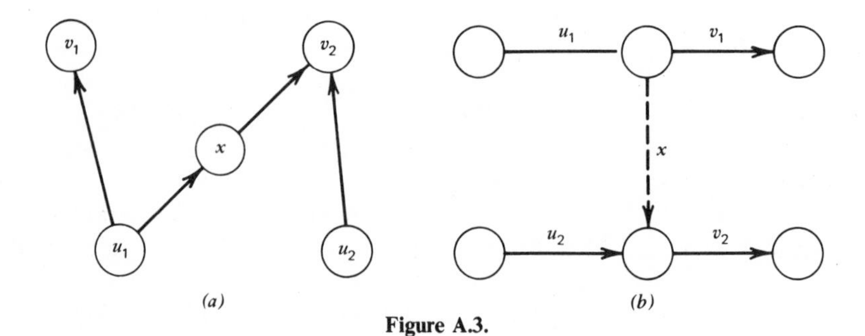

Figure A.3.

PROBLEMS

1. In an arbitrary network of N nodes and A arcs, what is the largest number of arcs in any path from origin to terminal? Prove your statement.
2. What are the three uses of dummy activities and dummy events in the A-on-A representation? Illustrate the uses by examples.
3. Construct a tabular format for the determination of the CP, the node slacks, and the activity floats. (*Hint*: Use Eq. (1.2) and the adjacency matrix in the case of A-on-A representation.) Apply it to the project of Problem 4.
4. a. Construct the activity network for the following "hospital moving" project. (Note that the activities may need renumbering to conform to the requirement of ascending node number of the arrowhead.)

Activity	Precedes	Duration
1. Meet with department heads	9, 11, 12, 13, 14, 15, 16	2
2. Consult with director	3	1
3. Tag furniture to be moved	9, 11, 12, 13, 14, 15, 16	1
4. Appoint move advisory committee	5	8
5. Establish move admission policies	6, 19, 21	10
6. Revise admission policies	32, 40, 41, 42, 43, 44, 45	10
7. Modify disaster plans for new hospital	31	10
8. Plan preliminary public relations activities	33, 46, 47, 48, 49	10
9. Meet with city traffic engineer	19, 21	5
10. Have preliminary meeting with police officials	17	5
11. Consult with department heads	19, 21	15
12. Negotiate contract with outside firms on specialized equipment	19, 21	20
13. Negotiate preliminary contract on ambulance buses	19, 21	16
14. Develop preliminary move plan	19, 21	20
15. Develop preliminary contacts with other hospitals	19, 21	17
16. Develop preliminary considerations of security needs for both old and new hospitals	19, 21	12

	Activity	Precedes	Duration
17.	Coordinate with police officials and develop policy move assistance plan	57, 58, 59	170
18.	Plan employee orientation at new hospital	54	55
19.	Develop final move plan	20, 23, 24, 29, 63	20
20.	Consult with building coordinator and contractor	62	20
21.	Establish packing material supply center	22	20
22.	Stock packing materials center	30	25
23.	Prepare final move tags	28	28
24.	Plan and arrange shuttle bus service between old and new hospital	35, 37, 38, 39, 40, 41, 42, 43, 44, 45, 52	54
25.	Finalize contacts with outside firms	35, 37, 38, 39, 40, 41, 42, 43, 44, 45, 52	33
26.	Publish plans	27, 34	3
27.	Modify plans	35, 37, 38, 39, 40, 41, 42, 43, 44, 45, 52	28
28.	Tag furniture & equipment	35, 37, 38, 39, 40, 41, 42, 43, 44, 45, 52	33
29.	Plan security for open houses & old hospital after move	35, 37, 38, 39, 40, 41, 42, 43, 44, 45, 52	38
30.	Plan premove activity and equipment	35, 37, 38, 39, 40, 41, 42, 43, 44, 45, 52	40
31.	Publish new disaster plan	55, 56	68
32.	Implement premove admissions policies	55, 56	28
33.	Hold reception for press Lay Board as hosts	20, 23, 24, 29, 63	1
34.	Construct cleaning station	37	22
35.	Establish parking plans	36	17
36.	Implement street & parking plans	—	28
37.	Clean equipment on arrival	—	37
38.	Clean in new hospital	55, 56	28
39.	Serve lunches in new hospital	55, 56	18
40.	Make final arrangements with other hospitals	55, 56	30
41.	Finalize ambulance arrangements	55, 56	30
42.	Contact army regarding ambulance buses	55, 56	30

	Activity	Precedes	Duration
43.	Move equipment, furnishings, and functions	—	56
44.	Prepare for patient move	55, 56	35
45.	Operate shuttle bus	57, 58, 59	28
46.	Prepare and distribute move newsletter to employees and medical staff	55, 56	85
47.	Plan open houses	50	60
48.	Prepare move fact book for employees	50	52
49.	Plan dedication of new hospital	53	75
50.	Train guides (guild volunteers)	51	10
51.	Hold open house	55, 56	15
52.	Provide for security as required	55, 56	15
53.	Plan dedication	55, 56	10
54.	Hold employee orientations at new hospital	55, 56	40
55.	Arrange for public relations to assist press with coverage	57, 58, 59	1
56.	Prepare for patient move	58	1
57.	Implement post-move admissions policies	—	8
58.	Provide for security at old hospital	—	12
59.	Move remaining equipment	60	10
60.	Prepare for sale of surplus material to employees	61	2
61.	Dispose of unsellable items	—	2
62.	Decide on move day	25, 26	10
63.	Publish instructions	30	20

b. Check for the absence of cycles.

c. Determine the earliest completion time of the project and the corresponding critical path(s).

d. Determine the four types of activity slacks for all activities and comment briefly.

5. Draw the network diagram of the following set of activities, which represent the design and implementation of software for a computer-based management information system. It is suggested that you draw an activity-on-node representation. Then determine the critical path and the *second* CP. Comment briefly on the "bottleneck" activities to be shortened to reduce the project duration to that of the second CP duration.

Activity Number	Immediate Precedors	Duration (days)
1	—	4
2	1	2
3	2	3
4	2	6
5	2	5
6	1	1
7	6	2
8	6	3
9	3,8	4
10	2,6,8	3
11	10	1
12	9	4
13	4,9	4
14	3	7
15	5	8
16	3	6
17	2	7
18	5	3
19	12,13	2
20	14,15	4
21	15	12
22	16	5
23	6	9
24	8	6
25	6	9
26	7	8
27	7	9
28	16,17,23	9
29	18	8
30	24,25	6
31	19,20,21	8
32	21	4
33	22	13
34	31,32	10
35	32	3
36	28,33	13
37	34,35	7
38	26,27	8
39	25	12
40	28,39	5
41	39	4
42	36	8
43	41	8
44	40,41	11

Activity Number	Immediate Precedors	Duration (days)
45	43	6
46	34, 35, 43	6
47	45, 46	3
48	44	7
49	10	14
50	11	3
51	11	7
52	50	10
53	52	9
54	52	6
55	51, 52	10
56	52	3
57	53	6
58	30, 38, 53	13
59	54, 55	6
60	56	10
61	57	4
62	57	3
63	61	3
64	61	8
65	62	8
66	58, 62	12
67	49, 59, 60, 61	7
68	49, 59, 60, 61	6
69	66	7
70	47, 63, 64, 67, 68	13
71	37, 42, 48, 65, 69	5
72	37, 42, 48, 65, 69	5
73	70, 71, 72	4

6. To any directed graph, whether cyclic or acyclic, there correspond three basic matrices:
   a. the *adjacency* matrix μ (given in the text): both row and column headings are the nodes 1 to n, and an entry $+1$ in position ij means that an arc leads from node i to node j; otherwise there is a 0 entry.
   b. The *incidence* (or vertex) matrix η: the row headings are the nodes 1 to n; and the column headings are the arcs of the graph; the entry is $+1$ if the arrow is incident *from* that node, it is -1 if the arrow is incident *on* that node, and it is 0 otherwise.
   c. The *circuit*† matrix γ: the row headings are the circuits (indepen-

†We define a circuit as a connected graph with every node of degree 2 (i.e., each node has two arcs incident on it).

dent of arrow direction), and the column headings are the arcs of the graph. An entry is $+1$ if the direction of the arrow coincides with the orientation of the circuit (which is arbitrary); it is -1 if it is opposite to the orientation; and it is 0 if the arrow does not lie on the circuit.

You are asked to construct all three matrices to the following AN and to verify the following properties:

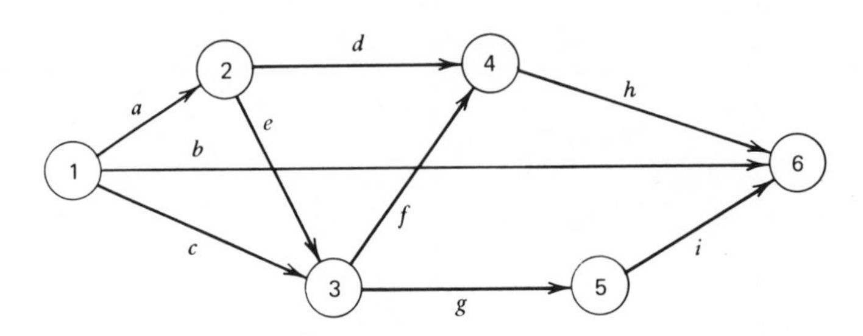

i. The rank of the incidence matrix η is $N-1$, where N is the number of nodes.

ii. If T is a tree of a connected directed graph G, then the $N-1$ columns of the incidence matrix corresponding to T constitute a nonsingular submatrix of the incidence matrix of G.

iii. The rank of the circuit matrix γ of a connected directed graph G of A arcs and N nodes is $A-N+1$.

iv. If the columns of the incidence matrix and the circuit matrix are arranged in the same arc order, then $\eta\gamma'=0$, where the prime denotes transpose.

7. Other texts emphasize the CP, which is the longest path in the network.

   Devise a procedure for determining the kth critical path (i.e., the kth ordered path), for $k=2,3,\ldots$. Apply this expression to the following AN to determine the second and third CPs. (The duration of each activity is indicated on the arc). To what use can you put this additional information?

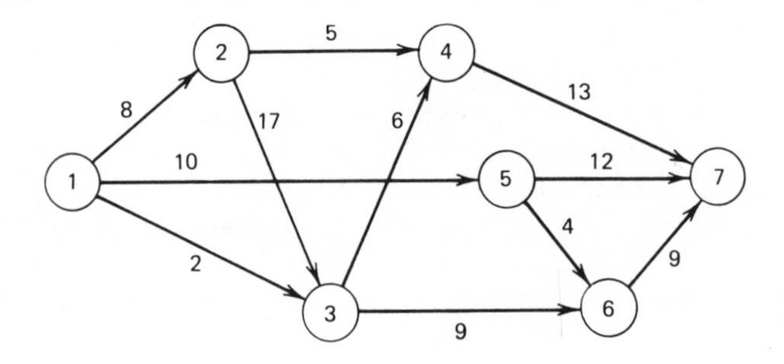

8. Discuss the LP model of Section 2.1 relative to the determination of: (a) all CPs; (b) $t_i(E)$ and $t_i(L)$ for all nodes of the network, (c) node slacks as well as activity floats, and (d) the kth CP.
9. In the network shown below, determine the second and third CPs, that is, the second and third longest paths from nodes *1* to *10*. Based on these three paths, which activities would you consider as "critical"?

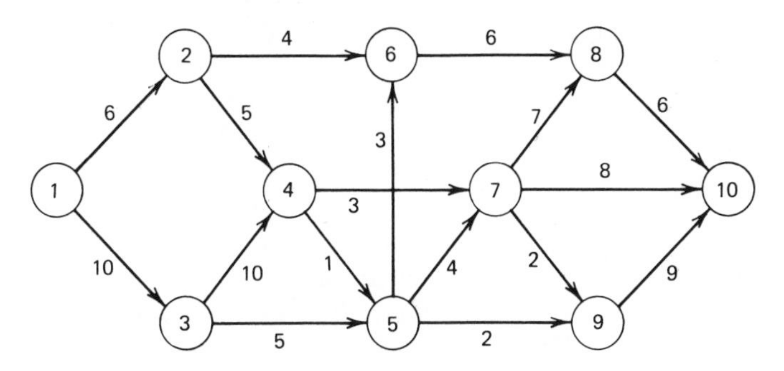

10. Consider the following AN (A-on-A representation):

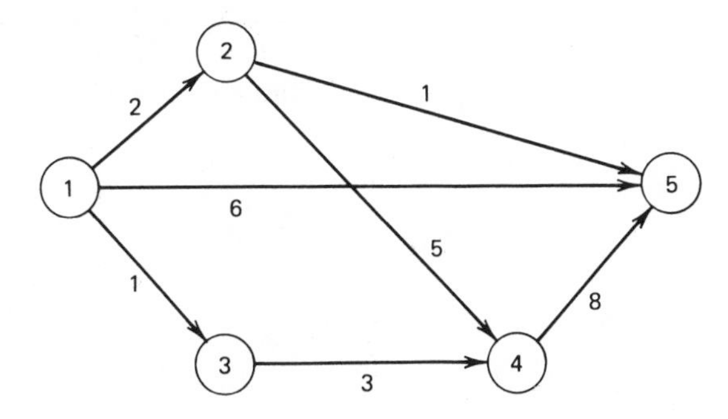

Determine all node slacks and all activity floats, and intepret $S_{1,3}^{(1)}$ and $S_{3,4}^{(1)}$. Can they be treated independently? Why?

11. Interpret the various measures of activity float discussed in Section 2, in particuular, relative to the impact of delaying the start of the activity by the total amount of the float (any one of the four values) on the preceding as well as succeeding activities.

    Is there any (algebraic) relationship between the activity floats and the node slacks? Substantiate your answers, either way.
12. Our discussion of "activity floats" was couched in terms of the A-on-A mode of representation. Several analysts (including this author) prefer the A-on-N mode of representation. But then one must use different entities to measure the floats since the end points of the activity are

now suppressed and the values $t_i(E)$, $t_i(L)$, $t_j(E)$, and $t_j(L)$ are no longer available. In their place we usually define the four entities: activity earliest start, ES; latest start, LS; earliest finish, $EF = ES + y$; and, finally, latest finish, $LF = LS + y$.

Let activity j be (immediately) preceded by activities $i_1, i_2, \ldots, i_m$, and be (immediately) succeeded by activities $k_1, k_2, \ldots, k_n$. Derive an expression for each of the four floats of activity j based on the values of ES, LS, EF, and LF of its preceding and succeeding nodes. Apply your formulas to derive the floats of each activity in the following A-on-N network. The duration of each activity is written next to the activity.

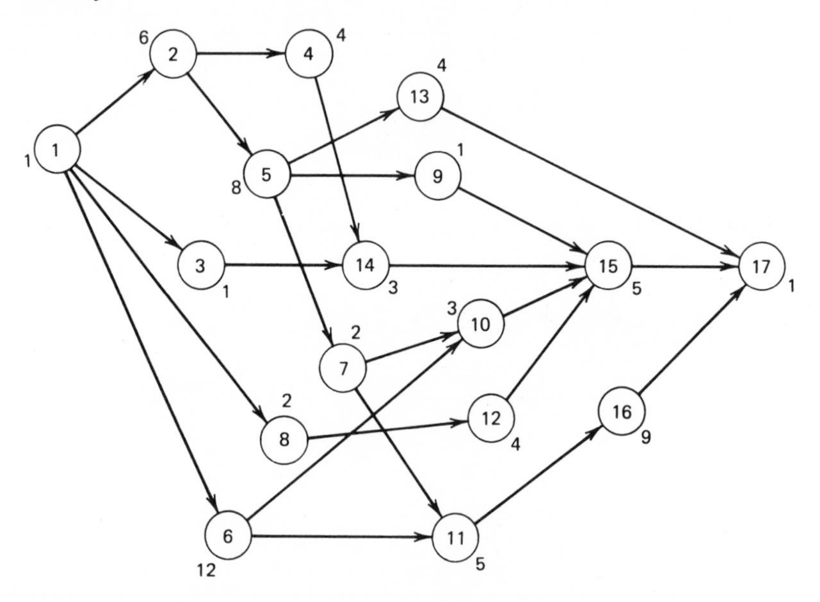

After performing these calculations translate the network into A-on-A representation and verify the values of the floats obtained.

13. The following are estimates (in weeks) of activity times. Opposite each activity are given three numbers that represent the optimistic, most likely, and pessimistic estimates, respectively.
    a. Draw the PERT network.
    b. Determine the CP.
    c. Determine the probability of completing the project five weeks earlier than the duration of the CP obtained in (b). Interpret such probability.
    d. Suggest a general algorithm for determining the kth CP in directed networks with no loops, such as in PERT. (*Hint*: Use either a combinatorial or Dynamic Programming approach.)

e. Determine the sequence of activities that constitute the second and third CPs, that is, the paths with the *second* and *third* longest duration.

f. Based on the first, second, and third CPs determined in (e), determine the four most critical activities. Explain your reasons for choosing these four activities.

Activity	Times	Activity	Times	Activity	Times
(1, 2)	3, 4, 8	(6,16)	1, 3, 8	(11,16)	1, 4, 8
(1, 3)	1, 2, 3	(6,20)	0, 3,10	(11,19)	1, 3,10
(1, 6)	0, 2,10	(7, 8)	1, 2, 3	(12,20)	2, 4, 5
(1, 8)	3, 3, 3	(7,11)	4,10,21	(13,17)	5,11,22
(2, 3)	6,15,16	(7,12)	3, 9,16	(13,18)	4,10,17
(3, 4)	5,10,20	(7,15)	4, 9,19	(14,15)	5,11,21
(3, 9)	3, 7,10	(8, 9)	2, 6, 9	(14,19)	3, 7,10
(3,16)	1, 8,14	(8,18)	1, 7,13	(15,17)	2, 8,15
(4, 7)	1, 3, 8	(9,10)	1, 2, 7	(15,18)	2, 5, 6
(4,11)	10,11,14	(9,13)	9,11,15	(15,19)	8,11,12
(4,16)	2, 5, 9	(9,19)	1, 4,10	(15,20)	0, 3,11
(5, 7)	10,12,17	(10,12)	8,10,15	(16,19)	7,11,17
(5,10)	2, 6, 8	(10,13)	1, 5, 7	(17,20)	2, 6, 8
(6, 7)	3, 4, 5	(10,20)	1, 3, 4	(18,19)	1, 5, 9
(6, 8)	2, 6, 8	(11,13)	2, 6, 8	(19,20)	1, 5,10

14. Consider the earliest completion time of a project $t_n(E)$, as determined by the length of the CP when the duration of each activity is taken to be the *average* duration [Eq. (1.9)]. Is $t_n(E)$ unbiased, optimistic, or pessimistic relative to the true average duration of the project?

15. The first assumption of the "classical" PERT approach is that the activities are independent.
    a. How valid is this assumption in practice?
    b. Demonstrate, using the following network, that even if the individual activities may be considered independent, the paths from node *1* to node *n* need not be independent (assume all activities to be exponentially distributed with the same mean μ).
    c. What is the impact of the conclusion in (b) on the PERT model?

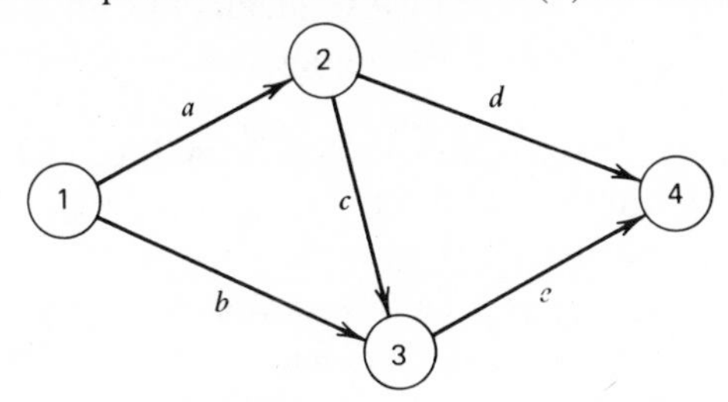

16. Consider the AN of Fig. 1.13a and assume that all activities possess the negative-exponential probability density function with means equal to the values indicated on the arc [e.g., activity (1,3) has a duration $Y_{1,3}$ whose density function is given by $0.1e^{-0.1t}$, $t \geqslant 0$,].

    Determine node numbers, $\{w_i\} \geqslant 0$, such that the probability of realization of each node i by time w_i is at least 0.95.

17. What is the sensitivity of the values $\{w_i\}$ obtained in the chance-constrained LP of Eq. (1.10) to the assumed PDF's of the activity durations $\{Y_u\}$?

    What is the sensitivity of the values $\{w_i\}$ to the stated probabilities $\{q_u\}$? (Note that all the probabilities need not be equal.) Alternatively, one may ask whether it would make a "great deal of difference" if the q_u values are low for all but the very last few activities.

    You are asked to respond to these two questions, which may be of interest from a research point of view. You may assume any "penalty function" you desire that is meaningful in this context.

18. Consider the small network of Fig. 1.13a, repeated below:

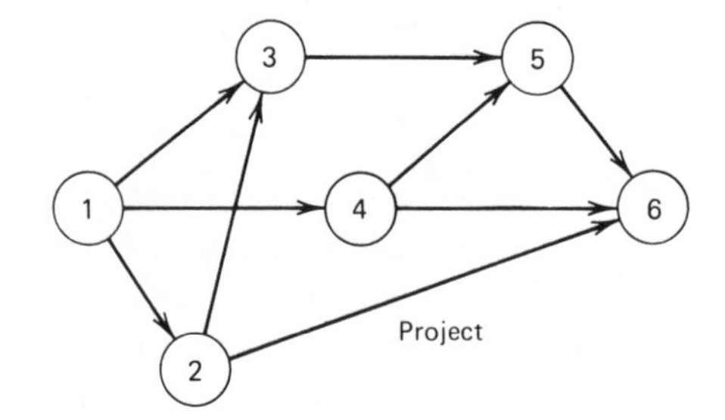

Project

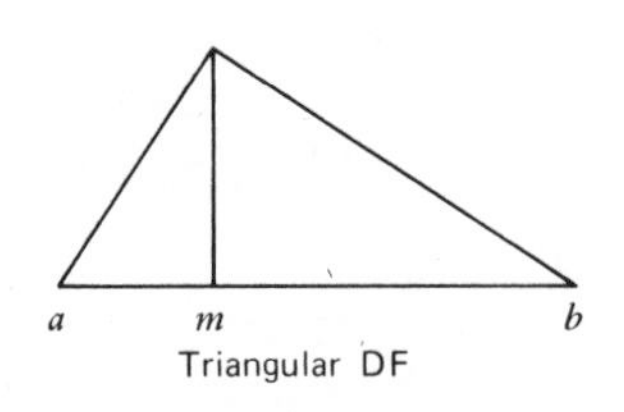

Triangular DF

    Suppose that all activity durations are "triangularly" distributed as shown above, with $a=5$, $m=9$, and $b=20$. It is desired to determine the node numbers $\{w_i\}$ so that $w_j - w_i$ will embrace the duration y_{ij} at least 95% of the time [i.e., $q_{ij}=0.95$ in constraints (1.11) for all (ij)].

19. Consider the network of Fig. 1.13a and assume that all activity durations possess the rectangular distribution function $f(x)=1/10$ for $1 \leqslant x \leqslant 11$.

    a. Following the "classical" PERT calculations [of Eq. (1.9)], estimate the time of realization of each node assuming a confidence level of 0.90.

    b. Determine the "node numbers", (w_i), of Eqs. (1.11) that yield the same confidence level of 0.90.

    c. Compare the results in (a) and (b).

20. Diagram the four lag restrictions exhibited in Fig. 1.8 (except the SF-lag restriction) in a manner similar to Fig. 1.9, in which you exhibit

all possible overlapping conditions. In each case, verify the expressions for the durations given in Fig. 1.8 and designate the various subactivities.

In each lag specification it is easy to recognize that for certain values of the parameters involved it would be more rational to reinterpret the lag as a different restriction and that such action would reduce the complexity of the expressions of the various subactivity durations. As an example, consider the SS lag; it is evident that if the SS lag has duration greater than y_u, then it is more appropriate to specify the restriction as an FS lag instead, as shown.

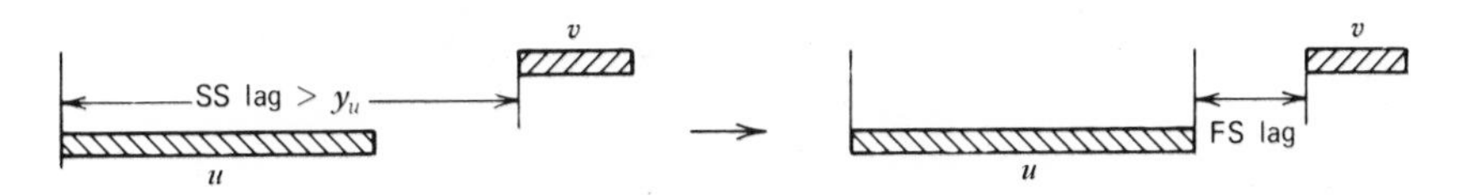

As a consequence, if the SS-lag is *always bounded by* y_u, then the expressions for the various subactivity durations are simplified to:

$$y_{u_1} = \min\{y_u; L_{SS}\}; \qquad L_{SS} \geqslant \text{SS lag}$$

$$y_{u_2} = y_u - y_{u_1}$$

$$y_{v_2} = \max\{0; \min(y_v, L_{SS} + y_v - y_u)\}$$

$$y_{v_1} = y_v - y_{v_2}$$

You are now asked to scrutinize the other four lag conditions, in terms of redefining the lag under certain conditions, and thereby simplifying the subactivity durations shown on Fig. 1.8.

21. You are asked to give the A-on-A representation of the various lag restrictions of Fig. 1.8. Compare with the A-on-N representation and comment briefly.
22. An important consideration in the analysis of ANs with lag restrictions, such as given in Fig. 1.8, is to verify the validity of these restrictions, especially relative to circularity (i.e., the existence of loops) and to infeasibility (i.e., impossibility of performing the activities in the time allowed with the given restrictions). Devise an algorithm to perform the requisite tests.
23. The specification of the "lags" (see Fig. 1.8) was stated in the context of DANs. The presence of an equivalent concept in PANs of the PERT model (i.e., in which the durations of the activities are RVs) has not been adequately investigated.

 By the very nature of the randomness in activity duration it is obvious that no restrictions can be imposed that entail the finish time of the activities. This leaves the start-to-start lag (SS lag) as the only valid restriction on the overlapping of activities.

Analyze the impact of this SS lag restriction on the two activities concerned, especially relative to the (possible?) division of the activities into subactivities, and on the durations of these subactivities.

24. For the following A-on-A network with specified "lags" between selected pairs of activities, determine the various temporal parameters of common interest (i.e., the CP, the activity floats, and the node slacks). The duration of each activity is indicated on the arc.

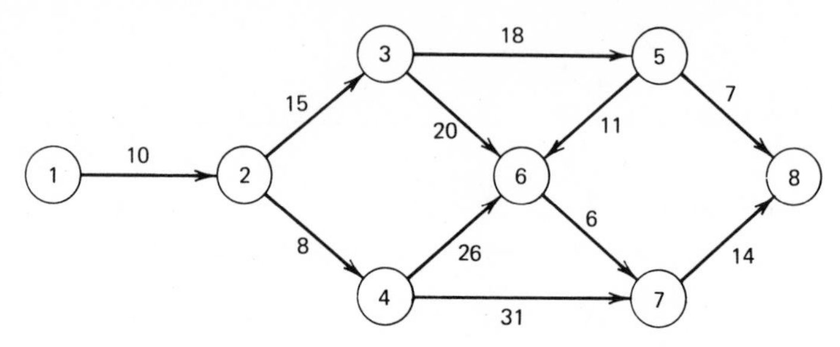

The "lags" are as follows:

Activity Pairs	Type of Lag	Duration of Lag
(1,2)–(2,3)	SS	5
(1,2)–(2,4)	FF	10
(3,5)–(5,6)	SF	8
(3,6)–(5,8)	FS	25

Compare the duration of the CP under these lag restrictions with the duration of the CP if the project were free of them. Devise a visual presentation scheme that summarizes your findings to management (who are assumed to be no "experts" in the field).

25. In Section 3.1 we argued that there exist two limit cost functions, (see Fig. 1.16), one corresponding to the activities started at their earliest start times, and the other corresponding to the activities started at their latest start times. Between these two bounds there exist infinitely many cost functions, corresponding to the infinitely many possible start times of activities with float.

It has been argued by some practitioners that there should be an "ideal" cost function to reflect the optimal expenditure of effort and money over the life of the project, and against which actual expenditures and work value are measured.

Discuss how you would determine an ideal cost function if you were asked to do so, and what type of control mechanisms you would impose on it.

26. The emphasis of the discussion of Section 3 is on cost. Yet, it is evident that the contractor is paid for his services not only to cover his costs but also his profit; thus, there are two streams of cash flow, namely, the expenditures and the income.

 It is possible to determine the expenditures on the basis of some "ideal" scheme as discussed in Problem 25. On the other hand, income is usually tied in to the completion of certain critical phases of the project. You may take it to be coincident with the time of realization of certain events in the project.

 Discuss the roles of the "ideal budget" and the "streams of payment" in bidding for, and control of, a project.

27. Read Refs. [Richardson, Butler, and Coup (1972) and Turban (1968)] on the "line of balance" approach and report on its merits and demerits, especially in comparison to the *PERT-Cost* procedure.

REFERENCES

Agard and Gamot, 1966.
Alsaker, 1962.
Antill and Woodhead, 1965.
Avots, 1962.
Battersby, 1967.
Becker, 1969.
Beckwith, 1962.
Berman, 1961.
Bildson and Gillespie, 1962.
Bird, Clayton, and Moore, 1973.
Boehm, 1962.
Boverie, 1963.
Burgess and Killebrew, 1962.
Carré, 1964.
Charnes, Cooper, and Thompson, 1964.
——— and ———, 1962.
Chaudhuri, 1968.
———, 1969.
Clark, 1954.
Codier, 1962.
Croft, 1970.
Crowston, 1971.
Davis, 1972.
———, 1973.
Digman, 1967.
Dimsdale, 1963.
DOD and NASA Guide, 1962.
Dogrusoz, 1961.
Dooley, 1964.
Fisher and Nemhauser, 1967.
Francis, 1962.
Freeman, 1960a.
Geller and Harary, 1968.
Gessford, 1966.
Goldberg, 1964.
Graham, 1965.
Gutsch, 1969.
Harary and Norman, 1961.
Hill, 1966.
Howard, 1971.
Jenett, 1969.
Joyce, 1974.
Kelley, 1961.
———, 1963.
Klass, 1960.
Levitt, 1968.
Levy, Thompson, and Wiest, 1963.
Lyden and Miller, 1967.

McNeill, 1964.
Malcolm, 1963.
———, Roseboom, Clark, and Fazar, 1959.
Marimont, 1959.
Martino, 1963.
———, 1965.
Mauchly, 1962.
Meyer, 1972.
Miller, 1962.
Minieka, 1975.
Minty, 1957.
Nevill and Fakoner, 1962.
Novick, 1965.
Paige, 1963.
Ponce-Campos and Kedia, 1976.
Richards, 1967.
Richardson, Butler, and Coup, 1972.
Robinson, 1965.
Roper, 1964.
Roy and Sussmann, 1964.
Russell, 1970
Schick and Maybell, 1968.
Schoderbek, 1966.
——— and Digman, 1970.
Schultis, 1962.
Sherrard and Mehlick, 1972.
Shier, 1974 *a* and *b*.
Soistman, 1962, 1965, 1966.
Thomas, 1969.
Thompson, 1966.
Turban, 1968.
Voraz, 1966.
Wilson, 1964.

2

MODELS OF TIME COST TRADE-OFFS IN DETERMINISTIC ACTIVITY NETWORKS

§1. PROLOGUE

Our concern in Chapter 1 was with temporal *analysis*: given the structure of the network and the activity durations, determine the earliest and latest times of realization of each event, the slack of events, the float of activities, the first, second,..., kth critical paths, and so on. Clearly, still within the context of temporal considerations there are important (from a managerial point of view) and interesting (from an analytic point of view) questions of *synthesis* that relate, in the main, to trade-offs between durations and cost. In other words, our concern is with the very meaningful questions of how long "should" an activity take to achieve certain objectives. The rationale for posing such a question is as follows.

It is commonly recognized that the majority of activities encountered in the real-life applications of projects can be accomplished in shorter or longer durations by increasing or decreasing the resources available to them, such as facilities, manpower, machinery, funds, and so on. For instance, if a hospital is urgently needed, the construction crew may be placed on overtime, a two-shift operation instituted, more mechanized equipment acquired, or some components purchased ready-made instead of on site construction (e.g., prestressed concrete slabs). Naturally, such acceleration of operation entails additional costs in the accomplishment of the activities. Such action would be rational only if these additional costs are more than offset by the anticipated "gains" from the completion of the project at an earlier date. Conversely, a more leisurely pace implies reduced costs of the activities but probably a higher "penalty" for late termination of the project. The determination of the "gains" from the earlier completion of the project is often a straightforward calculation—such as in the case of constructing a factory to satisfy a known demand. On the other hand, the determination of such "gains" or "penalties" may be quite a knotty problem involving hard-to-measure social or national values and objectives—such as in the early completion of a hospital or a defense system. In this case, one speaks of social benefits instead of monetary gains and losses, and of cost/benefit analysis.

In any event, it is meaningful to inquire into the utility of such trade-off between duration and cost. (In Problem 11 at the end of the present chapter we ask the reader to establish a criterion function for a project for the construction of a sanitarium.)

Our analysis gives form to such inquiry. But before delving into the mathematical formalism and the manipulation of the various forms that such a problem may assume, it is suggested that the reader take a "global" view of what is being discussed in order to be able to better assess the domains of applicability of the subsequent sections, as well as the avenues of research needed to further sharpen the available technology.

Throughout our discussions it is assumed that the complex interaction between the required resources and duration of the activity *can be summarized in a single functional relationship between cost and duration.* In other words, it is assumed that cost reflects the aggregate consumption of the requisite resources *in the most efficient manner* to complete the activity in the specified time. Or, putting it still differently, the individual resources lose their identity—for a specified duration y there exists an optimal (or near optimal) allocation of the various resources (the so-called factors of production in economic theory, composed of the various skills of labor, the various tools, machinery and equipment, capital, land, etc.) that yields the cost c. The determination of this optimal combination of the resources at various activity durations is *not* the subject of discussion—*it is assumed known*, as is its cost.

Herein lies what may be the most crucial assumption of all the OR technology that is applied to this problem.

From a managerial point of view, this assumption may be "reasonable" or "unreasonable" depending on the activity and on management's past experience. Lower and middle management with a good deal of experience in the activities within their jurisdiction have an uncanny feeling for the "right" mixture of the various resources and the time it takes to accomplish an activity using that mixture. However, if such experience is lacking, or if the environment of the activity precludes the a priori knowledge of the "right" mixture of resources and its cost (e.g., the case of unstable resource markets), the type of analysis suggested below would have conclusions that are suggestive of relative magnitudes rather than of precise optimal values.

The saving grace in what may seem at first blush as "demanding too much" from the practitioner is that the precise determination of parameters is not needed—only approximations to relative magnitudes and functional relations. It is the duty of the technically competent analyst to perform the various postoptimum and other studies that the current methodologies permit him to do in order to guide the search for the best mode of operation.

Given the full understanding of the implications of this basic premise, one may wonder what managerial questions are asked. There are several, for instance:

1. How much time should be allowed each activity to complete the project at a certain target date and with minimum cost?
2. Given a fixed amount of money to spend on the project, what is the shortest possible duration achievable?

3. For any given duration of the project, what is the marginal cost of reduction in that duration?
4. Are the same activities the "bottleneck" activities at all durations of the project—or are there different sets at different durations?

As may be suspected, these questions are not unrelated. This chapter is devoted to detailed discussion of the models that attempt to resolve these, and other, questions under varying assumptions.

In a more technical vein, the assumption of a time–cost trade-off is translated into a function $c_a = \phi(y_a)$, where y_a is the duration of the activity a and c_a is its cost. The approach to the resolution of the issues raised above is dependent on the form of the function ϕ. The following sections deal with these issues under different assumptions on ϕ: Sections 2 and 3 assume a linear function, Section 4 a convex function, Section 5 a concave function, and Section 6 a discontinuous nonincreasing function.

The key to answering all these questions is often to embed the problem in a more general setting, which essentially asks for the minimal total cost of the project *for all possible durations*. When this is not a (computationally) feasible task, then the managerial questions stated above are, indeed, answered piecemeal at a considerable computing cost.

One final remark is in order. Sections 2–6 deal with exact approaches to achieve the optimum answers under varying assumptions on the form of the function ϕ that relates cost to duration. The reader may wonder whether we are placing undue emphasis on achieving "the optimum," especially in view of the apparent extensive computations that are needed to achieve such optimum and whether there are no simpler and computationally faster and more expedient approaches that, although not achieving the optimum, still realize an acceptable closeness to it.

In a text aspiring to provide the more recent techniques for resolving these knotty problems, there is no substitute to "telling the story as it is," which necessitates treating the analytical approaches in their full detail. More importantly, the reader who is armed with such knowledge will then be able to devise his own heuristics to approximate any step, or steps, of the more rigorous procedures. In addition, he will be able to test the "goodness" of his or anybody's approximation.

The answer to the second question is a definite "yes." It is a common fact that almost everybody who has dealt with the problems of project time–cost trade-offs in a practical context has devised his own approximate procedure. The literature on project planning by network methods is replete with such examples. See the works by [Moder and Phillips (1964), Wiest and Levy (1969), Siemens (1971), and Goyal (1975)]. From a pedagogical point of view we prefer to leave the construction of such procedures to the practitioner, after having provided him with the more advanced approaches of analysis. This is the more appropriate due to the observed fact that heuristic procedures are tailored to a particular context that violates, in one way or another, the assumptions of the models presented herein (otherwise one would apply the models directly). Also, the degree and extent of such violation varies from situation to situation and cannot be foreseen. In Problem 13 we ask the reader to devise such an approximate procedure under specific conditions.

§2. LINEAR COST-DURATION FUNCTION: A NETWORK FLOW ALGORITHM

The most elementary cost-duration relationship is, indeed, the linear cost function, which assumes that a unit *decrease* in activity duration entails an *increase* in its cost by a_{ij} units. Apart from its simplicity, linearity is also appealing from another point of view, namely, that linear functions (or piecewise linear functions) often provide excellent approximations to nonlinear functions (as we see in Section 5). Hence, knowledge of the optimum under the assumption of linearity is the fundamental building block in achieving bounds on the optimum under nonlinear cost functions (or in achieving the optimum itself). Finally, the assumption of linearity leads directly to a LP model, and thus places at the disposal of the analyst the rich techniques of that field of OR.

Under the assumption of linearity, all of the managerial questions raised in the previous section can be answered in one sweep, the reason being that we can construct the cost-duration curve for *all* realizable durations of the project [the bounds on such duration are defined in Eq. (2.3)]. Unfortunately, such an achievement is not attainable under more general cost structures, except at a considerable computing cost.

Suppose the network has A activities and n nodes, and assume that for any activity $a \in A$ the cost of the activity, denoted by c_a, is a linear function of its duration between a lower and an upper bound, l_a and u_a, respectively; in other words, assume

$$c_{ij} = b_{ij} - a_{ij}y_{ij} \qquad \text{for} \qquad l_{ij} \leqslant y_{ij} \leqslant u_{ij}, \qquad a_{ij}, b_{ij} \geqslant 0$$

where l_{ij} is interpreted as the shortest possible duration (the "crash" duration) and u_{ij} the time the activity would take if it is desired to minimize the cost of that individual activity (see Fig. 2.1). If t_j represents, as always, the realization time of event j, then the precedence constraints imply that

$$t_j \geqslant t_i + y_{ij} \qquad \text{for all } (ij) \in A, \qquad j = 2, 3, \ldots, n; \; t_1 = 0$$

which simply means that if event i precedes event j then the latter cannot be realized before y_{ij} time units after event i has been realized.

The problem of optimal activity durations to accomplish the project in a specified time T can be succinctly stated as follows. Determine activity durations $\{y_{ij}\}$ and node-realization times $\{t_j\}$ that

$$\text{minimize } z = \sum_{(ij) \in A} c_{ij} = \sum_{(ij) \in A} (b_{ij} - a_{ij} y_{ij}) \tag{2.1}$$

s.t.

$$\begin{aligned} t_i - t_j + y_{ij} &\leqslant 0, \; (ij) \in A \\ y_{ij} &\leqslant u_{ij} \\ y_{ij} &\geqslant l_{ij} \\ t_n &= T \end{aligned} \tag{2.2}$$

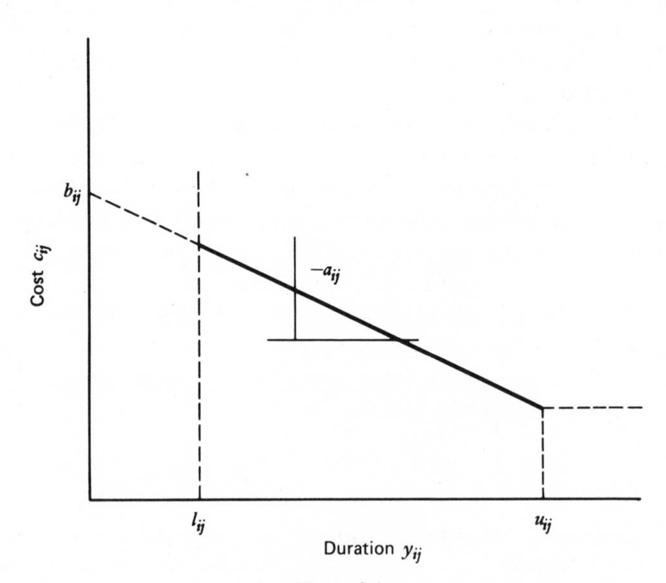

Figure 2.1.

The last equation defines the desired duration of the project as equal to $T > 0$. In particular, this is a parametric LP problem if T is permitted to range over its feasible region given by

$$\underline{T} \triangleq (t_n | y_{ij} = l_{ij}) \leqslant T \leqslant \bar{T} \triangleq (t_n | y_{ij} = u_{ij})$$

in which the lower bound, $\underline{T}$, signifies the length of the CP when all of the arcs of the network are at their lowest possible durations, and the upper bound, $\bar{T}$, signifies the length of the CP when all the arcs of the network are at their "normal" durations. Obviously, a duration $T < \underline{T}$ is infeasible, in the sense that it is impossible to realize, and a $T > \bar{T}$ is wasteful in the sense that the project is delayed with no economic advantage.

The lower and upper bounds on the activity durations lead immediately to the following theorem, which characterizes the node realization times as well as the activity durations.

Theorem 2.1: In an optimal schedule†, activities either do or do not possess float. Activities with no float form a "critical subgraph"; that is, a subgraph each path of which (from *1* to *n*) is a CP and hence of the same length. Activities with float are set at their maximum durations. These latter, together with any arborescence‡ of the critical subgraph, uniquely determine the earliest node realization times for all nodes.

The formal proof of this theorem is elementary and is left as an exercise to the reader. But its heuristic reasoning gives the key to its interpretation.

Consider any project duration T in the feasible range, $T \in [\underline{T}, \bar{T}]$, and consider the optimal activity durations which realize such a T. From the origin, node *1*, there must exist at lease one CP to the terminal node n (the so-called critical subgraph). Clearly, the activities along these CPs possess no slack, and their durations are $\leqslant u_a$, respectively, with *strict inequality* for some activities if $T < \bar{T}$. For nodes lying on these CPs, it is sufficient to have an arborescence in order to determine their earliest times of realization (by the definition of arborescence).

On the other hand, since the remaining activities do not fall on any CP and, in fact, possess some float, it is optimal to have their durations set equal to their respective upper limits. This completes the specification of all the activity durations of the network, and it is then an easy matter to

†Or, for that matter, in any schedule in which we assume that activities start at their earliest start times.

‡The word *arbre* in French means "tree." An arborescence from node s over n nodes (which include s) is a set of arcs $\mathscr{A}$ with the following properties: (a) $\mathscr{A}$ contains $n-1$ arcs, (b) $\mathscr{A}$ contains no cycles, and (c) $\mathscr{A}$ contains a unique directed path from node s to each node j; $j \neq s$.

determine the earliest times of realization of any nodes not lying on any CP.

We now turn our attention to the determination of the optimal activity durations of *those activities lying on the CPs,* assuming the above-designated cost structure.

As is well known from LP theory, the value z of the objective function [Eq. (2.1)] is a convex, piecewise linear function of T, as shown in Fig. 2.2. The optimal duration T^* corresponding to any fixed level of expenditure can be obtained directly from such a relationship. On the other hand, the optimal activity durations $\{y_{ij}^*\}$ for any fixed T are obtained from the solution of the LP of Eqs. (2.1)–(2.2).

The LP literature should be consulted for the techniques of solving such a parametric LP and generating the complete z–T curve. However, such an approach would be wasteful in computing effort and time; even a cursory glance at the LP of Eqs. (2.1)–(2.2) would reveal the very peculiar structure of the matrix of coefficients, which are 0, $+1$, or -1. Common sense and past experience indicate that before one resorts to general algorithms designed for arbitrary coefficients, such as parametric LP, one should first try to take advantage, as far as possible, of the special structure of the problem at hand.

Such special structure was capitalized on by Fulkerson, who developed a rather simple and elegant flow network algorithm for the determination of the complete time-cost (T–C) curve of Fig. 2.2.

For the sake of complete and full understanding of this clever algorithm the reader is urged at this juncture to familiarize himself with the basic concepts of flow network theory and primal–dual arguments. The reader is

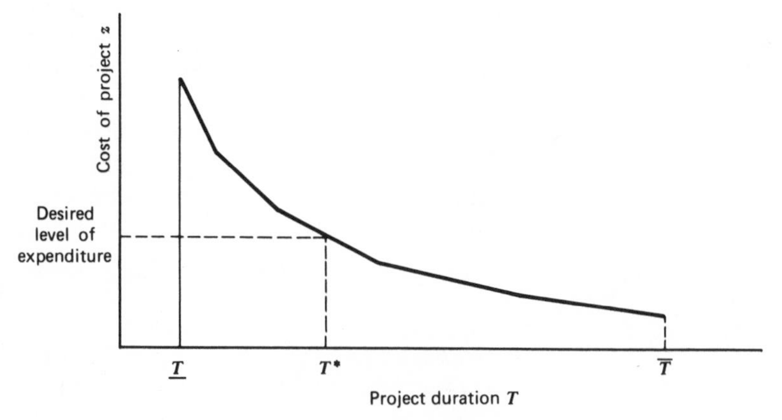

Figure 2.2. The z–T curve.

urged to consult the several excellent books on the subject, such as Hadley (1964), Dantzig (1963), Simonnard (1966), and Ford & Fulkerson (1962). To render this text self-contained in this respect, we include at the end of the chapter: Appendix B, which reviews the more elementary concepts in flow networks, especially the labeling procedure utilized herein, and Appendix C, which reviews the concepts of the primal–dual algorithm, especially as related to flow networks. In what follows we assume such knowledge.

First, we restate the parametric LP of Eqs. (2.1)–(2.2) in the following form:

$$\text{maximize } z(T) = \sum_{(ij)\in A} a_{ij} y_{ij} \tag{2.3}$$

s.t.

$$\begin{array}{rl} & \underline{\text{Dual Variables}} \\ t_i - t_j + y_{ij} \leq 0 & f_{ij} \\ -t_1 + t_n = T & v \\ y_{ij} \leq u_{ij} & g_{ij} \\ -y_{ij} \leq -l_{ij} & h_{ij} \end{array} \tag{2.4}$$

in which we also exhibited the dual variables. Then the dual LP is given by

$$\text{minimize } Tv + \sum_{(ij)} u_{ij} g_{ij} - \sum_{(ij)} l_{ij} h_{ij} \tag{2.5}$$

s.t.

$$f_{ij} + g_{ij} - h_{ij} = a_{ij}: \quad (ij) \in A \tag{2.6}$$

$$\sum_j \left[f_{ij} - f_{ji} \right] = \begin{cases} v, & i = 1 \\ 0, & i \neq 1, n \\ -v, & i = n \end{cases} \tag{2.7}$$

Here, all the variables are nonnegative, and equalities are imposed because the primal variables were not explicitly constrained in sign. Moreover, it is worth remarking that Eqs. (2.7) are the well-known "flow conservation constraints" when the dual variables (f_{ij}) are interpreted as the flows in the arcs.

Since the objective is to minimize Eq. (2.5), and considering the relation between g_{ij} and h_{ij} for any fixed f_{ij} expressed by Eq. (2.6), it immediately follows that *at the minimum*, either g_{ij} or h_{ij} must be equal to zero. In other words, it is not possible for both g_{ij} and h_{ij} to be simultaneously different from zero in the optimal solution to the dual LP. This fact can be seen with equal ease by reference to the primal LP and, in particular, to the last two sets of constraints of Eqs. (2.4). Considerations of duality and complementary slackness indicate that if, *at the optimal*, $g_{ij}>0$, then it must be true that $y_{ij}=u_{ij}$, hence $y_{ij}>l_{ij}$ which, in turn, indicates that h_{ij} must be $=0$. A similar argument applies if we assume that $h_{ij}>0$.†

Hence, we may assume from the outset that

$$g_{ij}=\max\left[0; a_{ij}-f_{ij}\right]$$

and (2.8)

$$h_{ij}=\max\left[0; f_{ij}-a_{ij}\right]$$

If we interpret f_{ij} as the "flow" in arc (ij), as suggested above, it seems reasonable to interpret a_{ij} as the "capacity" of arc (ij), from which it follows that g_{ij} can be interpreted as the residual capacity, and h_{ij} as the flow in excess of a_{ij}. Moreover, since a_{ij} is a given constant, it follows from Eqs. (2.8) that g_{ij} is linear in f_{ij} for $0\leqslant f_{ij}\leqslant a_{ij}$, while h_{ij} is linear in f_{ij} for $a_{ij}\leqslant f_{ij}$. The two variables can be sketched as in Fig. 2.3 as functions of the flow f_{ij}. Therefore, the objective function of Eq. (2.5) can be written as

$$\text{minimize } Tv+\sum_{(ij)} u_{ij}\max(0; a_{ij}-f_{ij})-\sum_{(ij)} l_{ij}\max(0; f_{ij}-a_{ij})$$

which clearly indicates that the objective function is linear in v and piecewise linear in f_{ij}. Consequently, denote by

$f_{ij}^{(1)}$ the flow f_{ij} for $0\leqslant f_{ij}\leqslant a_{ij}$, and

$f_{ij}^{(2)}$ the flow $(f_{ij}-a_{ij})$ for $a_{ij}\leqslant f_{ij}$

In other words, any flow $f_{ij}\geqslant 0$ is to be represented from now on as the sum of two flows

$$f_{ij}=f_{ij}^{(1)}+f_{ij}^{(2)}$$

†Of course, we are assuming that $u_{ij}>l_{ij}$, for otherwise the arc duration, y_{ij}, is a fixed constant and should be treated as such.

as defined above. Then we can substitute in the objective function

$$\sum_{(ij)} u_{ij} \cdot \max(0; a_{ij} - f_{ij}) = \sum_{(ij)} u_{ij}\left(a_{ij} - f_{ij}^{(1)}\right)$$

$$= \sum_{(ij)} u_{ij} a_{ij} - \sum_{(ij)} u_{ij} f_{ij}^{(1)}$$

$$= \text{const} - \sum_{(ij)} u_{ij} f_{ij}^{(1)}, \quad \text{and}$$

$$\sum_{(ij)} l_{ij} \cdot \max(0; f_{ij} - a_{ij}) = \sum_{(ij)} l_{ij} f_{ij}^{(2)}.$$

To obtain the dual LP in terms of $f_{ij}^{(1)}$ and $f_{ij}^{(2)}$ we write

$$\text{minimize } Tv - \sum_{(ij)} \left(u_{ij} f_{ij}^{(1)} + l_{ij} f_{ij}^{(2)}\right) + \text{const} \tag{2.9}$$

s.t.

$$\sum_{j \in \mathcal{A}(i)} \left(f_{ij}^{(1)} + f_{ij}^{(2)}\right) - \sum_{j \in \mathcal{B}(i)} \left(f_{ij}^{(1)} + f_{ij}^{(2)}\right) = \begin{cases} v, & i = 1 \\ 0, & i \neq 1, n \\ -v; & i = n, \end{cases} \tag{2.10}$$

$$0 \leqslant f_{ij}^{(k)} \leqslant a_{ij}^{(k)}, \quad k = 1, 2, \text{ all } (ij) \in A \tag{2.11}$$

where $a_{ij}^{(1)} = a_{ij}$ and $a_{ij}^{(2)} = \infty$, and $\mathcal{A}(i)$ is the set of nodes connecting with node i and occur after i, and $\mathcal{B}(i)$ is the set of nodes connecting with node i and occur before i.

Compare the primal LP of Eqs. (2.3)–(2.4) and the dual LP of Eqs. (2.9)–(2.11): the primal variables are, of course, the t_i and y_{ij} values, while the dual variables appearing *explicitly* in the dual LP are the $f_{ij}^{(k)}$ values and v. The g_{ij} and h_{ij} values have been removed from the explicit statement of the dual through the device of defining $f_{ij}^{(k)}$; $k = 1, 2$; and modifying the objective function accordingly [cf. formulation of Eqs. (2.5)–(2.7)]. However, they are implicitly there, always. In fact, from the primal–dual formulations we have:

1. $g_{ij} > 0 \Rightarrow$ (i) $h_{ij} = 0$, which we already know;
   (ii) $y_{ij} = u_{ij}$, by complementary slackness;
   (iii) this is the range of $f_{ij}^{(1)}$ (see Fig. 2.3).

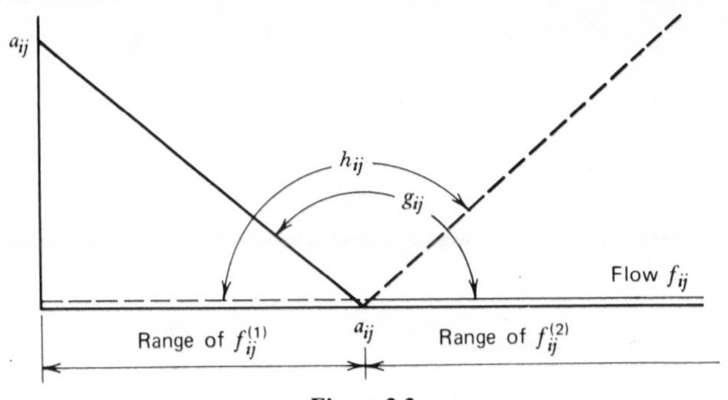

Figure 2.3.

It immediately follows, again by complementary slackness considerations, that

$$\text{if} \quad t_i - t_j + u_{ij} < 0 \quad \Rightarrow f_{ij}^{(1)} = 0;$$
$$\text{if} \quad 0 < f_{ij}^{(1)} \leqslant a_{ij} \quad \Rightarrow t_i - t_j + u_{ij} = 0$$

To simplify notation, let

$$s_{ij}^{(1)} \triangleq t_i - t_j + u_{ij} \tag{2.12}$$

Then we conclude that

$$\text{positive flow } f_{ij}^{(1)} > 0 \textit{ is permissible} \text{ only if } \quad s_{ij}^{(1)} = 0 \tag{2.13}$$

2. $h_{ij} > 0 \quad \Rightarrow$ (i) $g_{ij} = 0$, which we already know;
   (ii) $y_{ij} = l_{ij}$, by complementary slackness;
   (iii) this is the range of $f_{ij}^{(2)}$ (see Fig. 2.3).

It immediately follows, again by complementary slackness considerations, that

$$\text{if} \quad t_i - t_j + l_{ij} < 0 \quad \Rightarrow f_{ij}^{(2)} = 0;$$
$$\text{if} \quad f_{ij}^{(2)} > 0 \quad \Rightarrow t_i - t_j + l_{ij} = 0;$$
$$f_{ij}^{(1)} = a_{ij}$$

To simplify notation, let

$$s_{ij}^{(2)} \triangleq t_i - t_j + l_{ij} \tag{2.14}$$

Then we conclude that

$$\text{positive flow } f_{ij}^{(2)} > 0 \text{ is } \textit{permissible} \text{ only if } s_{ij}^{(2)} = 0; \quad \text{and} \quad f_{ij}^{(1)} = a_{ij} \tag{2.15}$$

Now, appealing to the duality theorem proper, it is sufficient for optimality (of both primal and dual) to discover node "numbers" (i.e., times) t_i, and arc "flows" $f_{ij}^{(1)}$ and $f_{ij}^{(2)}$, which satisfy the optimality conditions given by the statements (2.13) and (2.15).

A possible interpretation of the dual LP of Eqs. (2.9)–(2.11), and in particular of constraints (2.10), is that we now have a network in which each arc of finite marginal cost value a_{ij} is replaced by *two* arcs, the first of which has a "capacity" equal to a_{ij}, while the second arc has unlimited "capacity." Since $v = \sum_{j \in \mathcal{A}(1)} f_{1j} = \sum_{i \in \mathcal{B}(n)} f_{in}$ the problem is now translated into that of constructing flows $\{f_{ij}^{(k)}\}$; $k = 1,2$; in the new network in which the flows $\{f_{ij}^{(1)}\}$ are in the arcs of capacity a_{ij} and the flows $\{f_{ij}^{(2)}\}$ are in the arcs of unbounded capacity that satisfy the optimality conditions of (2.13) and (2.15).

The algorithm for achieving this for all values of project duration $T \in [\underline{T}, \overline{T}]$ is the next order of business.

The algorithm closely follows primal-dual arguments and procedures. In essence, we start at the longest possible project duration, $\overline{T}$, and proceed to shorten that duration in such a way that the resulting LPs are *always primal and dual feasible*. This is achieved by consistently satisfying the optimality conditions (2.13) and (2.15), which implies that the dual variables $\{f_{ij}^{(k)}\}$ are increased only along *permissible* arcs.

For simplicity of exposition we divide the presentation into three steps: (a) determining the feasibility of project duration $T < \overline{T}$, (b) labeling for minimum-cost activities, and (c) effecting the reduction in the project duration—the node time-change routine. As a vehicle for illustrating the different steps of the procedure we utilize the network shown in Fig. 2.4, in which we drew only a single arc between any two nodes to avoid cluttering the diagram, with the understanding that the second arc with infinite capacity is always present, to be brought into the picture when needed.

STEP 1. *Determining the Feasibility of $T < \overline{T}$*†. Put the activity durations $y_{ij} = u_{ij}$ for all activities. Label the origin, node *1*, with $(\infty, 0)$. Check each arc originating from node *1* for the relationship

$$s_{ij}^{(2)} = 0; \tag{2.16}$$

†The direct approach to answer this question of feasibility is to verify that $\underline{T}$ of Fig. 2.2 is, indeed, $< \overline{T}$. However, this network flow argument is helpful as an introduction to the general algorithm.

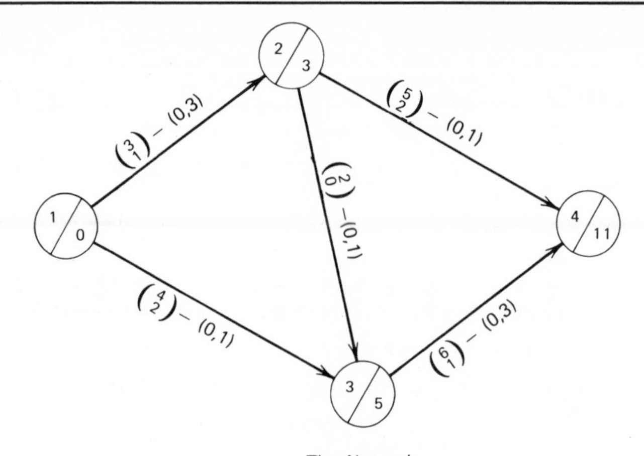

The Network

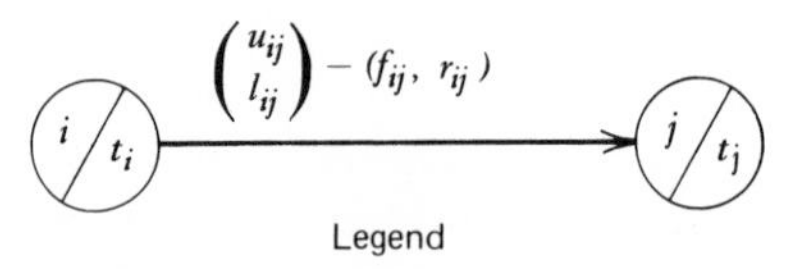

Legend

Figure 2.4.

and any node j for which Eq. (2.16) holds should be labeled $(\infty, 1)$. In general, for any node i labeled with $q_i = \infty$, check all nodes j connected with it for $s_{ij}^{(2)} = 0$ and label such j with (∞, i). Continue the labeling until one of two conditions is obtained: (a) the terminal n cannot be labeled (called *nonbreakthrough*) or (b) node n can be labeled with ∞ (called *breakthrough*). In the first eventuality it is possible to reduce $\overline{T}$; consequently, go to the labeling for minimum-cost activities subroutine below. In the second eventuality, it is not possible to reduce $\overline{T}$, and so terminate the analysis.

The reason for this latter conclusion is the following: if there is a breakthrough (to node n) with the label ∞, it means that there is a flow-augmenting path, and that such a path is a CP. Since each arc on this path is of duration l_{ij}, it is impossible to reduce its duration! Evidently, this can happen only if $u_{ij} = l_{ij}$ for a subset of activities that are the critical activities (i.e., that form a CP). This also explains the conclusion in the first eventuality since the absence of a flow-augmenting path when $y_{ij} = u_{ij}$ everywhere and the test of (2.16) is made indicates the possibility of shortening the project duration.

The application of this test to our illustrative example of Fig. 2.4 reveals that it is possible to shorten the duration to less than $\overline{T} = 11$.

STEP 2. *Labeling for minimum-cost activities subroutine.* Initialize all arc durations to $y_{ij} = u_{ij}$ and all flows to $f_{ij} = 0$. Start with the origin, node *1*, which is always labeled with $(\infty, 0)$. For any labeled node i check all unlabeled nodes j connected to it for which either

$$s_{ij}^{(1)} = 0 \qquad \text{and } f_{ij}^{(1)} < a_{ij} \tag{2.17}$$

or

$$s_{ij}^{(2)} = 0 \tag{2.18}$$

If conditions (2.17) hold, label node j with (q_j, i) where $q_j = \min(q_i; r_{ij})$, and $r_{ij} = a_{ij} - f_{ij}^{(1)}$. If condition (2.18) holds, label node j with (q_i, i).

At this junction the phenomenon called "reverse labeling" may occur. Here are the surrounding circumstances and the procedure to follow. Suppose that the arrow (ij) is directed $i \to j$, and that node j is labeled with q_j but node i is not labeled. Now perform the alternate check:

$$\text{if } s_{ij}^{(1)} = 0; \text{ then label } i \text{ with } (q_i, j); \text{ where } q_i = \min\left(q_j, f_{ij}^{(1)}\right); \tag{2.17a}$$

$$\text{and if } s_{ij}^{(2)} = 0; \text{ then label } i \text{ with } \left(\min\left(q_j, f_{ij}^{(2)}\right), j\right) \tag{2.18a}$$

This action simply "reverses" the amount of flow in arc (ij), in the direction $i \to j$, in preparation for subsequent action. Continue the labeling until one of the following three conditions obtain:

a. Breakthrough to n with q_n finite: increase the flow along the newly discovered flow-augmenting path by the amount q_n in the standard manner. Modify all residual capacities, erase all labels except those with ∞; and return to the labeling procedure with the latest flows.
b. Breakthrough to n with $q_n = \infty$; terminate the analysis.
c. Nonbreakthrough condition: go to the node time-change subroutine.

The realization of case (b) signals that the current project duration is the shortest possible ($= \underline{T}$), and hence the analysis is terminated (for the same reasons as explained in Step 1.)

If case (c) is realized—possibly after several realizations of case (a)—determine the cutset $\mathcal{C}$ that separates the set of labeled nodes from the set of unlabeled nodes, and go to the node time-change procedure.

As an example, the application of this step to the network of Fig. 2.4 yields the labeling shown in Fig. 2.5*a*, a breakthrough. The flow augmenting path is *1,2,3,4*. Increasing the flow by $q_4 = 1$ results in the flows shown in Fig. 2.5*b*, whose labeling results in a nonbreakthrough condition.

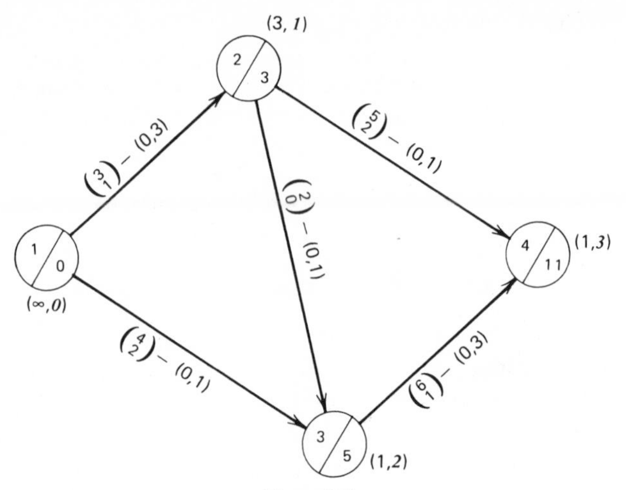

Figure 2.5*a*.

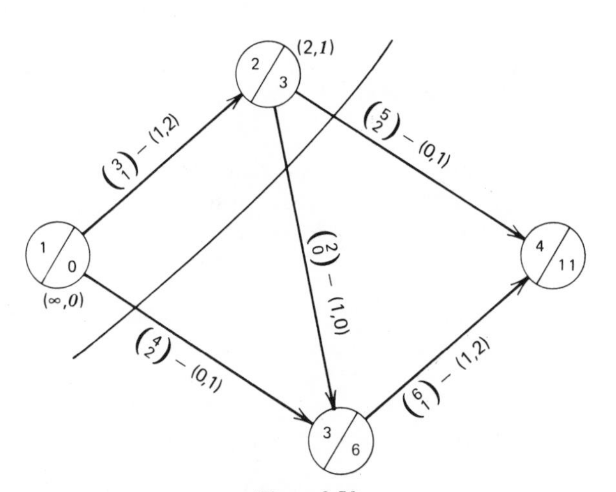

Figure 2.5*b*.

Since nodes *1* and *2* are labeled, but *3* and *4* are not, the cutset is as indicated on the figure, and is the set $\mathcal{C} = \{(1,3),(2,3),(2,4)\}$. Note that, true to its name, the labeling procedure did increase the flow to the point of "saturating" the minimal-cost activity, arc $(2,3)$. As is seen in the next step, this is precisely the activity that is shortened.

STEP 3. *The node time-change subroutine.* This subroutine is applied whenever a nonbreakthrough condition obtains. The nodes of the network can be divided into two mutually exclusive and totally exhaustive subsets: (a) the labeled nodes (including those labeled with $+\infty$) and the unlabeled nodes. Let $\mathcal{C}$ denote the cutset; then $\mathcal{C}$ contains two mutually exclusive (but not necessarily exhaustive) subsets of arcs, which we denote by Z_1 and Z_2 (either Z_1 or Z_2 may be empty), and define as follows:

$$Z_1 \triangleq \{(ij): i \text{ labeled}, j \text{ not labeled}, s_{ij}^{(k)} < 0\};$$

and

$$Z_2 \triangleq \{(ij): i \text{ not labeled}, j \text{ labeled}; s_{ij}^{(k)} > 0\}$$

In these definitions ignore values of $s_{ij}^{(k)} = 0$ for $k = 1, 2$. Let

$$\delta_1 = \min_{Z_1}\left(-s_{ij}^{(k)}\right) \quad \text{and} \quad \delta_2 = \min_{Z_2}\left(s_{ij}^{(k)}\right)$$

and put

$$\delta = \min(\delta_1, \delta_2)$$

For all nodes j that are *not* labeled, change the node times $\{t_j\}$ to $\{t_j - \delta\}$. Discard all labels other than ∞ and return to the labeling subroutine.

Applying Step 3 to the network of Fig. 2.5*b*, we discover that $Z_1 = \mathcal{C}$, $Z_2 = \phi$ (the empty set), $\delta_1 = \min(-s_{1,3}^{(1)}; -s_{2,3}^{(2)}; -s_{2,4}^{(1)}) = 1$, and $\delta = \delta_1 = 1$. Hence, t_3 and t_4 are changed to $t_3 = 4$ and $t_4 = 10$, and the labeling procedure is now applied to the network of Fig. 2.5*c*.

At this juncture it is important to perceive clearly what it is that the labeling and time-change subroutines really accomplish. We recall that the "capacity" a_{ij} is, in fact, the marginal cost of the primal problem. Moreover, the conditions $s_{ij}^{(k)} = 0$ in the labeling subroutine ensure dual feasibility, as has been already explained. Now, a breakthrough occurs only as a result of the discovery of a flow-augmenting path *that is also a CP*. According to the first subroutine, a breakthrough results in increasing the "flow" into n by the minimum "capacity" along that CP; in other words, it determines the minimum-cost activity along *one* CP. As long as breakthrough is possible, the flow-change subroutine saturates the "capacity" of one activity along a CP. The set composed of these saturated

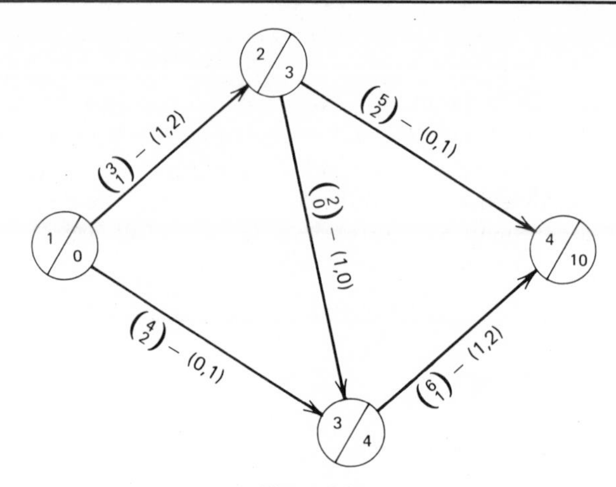

Figure 2.5*c*.

minimum-cost activities represents the arcs whose duration must be reduced, and the cost of such reduction is guaranteed to be minimal.

When a nonbreakthrough condition is obtained it must be that the CPs have been exhausted. The node time-change subroutine then proceeds to reduce the length of all CPs *simultaneously* by the maximum possible amount. This is the amount δ. It represents the possible reduction in all the CPs before either: (a) another path becomes critical or (b) an activity has been reduced to its lower bound, l_{ij}. Notice that the ∞-labeling permits the procedure to select the next-cheapest activity along a CP whenever the cheapest activity has reached its lower limit, l_{ij}.

From this point onward, and with the preceding remarks in mind, the successive application of the two subroutines yields the diagrams shown in Figs. 2.5*d–m*. The results are summarized in Table 2.1 and the complete cost-duration function is shown in Fig. 2.6, from which we see that the minimum duration is 3 at a cost of 27.

We draw the reader's attention to the following important remarks. First, whenever the duration of any arc has been reduced to its lower bound, l_{ij}, a second arc of infinite capacity is added in parallel with it. It is recalled that such arcs are always present to permit the flows $\{f_{ij}^{(2)}\}$, and that we just refrained from including them at the outset in order not to clutter the diagrams.

Second, it is evident that any shortening of the project duration must be effected through the reduction of the length of the CPs simultaneously. However, the process is not monotone for any particular activity, although it is monotone for any CP. For instance, the duration of activity *(2,3)*

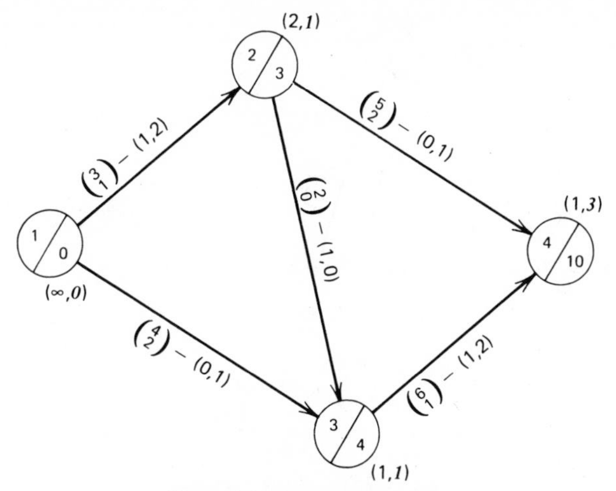

Figure 2.5*d*. Breakthrough.

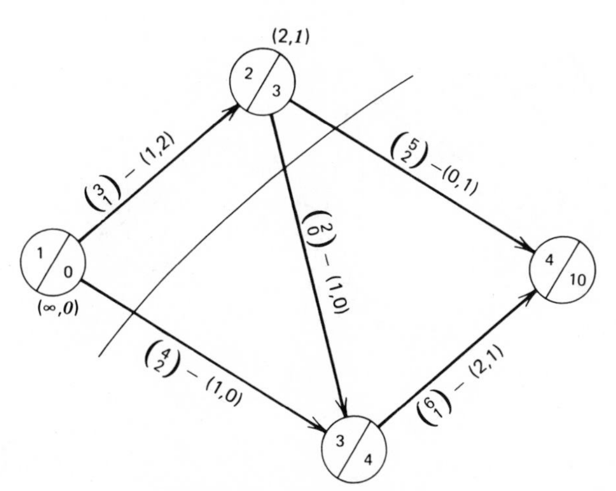

Figure 2.5*e*. Nonbreakthrough.

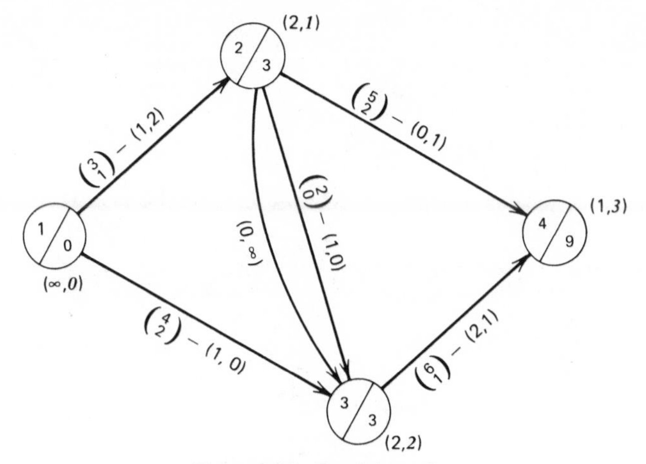

Figure 2.5*f*. Breakthrough.

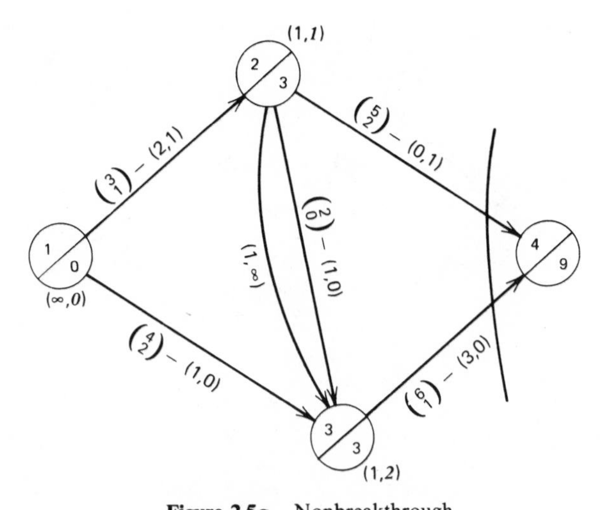

Figure 2.5*g*. Nonbreakthrough.

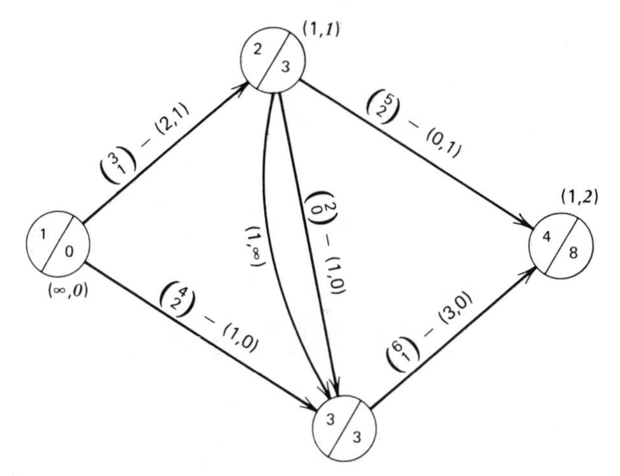

Figure 2.5*h*. Breakthrough.

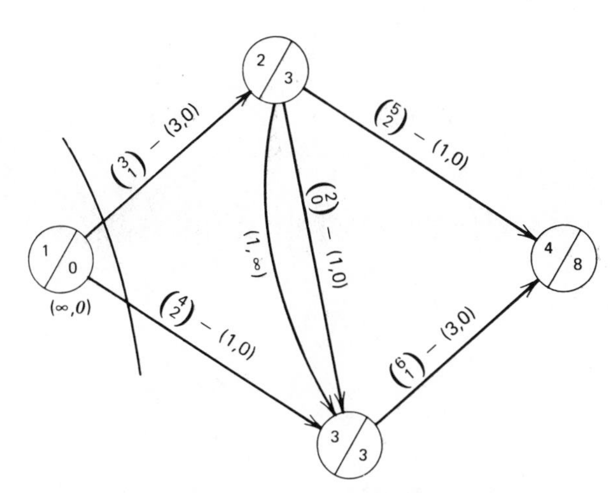

Figure 2.5*i*. Nonbreakthrough.

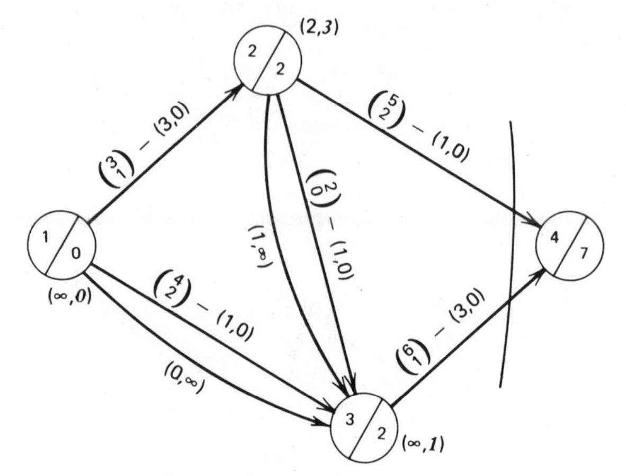

Figure 2.5*j*. Nonbreakthrough.

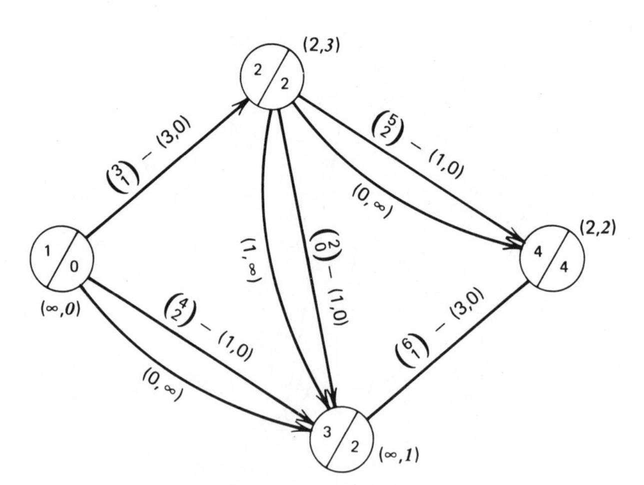

Figure 2.5*k*. Breakthrough.

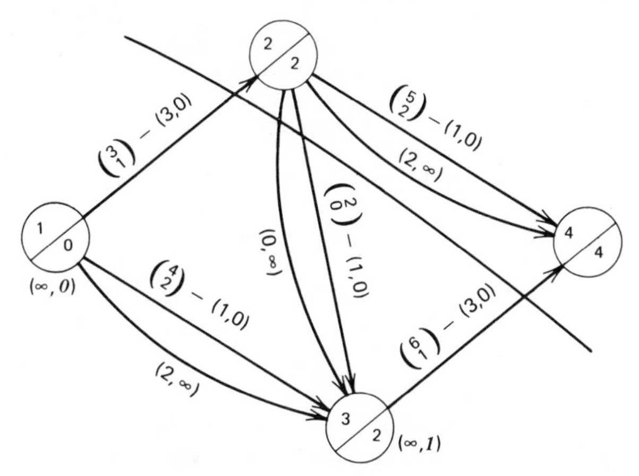

Figure 2.5*l*. Nonbreakthrough.

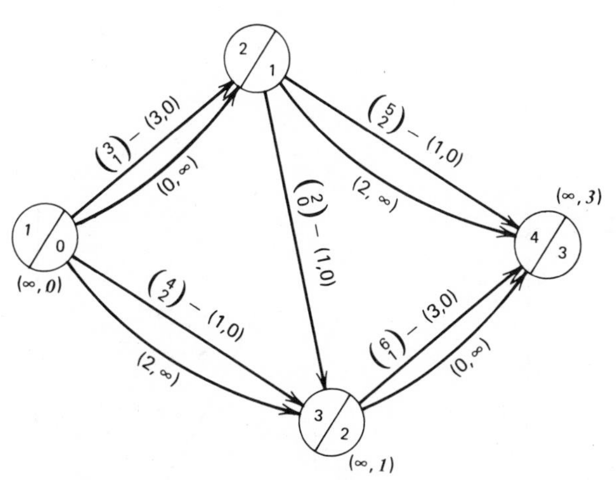

Figure 2.5*m*. Breakthrough with ∞ label. Terminal.

Table 2.1

Iteration Number	t_4	$F=\sum_i f_{i4}$	Reduction in Durations δ	Increase in Cost δF	Cumulative Increase in Cost
2	11	1	1	1	1
4	10	2	1	2	3
6	9	3	1	3	6
8	8	4	1	4	10
9	7	4	3	12	22
11	4	5	1	5	27
12	3	∞-terminate			

decreased from 2 to 0, stayed at 0 for a few iterations, and then increased to 1 in the last iteration. Such seemingly odd behavior is easily understood if one remembers that the procedure yields the *minimum-cost reduction in duration*. Hence, if at some iteration the procedure calls for the reduction in the duration of two activities which lie along the same CP, that particular CP would be shortened by 2δ and not just by δ. Consequently, to achieve minimum cost, an activity previously shortened can now be

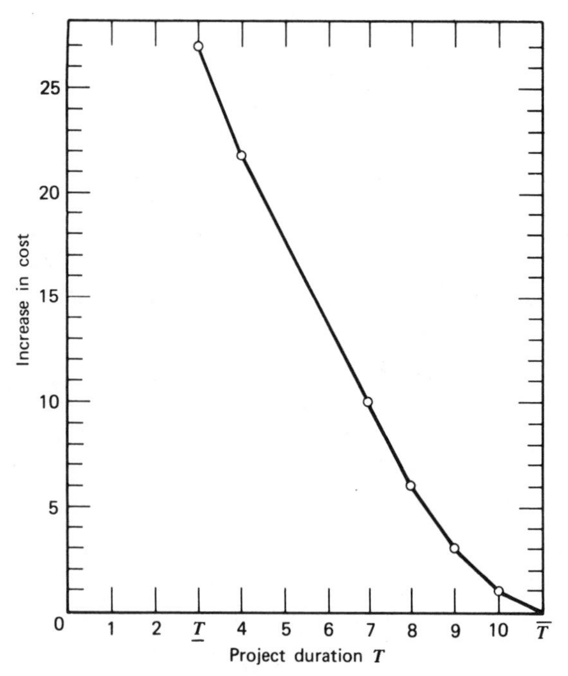

Figure 2.6. The optimal cost-duration curve.

lengthened by δ. This is precisely what happened to activity $(2,3)$ in the last iteration. Of course, if it is possible to lengthen more than one activity, the procedure would choose the most expensive to lengthen, thus ensuring the largest savings.

In naming the various sections of the procedure we have meticulously avoided giving the impression that the objective is to maximize flow. [Note that we called Step 2 *labeling for minimum cost activities* (p. 71), although the labeling procedure is, in fact, a convenient procedure for maximizing the *flow* in the network!]. Lest we have failed to convey this point and the reader is confusing the above procedure with flow maximization, we pause to reiterate that the procedure described above, in Steps 2 and 3, corresponds, in fact, to the primal–dual calculations of LP. In other words, had we elected to proceed with a frontal attack on the primal LP of Eqs. (2.3)–(2.4) and the dual LP of (2.5)–(2.7) using primal–dual arguments, we would have ended duplicating the steps of the flow algorithm. But this would have obscured the very special structure of the problem, and resulted in more elaborate calculations.

Returning to the managerial aspects of the problem (see p. 59), the reader should satisfy himself that one can respond positively to each question asked. For instance, consider the question as to what the marginal cost of project compression is at $T=7$. A glance at Fig. 2.6 reveals that the desired value is given by the slope of the line segment and is equal to -3. That is, a unit reduction in time will incur 3 units of cost. Incidentally, this marginal cost is applicable for any T in the range $4<T\leqslant 7$. At $T=4$ the marginal cost of compression changes to 5. It is left as an exercise to the reader to determine whether the set of "bottleneck activities" (i.e., those that must be shortened to achieve a reduction in the total project duration) were the same throughout the range of T.

§3. LINEAR COST-DURATION FUNCTION: AN ALTERNATIVE APPROACH

The approach discussed in the previous section may be viewed as a "dual" approach in which the mechanism used for discovering the least-cost activities at any iteration is the primal–dual considerations leading to the conditions (2.13) and (2.15). Furthermore, the arguments used were network-flow arguments throughout.

Such an approach is often criticized as being intuitively unappealing. The first impulse of any serious student of the problem of minimum-cost time compression is to utilize a "frontal" attack on the problem—which, incidentally, may be termed a "primal" algorithm. According to heuristic reasoning, it is evident that at any step of iteration we are concerned only with the CPs; thus, activities not lying on any CP are irrelevant to the issue of "time compression." Therefore, it seems logical to define these CPs, determine the least-cost activities on them, and proceed to shorten these activities in the most judicious manner to effect the desired reduction in total project duration.

The main pitfall in this line of reasoning is that *while* CPs *are always shortened, individual activities lying on these* CPs *do not exhibit such monotone behavior*. We witnessed a demonstration of this peculiar phenomenon on p. 74. Consequently, one should not only search for the *least* cost activities on the CPs, but also investigate the possibility of lengthening previously shortened activities, with the *most* expensive activity considered first! The network-flow arguments and the procedure based on them accomplish all these objectives automatically.

The following alternative approach, which is a longest-path algorithm in a "modified network," was proposed by Fulkerson (1964). The algorithm is best explained with reference to a numerical example.

Consider again the project network of Fig. 2.4, redrawn as Fig. 2.7, in which each arc bears three numbers only: the upper and lower limits of duration, u_{ij} and l_{ij}, and the marginal cost a_{ij}.

STEP 0. Compute the longest path(s) in the network assuming the durations $y_{ij} = u_{ij}$, all (ij); and determine the event times, $\{t_i\}$, and the CP(s). This constitutes an optimal solution for the completion time $T = \bar{T}$. This step is shown in Fig. 2.8, and $\bar{T} = 11$. The CP is shown in heavy arcs.

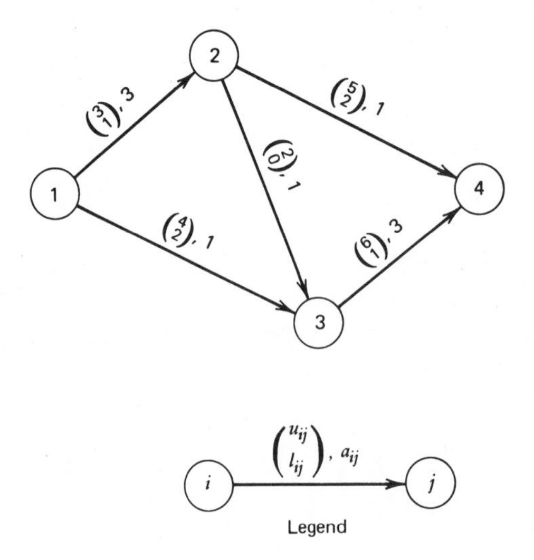

Figure 2.7.

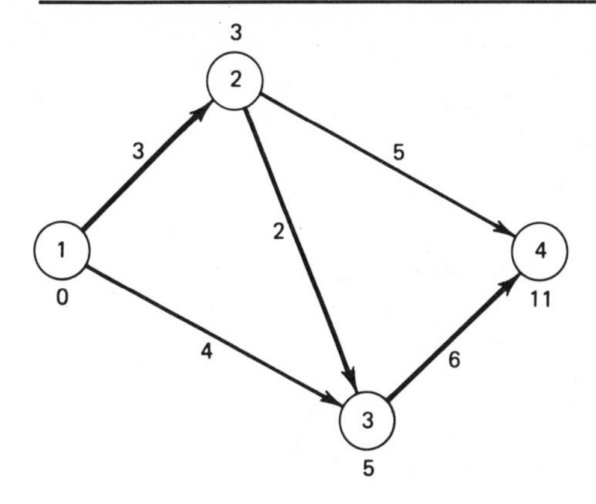

Figure 2.8. The *CP*.

STEP 1. Maximize the flow on the CP(s) *only*, in which arc capacities are taken to be the values of a_{ij}; using the standard labeling procedure. In the example the maximal flow is equal to one unit, determined by the capacity $a_{2,3}$. The flows in all arcs of the network are shown in Fig. 2.9. Clearly, this step emulates the determination of the least-cost activity (or activities) that are eligible for "compression" in the critical subnetwork.

STEP 2. Construct the *modified duration network* using the current flows in all arcs of the network, with arc "lengths" determined from the following Table 2.2. The longest distance from the origin to node i represents t_i, which is the current realization time of event i. The duration of any activity (ij) is given by

$$y_{ij} = \min(u_{ij}; t_j - t_i).$$

The rationale for Table 2.2 can only be gleaned with reference to the primal–dual programs of the previous section, and in particular to the

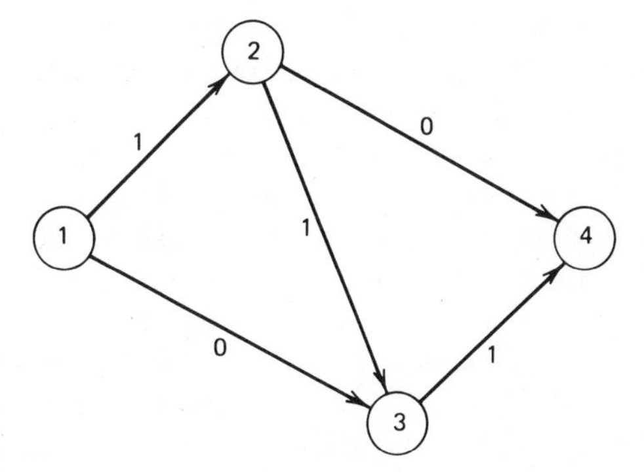

Figure 2.9. The corresponding flow.

Table 2.2

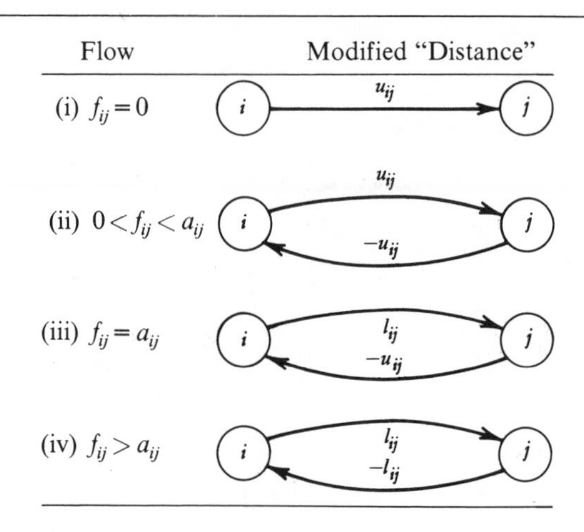

complementary slackness conditions of (2.13) and (2.15). For, a flow $f<a$ in an arc (which includes the case $f=0$) indicates that the arc has not been shortened at all. Hence, its duration is still fixed at its upper limit u. This explains the forward arc (i.e., $i \to j$) in cases (i) and (ii). The reverse arc (i.e., $j \to i$) in case (ii) is to permit reversal of flow in the arc while preserving complementarity—hence, the "length" is fixed at $-u$. Next, arcs with $f=a$ are the ones eligible for shortening (and may, in fact, have been shortened). The linearity of costs implies that we may shorten them as much as possible. Hence, their duration will be l. This explains the forward arc of (iii). On the other hand, if the arc is to be lengthened it should be lengthened to its upper bound u. This explains the reverse arc of (iii). Finally, if $f>a$, then the arc must be at its "crash" duration, and any increase in (forward) flow or any reversal of flow must take place at duration l; hence, the forward and reverse lengths of (iv) result. An important consequence of Table 2.2 is that the resulting modified network has cycles, all of which are of nonpositive lengths. In our example we have the modified network of Fig. 2.10, which also shows the t_i values. Clearly, $T=10$.

There are two pertinent remarks to make at this juncture. First, the determination of the node times $\{t_i\}$ is no longer the simple task of the standard CPM calculations. In fact, the presence of cycles in the modified network forces one to adopt some general longest-path algorithm in directed, cyclic networks, such as that proposed by Dijkstra (1959). This fact becomes manifest in the text that follows (see, in particular, Fig. 2.12(c)). Second, putting $y_{2,3}=l_{2,3}=0$ permitted the detection of the second

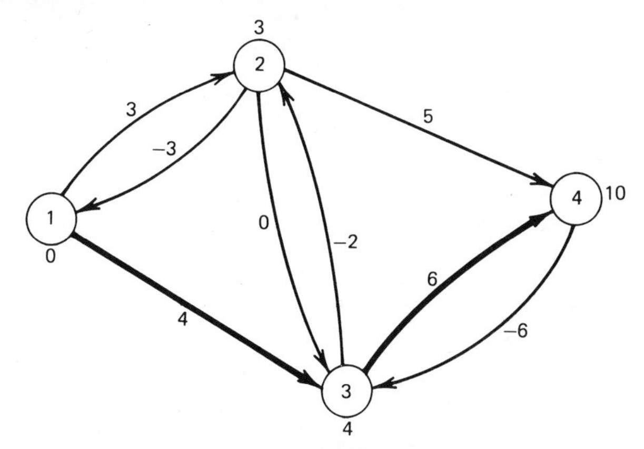

Figure 2.10. Modified network.

longest CP, namely, the path *1,3,4*. As a result, activity (*2,3*) has been shortened by only one time unit.

STEP 3. Find the longest path(s) from *1* to *n* in the modified network. Maximize the flow over the subnetwork of longest path(s), using the labeling procedure, where the arc capacities are defined as shown in Table 2.3. Return to Step 2.

In the example, the calculations of this step are illustrated in Fig. 2.11. As indicated above, the longest path in the modified network of Fig. 2.10 is

Table 2.3

(a) *Forward arcs*: arcs on the path whose direction is the same as the arrow of the original project network:	
State of Current Flow†	Capacity
(i), (ii)	$r_{ij} = a_{ij} - f_{ij}$
(iii), (iv)	∞
(b) *Reverse arcs*: arcs on the path whose direction is the opposite of the arrow of the original network:	
State of Current Flow	Capacity
(ii), (iii)	f_{ij}
(iv)	$f_{ij} - a_{ij}$.

†Refers to Table 2.2.

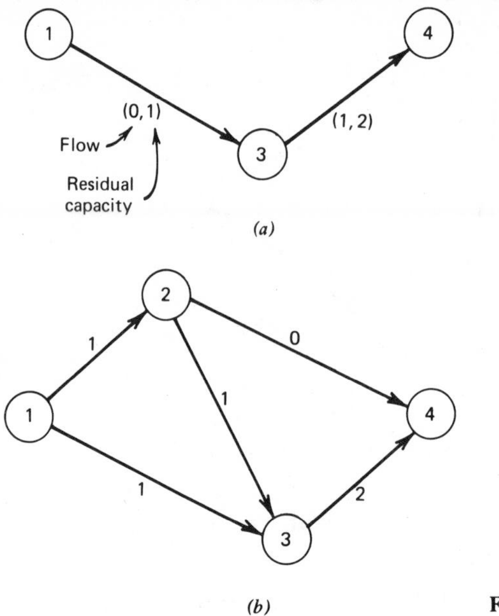

Figure 2.11.

the path *1*,*3*,*4*, of duration 10, shown in heavy lines. In Fig. 2.11*a*, the "as is" flow in this path, together with the capacity of each arc on the path according to Table 2.3, are given. Fig. 2.11*b* shows the resulting augmented flow in the network.

We return to Step 2 and repeat Steps 2 and 3 until the longest path(s) in the modified network has infinite capacity, which implies that all arcs on the CP(s) of the original network have reached the lower bounds of their durations, and it is no longer possible to reduce the total completion time of the project.

Taking off from Fig. 2.11, we obtain successively the results shown in Fig. 2.12*a*–*d*. The complete minimum-cost curve, $C(T)$, for the project can now be determined over the range of feasible T, $3 \leqslant T \leqslant 11$, using the optimal job times obtained from the modified network in each iteration. The result was previously given in Fig. 2.6.

Two pertinent remarks would assist the reader in fully comprehending the steps of iteration reported in Fig. 2.12. First, consider Fig. 2.12*c*. The realization time of node *2*, $t_2 = 2$, is obtained from the path *1*,*3*,*2*. Furthermore, the longest path in the modified network is *1*, *3*, *2*, *4*, shown in heavy lines. Here is an instance in which an arc on the longest path, namely, arc

Modified network

Flow network

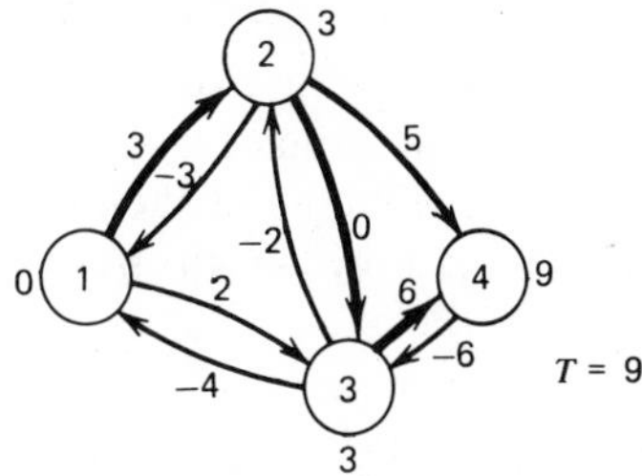

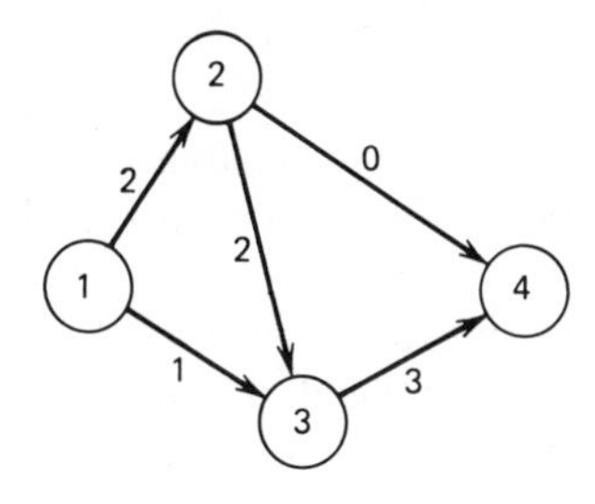

(a)

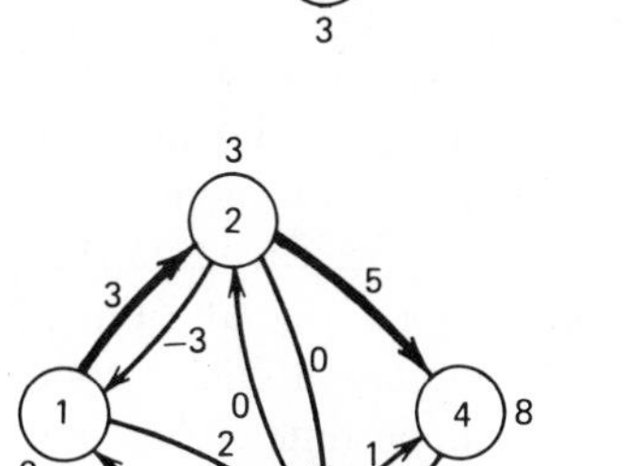

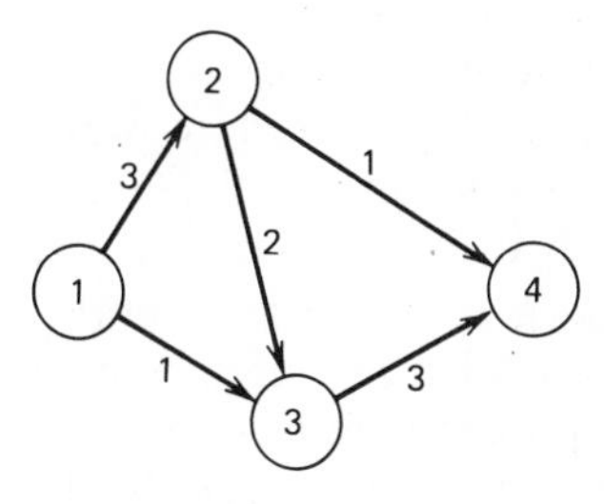

(b)

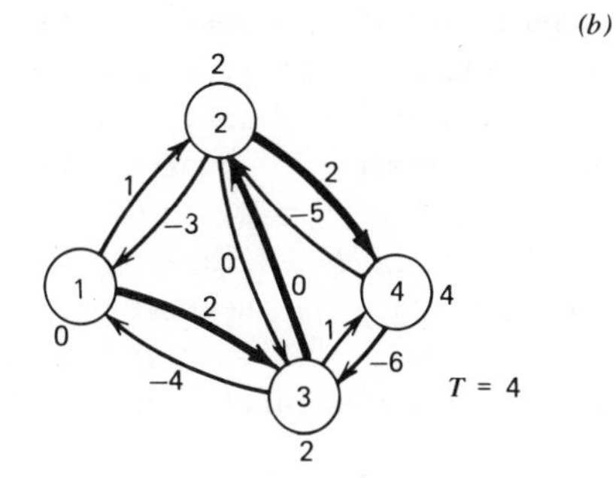

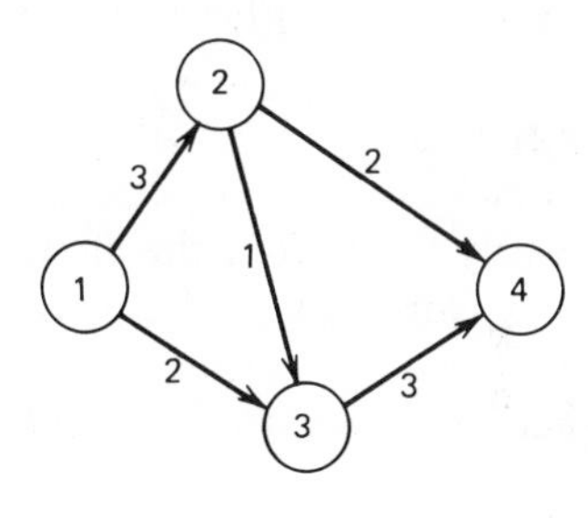

(c)

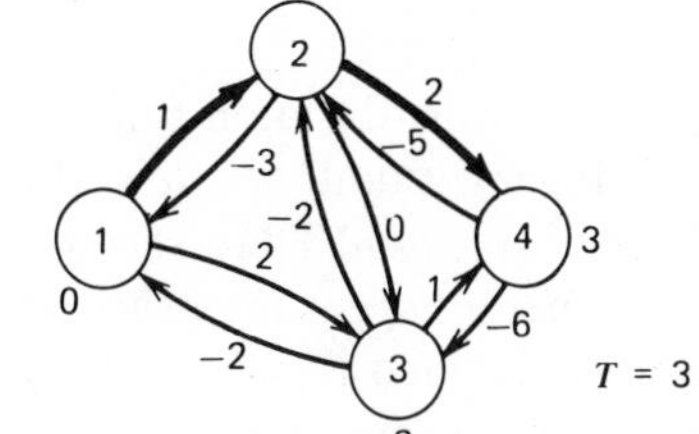

Here the capacity of the chain *1*, *2*, *4*, is infinite. Thus we have reached the lower bound of $\underline{T}$.

(d)

Figure 2.12.

$(3,2)$ is a *reverse* arc as defined in Table 2.3. Since the flow in the forward arc (shown in Fig. 2.12*b*) is $f_{2,3}=2>a_{2,3}=1$, then the capacity of the reverse arc $(3,2)$ is given by $f_{ij}-a_{ij}=2-1=1$, according to Table 2.3. The capacities of the other two arcs on the longest path, namely, arcs $(1,3)$ and $(2,4)$, are infinite (why?). Hence the maximum flow achievable is still unity. The unit flow in the direction $3 \to 2$ "reverses" some of the original flow in the direction $2 \to 3$, and the net resultant flow on all arcs is as shown in the flow network of Fig. 2.12*c*.

Second, the alert reader will notice that the "compression" of the project duration proceeded directly from $t_n=8$ to $t_n=4$ (see Figs. 2.12*b*, *c*); ignoring the pseudobreakpoint at duration $t_n=7$ obtained by the approach of the previous section (as reported in Table 2.1). Why did this happen? The answer lies in the process of detecting a minimal-capacity cutset under the labeling procedure; whenever there are two or more cutsets of the *same* minimal-capacity, the labeling procedure detects them one at a time, from "left to right," and each time indicates a "nonbreakthrough" condition that generates a pseudobreakpoint through the node time-change routine. Since all such cutsets have the same (minimal) capacity, the marginal cost of project reduction remains unchanged until they are all detected and the project duration accordingly reduced.

This is exactly what happens in determining the maximal flow in the longest path of Fig. 2.12*b*: the residual capacity of arc $(1,2)$ is equal to $a_{1,2}-f_{1,2}=3-2=1$, which is the *same* as the capacity of arc $(2,4)$. Hence, the flow of one unit saturates *both* arcs simultaneously, giving rise to *two* minimal-capacity cutsets: $\mathcal{C}_1=\{(1,2),(1,3)\}$ and $\mathcal{C}_2=\{(2,4),(3,4)\}$—as is easily seen from the flow network of Fig. 2.12*b*—of capacity 4. The reader can easily verify, by reference to Figs. 2.5*i*, *j* that these are precisely the cutsets detected by the labeling procedure, that gave rise to the pseudobreakpoint of duration 7 in the cost-duration function of Fig. 2.6.

§4. STRICTLY CONVEX COST FUNCTION

The assumption of strict convexity of the activity cost-duration function ϕ may be intuitively appealing from the following point of view. Long activity durations usually involve a good deal of freedom. Thus, shortening the activity at such durations involves small expenses spent on measures of economizing in the use of the available resources. Such is not the case, so goes the argument, at short activity durations, where one expects the activity to be tightly controlled, and a larger cost must be incurred for the same amount of time "compression," that is, reduction in the activity duration. Naturally, activities are encountered that do not exhibit such behavior, and then the other models discussed in this chapter may apply.

From an analytical point of view the assumption of strict convexity has great appeal since one may thereby derive some interesting and illuminating properties of the optimal solution. These form the subject matter of this section.

Thus, we assume[†] that each activity $a \in A$ has a cost $c_a = \phi(y_a)$ where ϕ is *continuous and strictly convex* in the duration of the activity y_a, as y_a ranges over the nonnegative values (i.e., $0 \leqslant y_a \leqslant \infty$). We further assume that $c_a \to \infty$ as $y_a \to 0$ and $c_a \to 0$ as $y_a \to \infty$, as shown in Fig. 2.13. More formally, denoting the first time derivative $\partial c_a / \partial y_a$ by c'_a, and the second time derivative by c''_a, we have

$$
\begin{aligned}
&c_a \to +\infty \quad \text{as } y_a \to 0^+ \\
&c_a \to 0^+ \quad \text{as } y_a \to +\infty \\
&-\infty \leqslant c'_a \leqslant 0, c''_a > 0, \quad \text{continuous.}
\end{aligned}
\tag{2.19}
$$

The problem posed is as follows: given a total project duration $T = t_n - t_1 = t_n$, since t_1 is usually put equal to zero, what are the optimal durations of the activities that complete the project in time T and which minimize the total project cost?

The main result of this section is Theorem 2.2, which provides necessary and sufficient conditions for the optimal schedule of activities. As a corollary to this theorem we derive Berman's main result for the special cost structure stated above. This section concludes with a discussion of the constrained problem; which is the same problem but with upper and lower bounds on the individual activity durations.

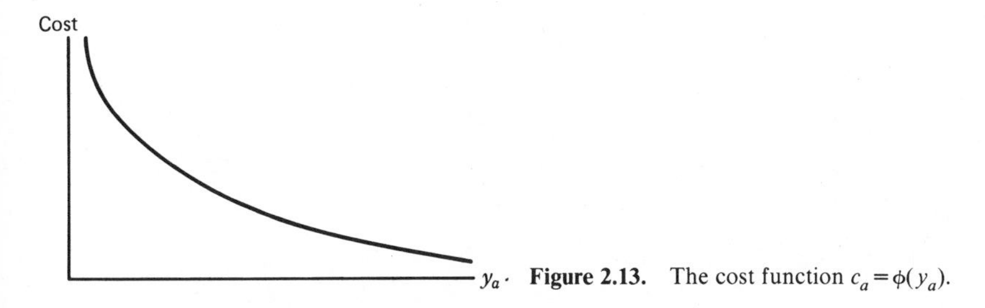

Figure 2.13. The cost function $c_a = \phi(y_a)$.

[†]Some of the assumptions concerning the form of c_a are modified later. Also, note that we use the symbol A to denote the *set* of arcs in the network as well as their *number*. The intent should be evident from the context.

Theorem 2.2: There is one and only one optimum schedule with duration T.

A necessary and sufficient condition for optimality is that:

1. The schedule has no slack.
2. If $D(\tau)=\Sigma_P \partial c_a/\partial y_a|_\tau$, where P is the set of all activities in progress at time τ, then $D(\tau)=$constant for all τ, $0 \leqslant \tau \leqslant T$.

Proof. A heuristic reasoning of the validity of the theorem can be gleaned from a mechanical analog suggested by Clark (1961) (Fig. 2.14); see also Minty (1957) and Prager (1963). In this model, the nodes are solid rings or plates and the arcs are elastic strings or springs. When node *1* is anchored at distance $x=0$ and node n at distance $x=T$, the tension in the springs will achieve an equilibrium position and the nodes will remain at rest. At equilibrium, the length of each spring corresponds to the duration of the activity, and the cost of the activity corresponds to the potential energy.

Since the formal proof is lengthy, albeit elementary, it is omitted here. The interested reader may consult the work by Elmaghraby (1968). However, in the course of proving the sufficiency of the condition of the theorem, two lemmas are established. We state them, without proof,

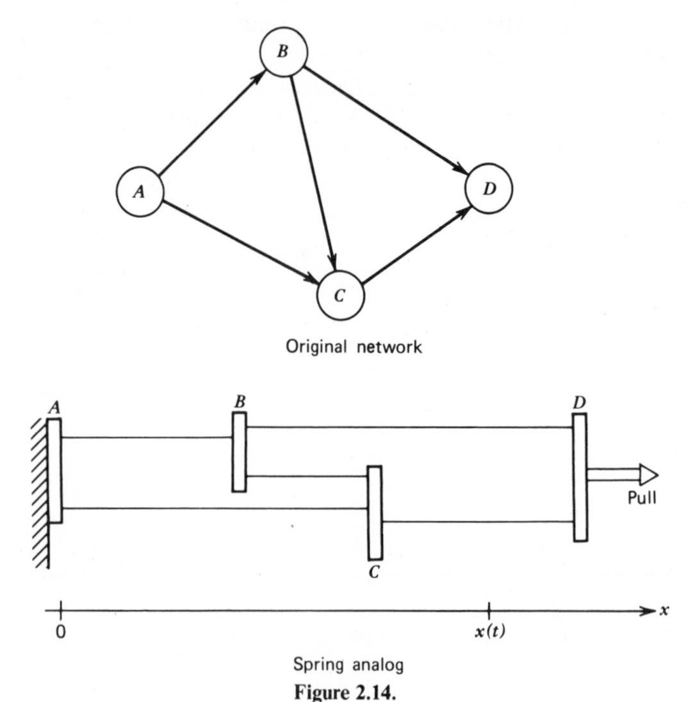

Figure 2.14.

because of their intrinsic value in adding insight into the inner workings of the model.

Lemma 2.1: In a directed acyclic network with no slack, the times of realization of all nodes as well as the durations of all activities are completely specified by specifying the duration of the arcs of any spanning tree.

Lemma 2.2: Under the cost structure of Eq. 2.19, the cost of an optimal schedule with no slack is strictly convex in the project duration T.

Lemma 2.1 asserts that *in the absence of any slack* in the network it is sufficient to know the durations of only $n-1$ activities that constitute a spanning tree in order to deduce the times of realization of *all nodes* of the network (since they lie on the tree) and the durations of *all other arcs* (equal to the difference between the times of realization of their respective defining nodes).

Lemma 2.2 asserts that the C–T curve is strictly convex. This is true because the total cost is the sum of the individual activity costs, which were assumed strictly convex.

Now, Theorem 2.2 asserts, first, that an optimal schedule possesses no slack, for if an activity (ij) has any float we can always increase its duration to be exactly equal to $t_j - t_i$ with no adverse effect on the total project duration. But by the strict convexity of the cost, such prolongation of its duration would *reduce* the total cost, which contradicts the optimality of the original schedule. Hence, no activity can have any float. This immediately leads, by invoking Lemma 2.1, to the conclusion that knowledge of the arc durations on a tree is sufficient to determine all node times and all arc durations.

Second, the theorem asserts that the sum of the derivatives, $D(\tau)$, is constant over the duration of the project. This property is the key to the actual determination of the activity durations, as we see below. It asserts that if one considers two distinct points in time, say τ_1 and τ_2, with $0<\tau_1<\tau_2<T$, and if at time τ_1 activities $a_1, a_2, \ldots, a_k$ are in progress, whose durations are $y_{a_1}, y_{a_2}, \ldots, y_{a_k}$ and costs are $c_{a_1}, c_{a_2}, \ldots, c_{a_k}$; while at time τ_2 activities $b_1, b_2, \ldots, b_r$ are in progress, with respective parameters y_{b_1}, $y_{b_2}, \ldots, y_{b_r}$ and $c_{b_1}, c_{b_2}, \ldots, c_{b_r}$; then

$$D(\tau_1) = \sum_{i=1}^{k} \frac{\partial c_{a_i}}{\partial y_{a_i}} = \sum_{j=1}^{r} \frac{\partial c_{b_j}}{\partial y_{b_j}}$$

The equality of the marginal costs at all points of time is intuitively appealing when one realizes that individual activity durations were assumed to vary between 0 and $+\infty$.

The utility of Theorem 2.2 is in its application to the determination of optimal arc durations. The development of the argument is as follows. Consider an optimal schedule, S^*, and assume for the time being that all node realization times are different (this assumption will be relaxed below). Then the t_i values can be arranged such that

$$0 = t_1 = t_{i_1} < t_{i_2} < \cdots < t_{i_n} = t_n = T \tag{2.20}$$

Furthermore, since

$$D_k(\tau) = \text{constant for } t_{i_k} < \tau < t_{i_{k+1}} \tag{2.21}$$

then we conclude, by condition 1 of Theorem 2.2 and Lemma 2.1, that there are exactly $n-1$ unknowns, specifically, the arc durations of the spanning tree. But condition 2 of Theorem 2.2 and Eq. (2.21) yield the conclusion that there are $n-2$ equations derived from the relations

$$D_1(\tau) = D_2(\tau) = \cdots = D_{n-1}(\tau) \tag{2.22}$$

In addition, there is the condition on the total duration of the project,

$$t_n = T \tag{2.23}$$

Thus, S^* must be determined by the solution of this system of $n-1$ equations in $n-1$ unknowns.

Now we may relax the assumption of strict inequalities in (2.20) and assert that the result derived above still holds. For suppose that two nodes, say i and j, occur at the same time; in other words, suppose that Eq. (2.20) is of the form

$$0 = t_1 = t_{i_1} < t_{i_2} < \cdots < t_i = t_j < t_{i_{j+1}} < \cdots < t_{i_n} = T$$

If i and j are connected by an arc, then activity (ij) may be regarded as a "dummy" activity of duration zero. Thus, the loss of one equality relationship of the form (2.22) is offset by the knowledge of one arc duration, and we still have $n-2$ equations in $n-2$ unknowns. If, on the other hand, nodes i and j are not connected, then there are only $n-3$ relationships derived from Eq. (2.22). However, in this case there is the added condition that $t_i = t_j$ which, together with conditions (2.22) and (2.23) result in the desired number of $n-1$ equations in $n-1$ unknowns.

Unfortunately, the solution of this set of equations, assuming that a solution is tried, is not an operational approach to the determination of the optimal activity durations, since one can write the set of equations only *after* one has already specified the optimal schedule! Evidently, what is needed is an iterative scheme by which a "guess" is made at the optimal schedule, the conditions for optimality are checked and, if not satisfied, the guess is updated, and the cycle repeats until a satisfactory schedule is achieved. Such an iterative procedure was proposed by Berman (1964) based on sufficiency conditions on the node times (rather than on the arc durations). We derive his result next.

§4.1 A Condition on Arc Durations

Theorem 2.2 specifies the necessary and sufficient conditions on the *arc* durations of the network. It is possible to deduce (equivalent) conditions on the *node* times.

Consider any node j whose time of realization in the optimal schedule S^* is t_j. Suppose that for some finite time span before and after t_j no other node has been realized. Denote the interval before j by β_j and the interval after j by α_j (see Fig. 2.15). The condition of the theorem that

$$D(\tau | \tau \in \beta_j) = D(\tau | \tau \in \alpha_j) \tag{2.24}$$

can now be translated relative to the time of realization of node j through

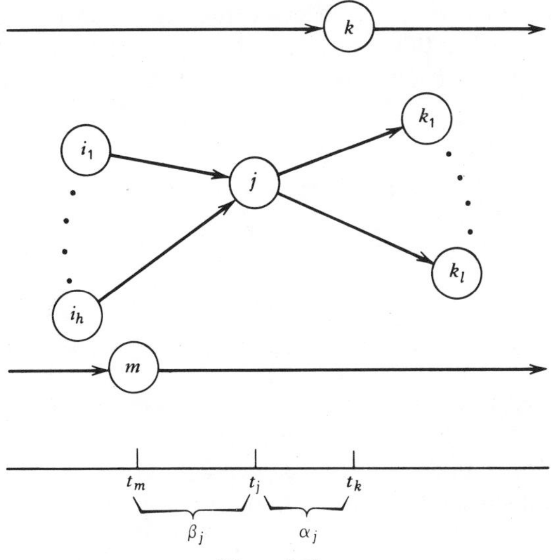

Figure 2.15.

a variational argument as follows. Suppose that nodes $i_1, \ldots, i_h$ connect *to* node j, while nodes $k_1, \ldots, k_l$ connect *from* node j. Then it is easy to see that an infinitesimal shift in node j to the right, that is, to time $t_j + \Delta t_j$, "stretches" the arrows into j and "contracts" the arrows out of j, all by the same infinitesimal amount, provided that nodes $i_1, \ldots, i_h$ and $k_1, \ldots, k_l$ are kept fixed. All nodes and arcs not connected to j remain unaffected. In particular, the $(i_u j)$ arcs become of duration $y_{i_u j} + \Delta t_j$ and the (jk_v) arcs become $y_{jk_v} - \Delta t_j$. The total change in the cost of arrows into j is

$$\sum_{i \to j} - \Delta c_{ij}$$

and the total change in the cost of arrows out of j is

$$\sum_{j \to k} \Delta c_{jk}$$

By Eq. (2.24) one has that, at the optimal, the total variation is zero; in other words,

$$K - \sum_{i \to j} \Delta c_{ij} = K + \sum_{j \to k} \Delta c_{jk}, \tag{2.25}$$

where K is the same constant on both sides of the equation, representing the value of $D(\tau)$ for the schedule before perturbation. Equation (2.25) yields, after canceling K from both sides and dividing throughout by Δt_j then letting $\Delta t_j \to 0$,

$$\sum_{i \to j} \frac{\partial c_{ij}}{\partial t_j} + \sum_{j \to k} \frac{\partial c_{jk}}{\partial t_j} = 0 \tag{2.26}$$

Equation (2.26), which is derived here from the theorem, is identical to the necessary condition for "local" optimality derived by Berman.

Example 2.4.[†] As an illustration of the concepts presented above, consider the network of Fig. 2.16, in which the numbers on the arcs are their "minimal cost" durations (to be explained below) and the numbers next to the nodes are their earliest realization times, $\{t_i\}$. We assume the cost

[†]Adapted from Berman (1964).

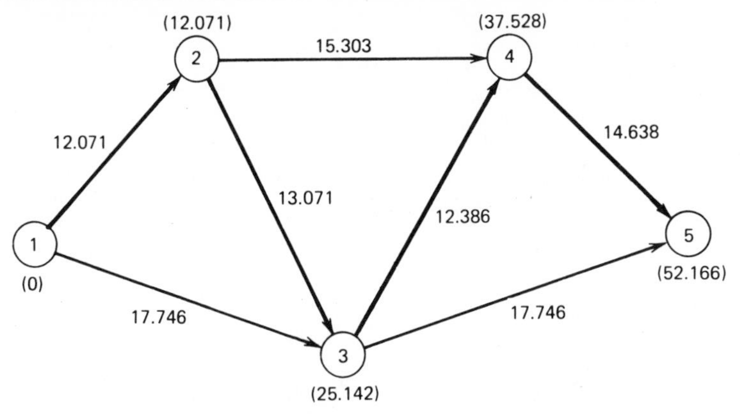

Figure 2.16. Network and its minimal-cost durations.

function is given by

$$\phi(y_{ij}) = \begin{cases} a_{ij} + b_{ij}y_{ij} + \dfrac{c_{ij}}{y_{ij} - l_{ij}}; l_{ij} \leqslant y_{ij} \leqslant l_{ij} + \left(\dfrac{c_{ij}}{b_{ij}}\right)^{1/2} \\ \phi\left[l_{ij} + \left(\dfrac{c_{ij}}{b_{ij}}\right)^{1/2}\right]; y_{ij} > l_{ij} + \left(\dfrac{c_{ij}}{b_{ij}}\right)^{1/2} \end{cases} \tag{2.27}$$

Here $\{a_{ij}\}$, $\{b_{ij}\}$, $\{c_{ij}\}$, and $\{l_{ij}\}$ are constants, in which l_{ij} represents a lower bound on the duration. The dichotomy in the function ϕ stems from the fact that ϕ achieves its minimum at $y = l + (c/b)^{1/2}$; hence, for $y > l + (c/b)^{1/2}$ the cost remains at the minimum of the function, representing the fact that slack is permissible in the system.† The interpretation of the cost function ϕ may be that the constant a_{ij} represents the fixed cost of undertaking the activity, with the second term roughly representing the impact on capital cost (which increases with duration) and the third term, the cost of resources (which must increase with shorter durations). The parameters of the sample problem are given in the first four columns of Table 2.4.

Column 5 of Table 2.4 is obtained by drawing the network assuming $y_{ij} = l_{ij} + (c_{ij}/b_{ij})^{1/2}$ for all (ij) as shown in Fig. 2.16. It is easy to see that the minimum-cost duration of the project is equal to 52.166 days, at a cost of

†Furthermore, the function ϕ is strictly convex only in the interval $[l, l + (c/b)^{1/2}]$. This relaxation of the assumptions of the model is illusory since all arc durations will be in this range to achieve the desired total project duration.

Table 2.4

(1)	(2)	(3)	(4)	(5)	(6)
Activity	b_{ij} \$/day	c_{ij} \$/day	l_{ij} days	Minimal-cost Duration	Initial Trial Duration y_{ij}^0
(1,2)	100	5000	5	12.071	6
(1,3)	50	3000	10	17.746[a]	13
(2,3)	80	4000	6	13.071	7
(2,4)	150	8000	8	15.303[a]	13
(3,4)	110	6000	5	12.386	6
(3,5)	100	6000	10	14.746[a]	17
(4,5)	120	7000	7	14.638	11

$a_{ij} = a = \$1000$ for all activities

[a] Float exists in these arcs.

$\$\Sigma\phi_{ij}[l_{ij} + (c_{ij}/b_{ij})^{1/2}] = \$22{,}588.4$. To be sure, $D(\tau) = 0$ for all τ in the interval $0 \leqslant \tau \leqslant T = 52.166$.

Now suppose that the desired completion time is $T = t_5 = 30$ days only; what is the optimal "compression" of the various activities? Assume initial trial durations, y_{ij}^0, as shown in column 6 of Table 2.4; also shown in Fig. 2.17. The network possesses no slack. The number in parentheses on each

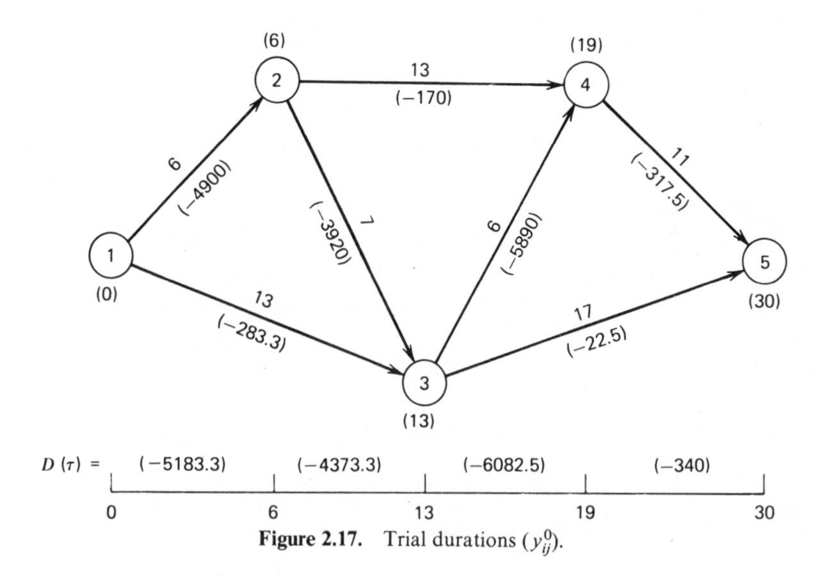

Figure 2.17. Trial durations (y_{ij}^0).

Table 2.5

Interval	$D(\tau)$
$0 \leqslant \tau \leqslant 6$	-5183.3
$6 \leqslant \tau \leqslant 13$	-4373.3
$13 \leqslant \tau \leqslant 19$	-6082.5
$19 \leqslant \tau \leqslant 30$	$-\ 340.$

arc is $\partial\phi_{ij}/\partial y_{ij}$; from which we see that $D(\tau)$ varies as shown in Table 2.5 (also shown along the time scale in Fig. 2.17).

Clearly, the greatest discrepancy in $D(\tau)$ is between the last two intervals of τ. We start the iterations by perturbing node *4* one unit of time to the right, which is achieved by shortening arc $(4,5)$ by one unit and lengthening arcs $(2,4)$ and $(3,4)$ by one unit each, leaving all other arcs undisturbed. The outcome is the network of Fig. 2.18, with the corresponding $D(\tau)$ as shown on the time scale.

Now the greatest discrepancy in $D(\tau)$ is between the second and third intervals of τ, and so on. Continuing in this fashion we finally arrive at the balanced (except for small rounded-off errors) network of Fig. 2.19, in which $D(\tau) \cong 1793.3$ for all τ; $0 \leqslant \tau \leqslant 30$. The total cost of the shortened

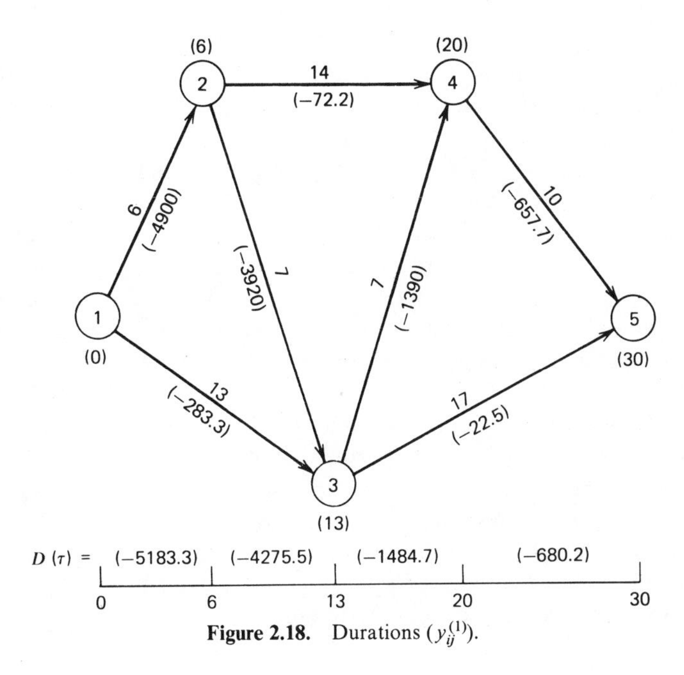

Figure 2.18. Durations ($y_{ij}^{(1)}$).

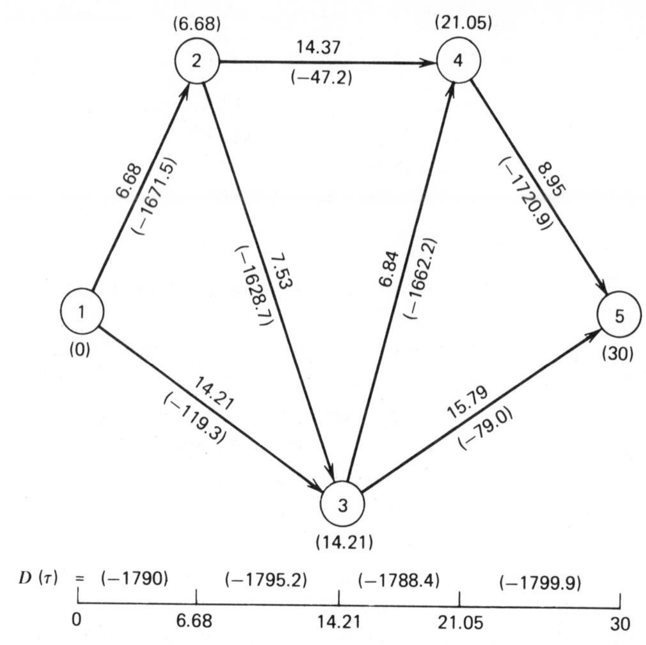

Figure 2.19. Optimal time compression for total duration $T=30$.

project is \$27,372.5. Thus, the reduction in duration by 42.49% [=(52.166 − 30)/52.166 × 100] increased the cost by 30.69%[= (20,372.5 − 15,588.4)/15,588.4 × 100] of the *variable costs.*

When the problem was formulated at the beginning of this section we adhered to the cost structure suggested by Clark. This was done deliberately to arrive at the same conclusions by a different reasoning. However, since none of the results depends on the assumption that the cost $c_a \to +\infty$ as $y_a \to 0$, we are in a position to drop this assumption completely and require only continuity and strict convexity. This permits the utilization of cost functions of the exponential form, which have a finite intercept at $y_a = 0$.

Harking back to the managerial questions posed on p. 59, it becomes immediately evident that the departure from linearity, presumably to be more "realistic" in modeling the cost function, takes its toll in our capability to answer these questions. Thus, it is no longer easy to construct the complete project cost-duration function, as we did in Fig. 2.6. To be sure, we can still determine the minimal total cost given a desired project completion date, that is, we can determine $C(T)$, but such determination is achieved at the expense of considerably more computing effort (as to be expected). The marginal cost, $\partial C^*/\partial T$ at such duration is easily determined. But the reverse problem [question (2) on p. 59] is much more difficult to resolve: given a fixed amount of financial resources, say C, what is the shortest realizeable project duration $T^*(C)$? The only approach we can offer the reader is to apply the procedure described above for different values of T, preferably in an optimally laid-out search scheme, until the value $T^*(C)$ is determined. Finally, the question on bottleneck activities [see (4) on p. 60] can be immediately and easily answered; under the assumptions of this model, all activities must be shortened—hence, they are all bottleneck activities.

§4.2 The General Case*

Turning to the general case in which activities have convex (but not strictly convex) cost functions, $\{c_{ij}(y_{ij})\}$, and the duration activity (ij) is bound from above and below,

$$l_{ij} \leqslant y_{ij} \leqslant u_{ij}$$

the problem can be posed as follows:

$$\text{minimize } C = \sum_{(ij) \in A} c_{ij}(y_{ij}) \tag{2.28}$$

subject to the linear constraints

$$t_j - t_i - y_{ij} \geqslant 0, \qquad (ij) \in A \tag{2.29}$$

$$y_{ij} \geqslant l_{ij} \qquad (ij) \in A \tag{2.30}$$

$$y_{ij} \leqslant u_{ij} \qquad (ij) \in A \tag{2.31}$$

$$t_n - t_1 = T \tag{2.32}$$

There are exactly $M = 3A + 1$ constraints: A constraints in each of Eqs. (2.29)–(2.31) and Eq. (2.32), which specifies the duration of the project.

Denote the coefficient of t_j in the rth equation of (2.29) by α_{rj}, $r = 1, \ldots, A$, and the coefficient of y_{ij} in the rth equation of (2.29)–(2.31) by β_{rj}, $r = 1, \ldots, 3A$. Note that α_{rj} and β_{rj} are either $+1$ or -1.

*Starred sections may be omitted on first reading.

Equation (2.28) is a convex separable function, explicit only in the arc durations $\{y_{ij}\}$. Consequently,

$$\frac{\partial C}{\partial y_{ij}} = \frac{dc_{ij}}{dy_{ij}}, \qquad (ij) \in A,$$

and

$$\frac{\partial C}{\partial t_j} = 0, 1 \leqslant j \leqslant n$$

The constraints of Eqs. (2.29)–(2.32) are linear and hence can be considered either convex or concave. It follows, by the Kuhn–Tucker theorem (1951) that a necessary and sufficient condition that C takes on its global minimum at $\{t_j^*, y_{ij}^*\}$ is that there exists a set of values $\lambda_1^*, \ldots, \lambda_M^*$, with

$$\begin{aligned} &\lambda_r^* \leqslant 0 \qquad \text{for } r = 1, \ldots, 2A \\ &\lambda_r^* \geqslant 0 \qquad \text{for } r = 2A+1, \ldots, 3A \\ &\lambda_M^* \text{ unrestricted in sign,} \end{aligned}$$

such that

$$\begin{aligned} &\frac{\partial C}{\partial t_j} = 0 \leqslant \sum_{r=1}^{M} \alpha_{rj} \lambda_r^*, \qquad j = 1, \ldots, n; M = 3A+1, \\ &\frac{\partial C}{\partial y_{ij}} = \frac{dc_{ij}}{dy_{ij}} \leqslant \sum_{r=1}^{M} \beta_{rj} \lambda_r^*, \qquad (ij) \in A \end{aligned} \tag{2.33}$$

and

$$\sum_{j=1}^{M} t_j^* \left[-\sum_{r=1}^{M} \alpha_{rj} \lambda_r^* \right] + \sum_{(ij) \in A} y_{ij}^* \left[dc_{ij}/dy_{ij} - \sum_{r=1}^{M} \beta_{rj} \lambda_r^* \right] = 0 \tag{2.34}$$

and

$$\sum_{r=1}^{A} \lambda_r^* [t_i + y_{ij} - t_j] + \sum_{r=A+1}^{2A} \lambda_r^* [l_{ij,r} - y_{ij,r}] + \sum_{r=2A+1}^{3A} \lambda_r^* [u_{ij,r} - y_{ij,r}]$$
$$+ \lambda_M^* [T + t_1 - t_n] = 0 \tag{2.35}$$

An alternative formulation to Eq. (2.34) is obtained as follows. Grouping the terms in λ_r^* in Eq. (2.34) one obtains

$$\sum_{(ij)\in A} y_{ij}^* \frac{dc_{ij}}{dy_{ij}} = \sum_{r=1}^{M} \lambda_r^* \left[\sum_{j=1}^{N} \alpha_{rj} t_j^* + \sum_{(ij)\in A} \beta_{rj} y_{ij}^* \right] \tag{2.36}$$

From Eqs. (2.29)–(2.32), the expression between square brackets in Eq. (2.36) is equal to 0 for $r=1,\ldots,A$; equal to $l_{ij,r}$ for $r=A+1,\ldots,2A$; equal to $u_{ij,r}$ for $r=2A+1,\ldots,3A$, and equal to T for $r=M$. Therefore,

$$\sum_{(ij)\in A} y_{ij}^* \frac{dc_{ij}}{dy_{ij}} = \sum_{A+1}^{2A} l_{ij,r}\lambda_r^* + \sum_{2A+1} u_{ij,r}\lambda_r^* + T\lambda_M^* \tag{2.37}$$

Equations (2.33), (2.35), and (2.37) are the necessary and sufficient conditions for the optimal duration of the project. It is not immediately obvious how these conditions can be utilized in a computing scheme (other than trial and error) for determining the optimal $\{t_j^*\}$, $\{y_{ij}^*\}$, and $\{\lambda_r^*\}$. This issue is still open to investigation.

Finally, this problem has been treated by Lamberson and Hocking (1970) by an approach based on the work of Hartley and Hocking (1963) on the minimization of a convex function over a convex space. The gist of the approach is as follows. The problem is first "linearized" by the introduction, at least conceptually, of a large number of tangent planes to approximate the nonlinear function. (In practice, iteration is carried out with only a few such tangent planes.) Then the dual of this LP is considered. By using a special pricing operation based on decomposition considerations, the algorithm indicates whether a tangent plane should be constructed and if so, which one, in order to improve the current solution through a standard LP pivoting operation. Iteration proceeds to termination, which occurs in a finite number of steps.

The relative (computing) efficiency of this approach to other approaches, especially the separable programming algorithms (e.g., that of IBM) is not evident. Space does not permit us to give a detailed account of the procedure, so we must be content with this brief discussion.

§5. CONCAVE COST-DURATION FUNCTION

As long as there is no empirical evidence to justify the assumption of strict convexity, or concavity, of the cost function, it seems reasonable to study both conditions. On the other hand, concavity of costs may be rationalized on the basis of accrued economies due to longer contractural periods, that is, for longer activity durations. Thus, while the cost of the activity decreases with its duration, the largest savings are realized, so to speak, at the longer durations (see Fig. 2.20). Denoting the duration of activity (ij) by y_{ij}, and letting its upper and lower limits be u_{ij} and l_{ij}, respectively, the problem may be formally stated as the following program:

$$P\text{: minimize } C = \sum_{(ij)\in A} c_{ij}(y_{ij})$$

$$\text{s.t.} \quad \left.\begin{aligned} t_i + y_{ij} &\leqslant t_j; && (ij)\in A \\ 0 \leqslant l_{ij} \leqslant y_{ij} &\leqslant u_{ij}; && (ij)\in A \\ t_1 = 0,\ t_n &= T; && \\ y_{ij} &\geqslant 0, && (ij)\in A \end{aligned}\right\} F \qquad (2.38)$$

Here, t_i is, as before, the time of realization of node i; and $c_{ij}(y_{ij})$ is assumed concave as shown in Fig. 2.20.

This problem was treated by Falk and Horowitz (1972), whose approach is the one detailed below. It is, in turn, an adaptation of the work of Falk and Soland (1969) and Rech and Barton (1970).

The approach is simply the following. Suppose that the concave cost function of Fig. 2.20 is approximated by a linear function as shown dotted in Fig. 2.21, which is the highest linear function to *underestimate* $c_{ij}(y_{ij})$. Denote such linear cost function by $c_{ij}^1(y_{ij})$. Then we may take $C^1(Y)$ as a

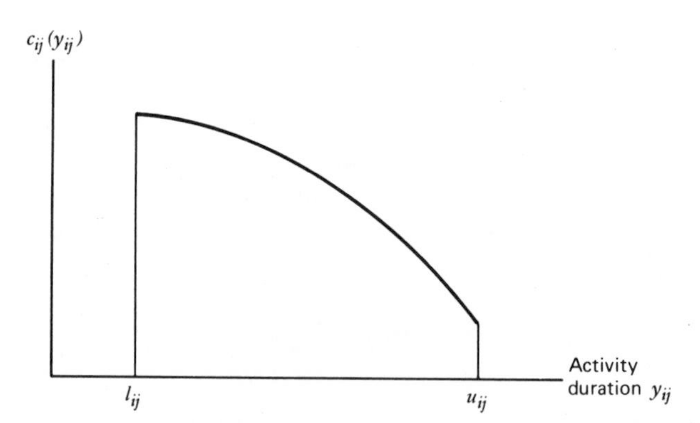

Figure 2.20. Concave cost-duration function.

first (lower) approximation to the objective function of the program P of (2.38), and formulate the first "estimating problem, Q^1" of the form:

$$Q^1\text{: minimize } C^1(Y) = \sum_{(ij)\in A} c_{ij}^1(y_{ij});$$

$$\left.\begin{aligned} \text{s.t.} \quad & t_i + y_{ij} \leqslant t_j; && (ij)\in A \\ & l_{ij}^1 \leqslant y_{ij} \leqslant u_{ij}^1; && (ij)\in A \\ & t_1 = 0;\ t_n = T; && \end{aligned}\right\} F^1 \tag{2.39}$$

Here, the upper and lower bounds on the duration y_{ij} are $l_{ij}^1 = l_{ij}$ and $u_{ij}^1 = u_{ij}$, as shown in Fig. 2.21.

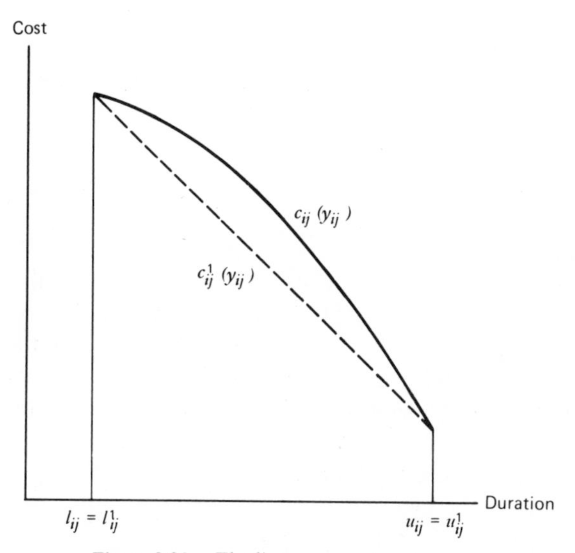

Figure 2.21. The linear approximation.

Interestingly enough, the program Q^1 is a *linear program* that can be solved by the Fulkerson approach detailed in Section 2 (or 3).

For the sake of simplicity of notation, denote the feasible space of the program P by F, and denote the feasible space of the program Q^1 by F^1. Clearly, $F^1 \equiv F$. Consequently, if P is feasible, so is Q^1. Let Y^1 denote the optimal solution of Q^1; then $C^1(Y^1) = \Sigma c_{ij}^1(y_{ij}^1)$ is a *lower bound* on the optimal value of the program P, which we denote by C^*. Furthermore, since Y^1 is a feasible solution for P, $C(Y^1)=\Sigma c_{ij}(y_{ij}^1)$ is an upper bound on C^*. Thus, we have succeeded in bounding the optimal value C^* based on the optimal solution of Q^1 as follows:

$$C^1(Y^1) \leqslant C^* \leqslant C(Y^1) \tag{2.40}$$

Clearly, if equality holds throughout (2.40), then the current trial solution is optimal. This remark holds for all subsequent iterations and hence is not repeated.

Now suppose that strict inequality holds in (2.40); then there is room for improvement. This is accomplished by producing a closer (albeit still an underestimate) linear approximation to the cost function $c(y)$. Consider the difference.

$$c_{ij}(y_{ij}^1) - c_{ij}^1(y_{ij}^1) \tag{2.41}$$

which is always $\geqslant 0$; and suppose that activity $(i_1 j_1)$ yields the maximum such difference; in other words,

$$c_{i_1j_1}(y_{i_1j_1}^1) - c_{i_1j_1}^1(y_{i_1j_1}^1) = \max_{(ij)\in A}\left[c_{ij}(y_{ij}^1) - c_{ij}^1(y_{ij}^1)\right] > 0$$

Such an activity must exist, for otherwise the differences of (2.41) are $=0$ for all activities, implying equality in (2.40); a contradiction. (In case of ties, any tied activity will do.) Divide the feasible domain of $y_{i_1j_1}$ into two subintervals: $[l_{i_1j_1}, y_{i_1j_1}^1]$ and $[y_{i_1j_1}^1, u_{i_1j_1}]$. [Recall that $y_{i_1j_1}^1$ is the value of the duration of activity $(i_1 j_1)$ obtained from the optimal solution of the program Q^1.] Construct the two cost functions: $c_{i_1j_1}^2(y_{i_1j_1})$ and $c_{i_1j_1}^3(y_{i_1j_1})$ as the maximum linear approximations that underestimate the original cost function $c_{i_1j_1}(y_{i_1j_1})$ in the two subintervals $[l_{i_1j_1}, y_{i_1j_1}^1]$ and $[y_{i_1j_1}^1, u_{i_1j_1}]$ (see Fig. 2.22). Now define the two linear programs:

$$Q^2\text{: minimize } \sum_{(ij)\neq(i_1j_1)} c_{ij}^1(y_{ij}) + c_{i_1j_1}^2(y_{i_1j_1})$$

$$\text{s.t.}\quad \left.\begin{array}{ll} t_i + y_{ij} \leqslant t_j; & \text{all } (ij)\in A \\ l_{ij} \leqslant y_{ij} \leqslant u_{ij}; & (ij)\neq(i_1j_1) \\ l_{i_1j_1} = l_{i_1j_1}^2 \leqslant y_{i_1j_1} \leqslant u_{i_1j_1}^2 = y_{i_1j_1}^1; & \\ t_1 = 0, t_n = T; & \end{array}\right\} F^2 \tag{2.42}$$

$$Q^3\text{: minimize} \quad \sum_{(ij)\neq(i_1 j_1)} c_{ij}^1(y_{ij}) + c_{i_1 j_1}^3(y_{i_1 j_1})$$

$$\text{s.t.} \quad \left.\begin{array}{ll} t_i + y_{ij} \leqslant t_j; & \text{all } (ij) \in A \\ l_{ij} \leqslant y_{ij} \leqslant u_{ij}; & (ij) \neq (i_1 j_1) \\ y_{i_1 j_1}^1 = l_{i_1 j_1}^3 \leqslant y_{i_1 j_1} \leqslant u_{i_1 j_1}^3 = u_{i_1 j_1} & \\ t_1 = 0, t_n = T & \end{array}\right\} F^3 \tag{2.43}$$

The logic of these two programs rests on the observation that the duration of the activity (i_1,j_1) must lie in either of the two subintervals of Fig. 2.22. Note that the only difference between these two programs is in the definition of the range of the duration $y_{i_1 j_1}$. Both programs are feasible, since the point Y^1 is still a feasible point of either of them.

Let F^2 denote the space of feasible solutions for the program Q^2 and F^3 denote the space of feasible solutions for the program Q^3. Clearly, F^2 and F^3 are defined by

$$F^2 = F^1 \cap \{(Y,t) : l_{i_1 j_1}^1 \leqslant y_{i_1 j_1} \leqslant y_{i_1 j_1}^1\}$$

$$F^3 = F^1 \cap \{(Y,t) : y_{i_1 j_1}^1 \leqslant y_{i_1 j_1} \leqslant u_{i_1 j_1}^1\} \tag{2.44}$$

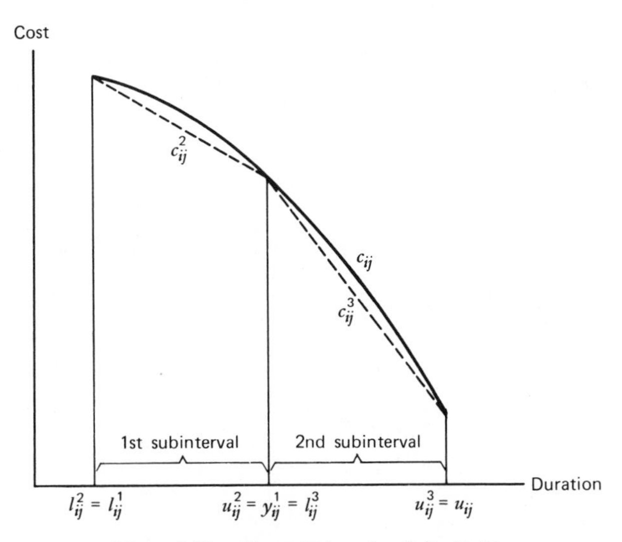

Figure 2.22. The splitting of activity $(i_1 j_1)$.

Furthermore,

$$c_{ij}^2(y_{ij}) = c_{ij}^3(y_{ij}) = c_{ij}^1(y_{ij}) \quad \text{for } (ij) \neq (i_1 j_1) \tag{2.45}$$

Therefore, problems Q^2 and Q^3 may be succinctly stated as:

$$Q^2: \text{minimize } C^2(Y), \quad \text{s.t. } (Y,t) \in F^2;$$
$$Q^3: \text{minimize } C^3(Y), \quad \text{s.t. } (Y,t) \in F^3$$

Now both problems Q^2 and Q^3 may be solved by the Fulkerson algorithm, yielding optimal durations Y^2 and Y^3, respectively. By virtue of the fact that the cost functions C^2 and C^3 serve as tighter underestimates of C over their domains, and that the feasible space $F \equiv F^1 = F^2 \cup F^3$, we have that

$$C^1(Y^1) \leqslant \min\{C^2(Y^2), C^3(Y^3)\} \leqslant C^*$$
$$\leqslant \min\{C(Y^1), C(Y^2), C(Y^3)\} \leqslant C(Y^1)$$

The rightmost inequality follows from (Y^2, t^2) and (Y^3, t^3) being feasible solutions to problem P. We have thus achieved improved bounds on the optimal value C^*.

From this point onward the algorithm proceeds in a series of stages. The zeroth stage detailed above consisted of problem Q^1 and its solution Y^1; and the first stage consisted of problems Q^2 and Q^3 and their solutions, Y^2 and Y^3. The kth stage consists of problems Q^{2k} and Q^{2k+1} and their solutions. The process is conveniently depicted by a typical binary-search tree whose nodes correspond to the problems $\{Q^s\}$. Figure 2.23 depicts such a tree with four stages and nine nodes. Branching occurs when a particular activity is selected to have one of its (duration) intervals divided into two subintervals, as exemplified in Fig. 2.22, with costs and bounds redefined as in (2.42) and (2.43). A branching node s may be selected according to some heuristic rule, for example, by choosing that problem Q^s whose optimal value, $C^s(Y^s)$, is minimal over all cost values associated with intermediate nodes (i.e., nodes from which no branching has occurred). The rationale here is that problem Q^s has the smallest lower bound on C^* and, hopefully, the feasible space F^s will contain a point yielding a value to C approximately equal to $C^s(Y^s)$. Another possible heuristic rule is to choose that problem Q^s whose optimal solution, Y^s, yields the smallest interval of uncertainty on C^*. The reader is invited to state the possible rationale behind this rule and to suggest other branching rules.

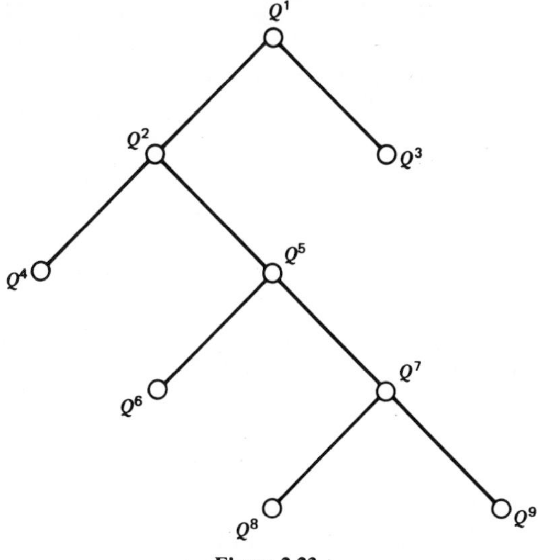

Figure 2.23.

In any event, having chosen a problem Q^s at stage k from which to branch, we create the problems Q^{2k} and Q^{2k+1} in a manner exactly analogous to the manner in which Q^2 and Q^3 were created from problem Q^1. The feasible spaces F^{2k} and F^{2k+1} are thus defined by

$$F^{2k} = F^s \cap \{(Y,t) : l^s_{i_s j_s} \leqslant y_{i_s j_s} \leqslant y^s_{i_s j_s}\}$$

$$F^{2k+1} = F^s \cap \{(Y,t) : y^s_{i_s j_s} \leqslant y_{i_s j_s} \leqslant u^s_{i_s j_s}\}$$

Furthermore, $c^{2k}_{i_s j_s}$ is the highest linear function underestimating $c_{i_s j_s}$ over the subinterval $[l^s_{i_s j_s}, y^s_{i_s j_s}]$, while $c^{2k+1}_{i_s j_s}$ is the highest linear function underestimating $c_{i_s j_s}$ over subinterval $[y^s_{i_s j_s}, u^s_{i_s j_s}]$. The sets F^{2k} and F^{2k+1} are feasible, since the point (Y^s, t^s) lies in both sets. The new problems Q^{2k} and Q^{2k+1} are linear problems with network constraints similar to those of problem P and thus may be solved by the Fulkerson algorithm. The optimal value C^* is bound at stage $k+1$ by

$$v^{k+1} = \min_{s=1,2,\ldots,2k+1} \{C^s(Y^s)\} \leqslant C^* \leqslant \min_{s=1,2,\ldots,2k+1} \{C(Y^s)\} = w^{k+1}$$

The process continues until either $v^k = w^k$ at some stage k or when the interval bounding C^* is deemed sufficiently small for practical purposes.

There remains to make the following three remarks. First, the choice of the subintervals on the duration $y_{i_s j_s}$ of problem Q^s is arbitrary—the division indicated above seems to be a reasonable heuristic. Second, the above approach is clearly applicable to piecewise linear functions, whether concave or convex. In fact, the numerical example worked out in the original paper of Falk and Horowitz (1972) contained arcs in both categories. Third, the proof that the algorithm is finite rests on the fact that the function C is concave and defined over a convex polytope; hence, it must assume its minimum at one of the finite vertices of the feasible space F. Most of the subproblems Q^s have their solutions at vertices of F; but since new vertices are created when new upper and lower bounds u_{ij}^s and l_{ij}^s are added, some Q problems may have solutions at points which are not vertices of F. The proof that only a finite number of such problems exist is given in the work by Falk and Soland (1969).

§6. DISCONTINUOUS NONINCREASING FUNCTION: A DYNAMIC PROGRAMMING APPROACH*

As before, we assume that to each activity there corresponds a cost-duration function $c_i = \phi(y_i)$ that reflects the optimal allocation of resources at different expenditures. Unlike the previous models, however, we impose the following milder restrictions on ϕ: it is assumed nonincreasing with possible discontinuities at cost points $\Delta, 2\Delta, 3\Delta, \ldots,$ starting from the minimal value, $\min c_i$, of the investment c_i in activity i (which is the cost corresponding to the "normal" duration of the activity), to the maximal value, $\max c_i$ (which corresponds to the cost of the "crash" duration).

Such a cost-duration function may represent a reasonable approximation to a continuous function that is neither convex nor concave, as shown in Fig. 2.24*a*. More importantly, it may reflect the discreteness of the cost estimates ab initio. In other words, it may be possible to secure cost estimates (or durations) at a few selected durations of (investments in) the activity and assume that the cost (duration) remain constant over a range about these values; the resultant function is as shown in Fig. 2.24*b*.

Since an investment x_i in activity i may "purchase" a variety of combinations of resources, we assume that the investment is made optimally, and let $g_i(x_i)$ denote the corresponding minimum duration of the activity. For notational convenience the domain of each x_i is transformed to $[0, \max x_i - \min x_i]$, so that now the "investment x_i" really denotes the *additional* investment used to shorten the duration of activity i. Furthermore, we

*Starred sections may be omitted on first reading.

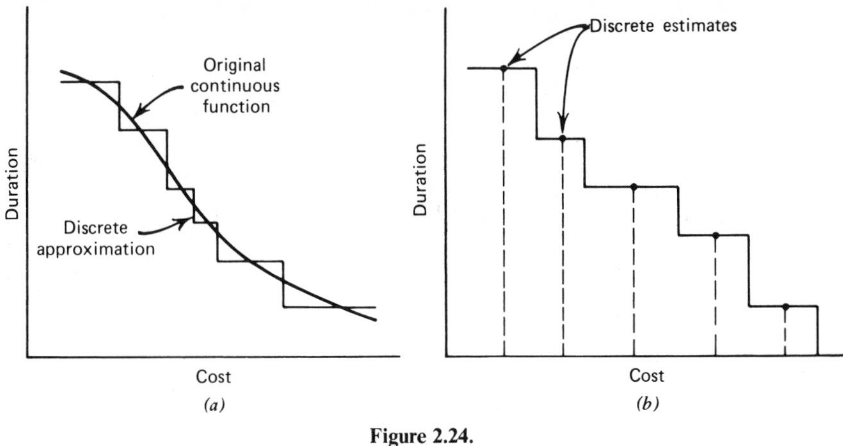

Figure 2.24.

consider the "duration" of the activity to be its actual duration minus its crash duration, l_i; in other words, we shift the duration origin to the point l_i. The function $g_i(x_i)$ is shown schematically in Fig. 2.25. Naturally, we always assume that a given duration is realized at the smallest value of investment x.

Parenthetically, we wish to alert the reader that in this section we are trying to determine what the minimal project duration is for a specified cost (or investment) level c [see also question (2) on p. 59]. The nature of the solution procedure (through dynamic programming) facilitates derivation of the solution for *all* values of investment c. Hence, the complete project cost-duration function is available, which greatly facilitates answer-

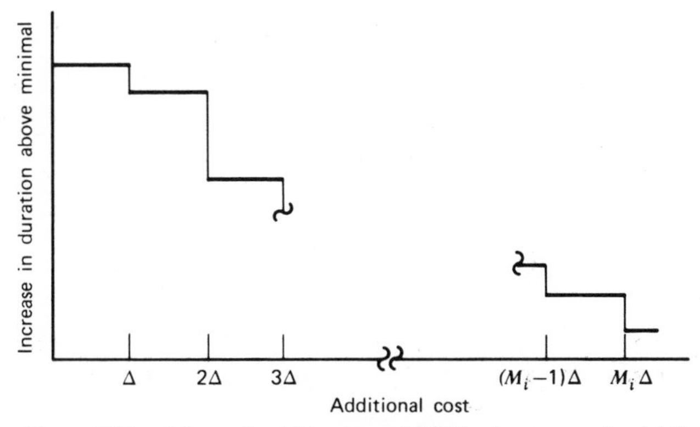

Figure 2.25. Schematic of function $g_i(x_i)(M_i = (\max x_i - \min x_i)/\Delta)$.

ing the other managerial questions of concern. The main drawback of this approach, however, is its computational difficulty except in some very special cases. Its application to large-scale networks must be done judiciously and with the benefit of insight into the structure of the network, or else it will quickly bog down into the quicksand of multidimensional state vectors.

Let a path in the network be denoted by π_k, $k=1,2,\ldots,K$, and let S_k denote the set of activities in path π_k. The duration, or length, of path π_k for a given level of investment $X=(x_i)$ is denoted by $L_k(X)$ and is given by

$$L_k(X)=\sum_{i\in S_k} g_i(x_i)$$

The duration of the project is determined by the length of the longest path; $\max_k L_k(X)$. The objective is to determine

$$\min_X \max_k L_k(X) \triangleq f_N(c)$$

$$\text{s.t.} \qquad \sum_{i=1}^{N} x_i \leqslant c \tag{2.46}$$

$$\text{and } 0\leqslant x_i \leqslant \max x_i - \min x_i = M_i\Delta$$

where c is the total available investment and N the number of "stages" of the DP formulation to be determined presently.

To achieve a DP formulation of the problem, let σ denote the set of activities common to *all* paths;

$$\sigma=\bigcap_{k=1}^{K} S_k$$

Then we may decompose the objective function of (2.46) as follows:

$$f_N(c)=\min_X\left[\sum_{i\in\sigma} g_i(x_i)+\max(L_1',\ldots,L_K')\right]$$

where (2.47)

$$L_k'=\sum_{i\in S_k'} g_i(x_i);\ S_k'=S_k-\sigma$$

Indeed, the activities in the set σ must be performed sequentially; hence, the sum of their durations must be added to the critical path of the residual network (i.e., after these activities have been removed).

The formulation of (2.47) leads one to investigate the (sufficient) conditions under which the minimand is completely decomposed into a sequence of one-dimensional optimization problems with c as the state variable. We have the following interesting sufficient condition:

Theorem 2.3: Let P_i be the set of all paths to which activity i belongs. If for all i and j, with $i \prec j$, either $P_i \subseteq P_j$ or $P_j \subseteq P_i$, then Eq. (2.47) decomposes into a sequence of one-dimensional optimization problems.

Before proving the assertion, we explain its content and implications by an example.

Consider the network of Fig. 2.26, in which we adopt the A-on-N convention. Since every path contains activities *1* and *6*, the set $\sigma = \{1, 6\}$. Furthermore, $2 \prec 5$ and $3 \prec 5$, and the path containing *2* also contains *5*, so does the path containing *3*. Hence the conditions of the Theorem 2.3 are satisfied and the function $f_N(c)$ of (2.47) is completely decomposable.

We illustrate how this decomposition is accomplished. For the sake of brevity we write g_i for $g_i(x_i)$. Then, capitalizing on the set σ, we have

$$f_N(c) = \min_X \left[g_1 + g_6 + \max(g_2 + g_5; g_3 + g_5; g_4) \right]$$

which, in turn, decomposes into

$$f_N(c) = \min_X \left[g_1 + g_6 + \max\{ g_5 + \max(g_2, g_3); g_4 \} \right]. \tag{2.48}$$

The DP recursive scheme is given by the movement from the innermost maxoperator to the outermost minoperator:

$$\text{put} \quad f_1(c) = g_3(c)$$

$$\text{then} \quad f_2(c) = \min_{x_2} \left[\max\{ g_2(x_2), f_1(c - x_2) \} \right]$$

$$f_3(c) = \min_{x_5} \left[g_5(x_5) + f_2(c - x_5) \right]$$

$$f_4(c) = \min_{x_4} \left[\max\{ g_4(x_4), f_3(c - x_4) \} \right]$$

$$f_5(c) = \min_{x_6} \left[g_6(x_6) + f_4(c - x_6) \right]$$

$$f_6(c) = \min_{x_1} \left[g_1(x_1) + f_5(c - x_1) \right]$$

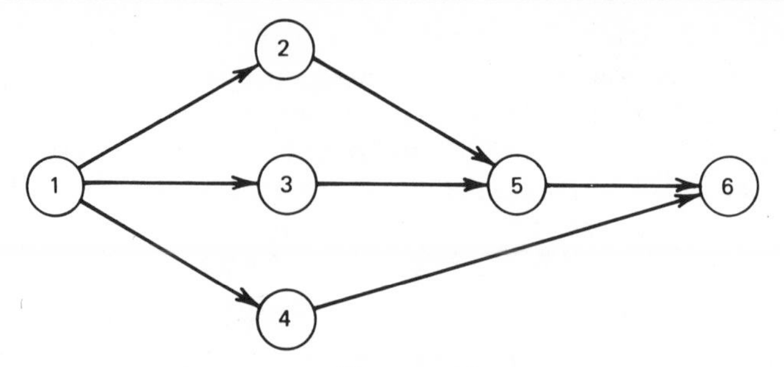

Figure 2.26.

Evidently, with complete decomposition of $f_N(c)$ the number of "stages," N, is equal to the number of activities.

Proof. To prove Theorem 2.3, suppose that the condition of the assertion is satisfied; that is, that each pair of nonconcurrent activities is s.t. every path containing one of the pair also contains the other. Determine the activity (or activities) that appears in the most paths π_i', and let it be activity k (activity *5* in the above example). If a tie exists, select from among the tied activities the one with the smallest index (which leads to an "earlier" activity).† Renumber the paths π_i' so that activity k is in paths $1,2,\ldots,j$ and k is not in paths $j+1,\ldots,K$. Renumber the sets S_i' and L_i' accordingly. The maximization portion of (2.47) may be written as

$$\max(L_1', L_2', \ldots, L_j'; L_{j+1}', \ldots, L_K') \tag{2.49}$$

Now we assert that $(\cup_{i=1}^{j} S_i') \cap (\cup_{i=j+1}^{K} S_i') = \phi$; that is, an activity appearing in any set S_i', $i=1,\ldots,j$, does not appear in any set S_i', $i=j+1,\ldots,K$. (Compare $S_1' = \{2,5\}$, $S_2' = \{3,5\}$ with $S_3' = \{4\}$ in the above example.) The proof is by contradiction; if such an activity exists, say activity l, then k and l are concurrent and there must exist two paths containing l, one of which violates the condition of Theorem 2.3 (viz., that one set of paths is wholly contained in the other); which is a contradiction.

This separability between $\cup_{i=1}^{j} S_i'$ and $\cup_{i=j+1}^{K} S_i'$ leads to the decomposition of (2.49) into:

$$\max\left[\max(L_1', \ldots, L_j'); \max(L_{j+1}', \ldots, L_k')\right] \tag{2.50}$$

†We assume that the convention is maintained, as advocated on p. 4, that the activities are numbered so that an arrow leads from a small-numbered activity to a larger-numbered activity.

Now, $\max(L'_1, \ldots, L'_j)$ can be further decomposed into the form of (2.47), with activity k (or the set of such activities common to *all* paths in this subset) playing the role of the set σ, and with the residual paths permitting further decomposition in the form of Eqs. (2.49) and (2.50). The decomposition process is continued since each step results in subsets satisfying the sufficient condition of Theorem 2.3. This process eventually terminates in an objective function $f_N(c)$ consisting of mutually exclusive sets of g functions linked together by combinations of maximization and addition operations, to which the recursive formalism of DP is applied.

An immediate question that comes to mind at this juncture is, what if the sufficient condition of Theorem 2.3 is not satisfied? The answer is twofold: first, certain network structures are amenable to partial decomposition in spite of the fact that the sufficient condition is not satisfied. Unfortunately, one may then be obliged to conduct multidimensional optimization. Second, one may appeal to *conditional optimization* to render the network decomposable. The major drawback of this latter approach is that is transforms the problem into a multidimensional state vector that is computationally unwieldy. We discuss these two approaches next.

§6.1 Structural Considerations

The "wheatstone bridge" structure is perhaps one of the simplest networks that do not satisfy the sufficient condition of Theorem 2.3 (see Fig. 2.27). Apart from activities *1* and *6*, which are common to all paths [and hence constitute the set σ of (2.47)], the residual pairs of activities do not meet the sufficient condition. Writing the objective function, we have

$$f_N(c) = \min_X \left[g_1 + g_6 + \max(g_2 + g_4, g_2 + g_5, g_3 + g_4, g_3 + g_5) \right]$$

However, it is easy to see that

$$\max(g_2 + g_4, g_2 + g_5; g_3 + g_4, g_3 + g_5)$$

$$= \max\left[g_2 + \max(g_4, g_5); g_3 + \max(g_4, g_5) \right]$$

$$= \max(g_2, g_3) + \max(g_4, g_5)$$

Hence, we may write the objective function as

$$f_N(c) = \min_X \left[g_1 + g_6 + \max(g_2, g_3) + \max(g_4, g_5) \right]$$

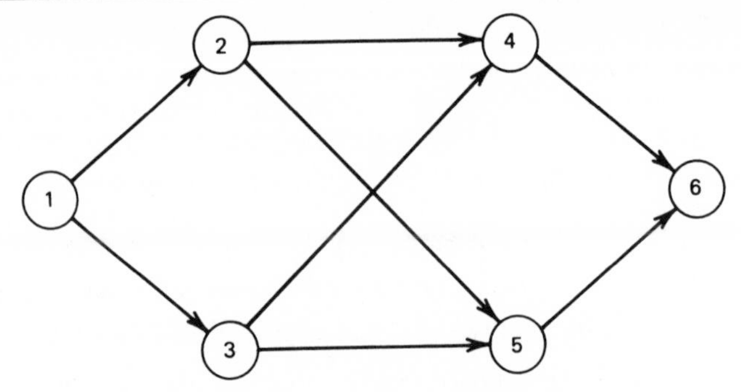

Figure 2.27. The "Wheatstone bridge," A-on-N.

The recursive procedure of DP may be written as:

$$f_1(c) = g_2(c)$$

$$f_2(c) = \min x_3 \left[\max(g_3(x_3), f_1(c - x_3)) \right]$$

$$f_3(c) = \min_{\substack{x_4, x_5 \\ x_4 + x_5 \leq c}} \left[\max\{ g_4(x_4), g_5(x_5)\} + f_2(c - x_4 - x_5) \right] \tag{2.51}$$

$$f_4(c) = \min_{x_6} \left[g_6(x_6) + f_3(c - x_6) \right]$$

$$f_5(c) = \min_{x_1} \left[g_1(x_1) + f_4(c - x_1) \right]$$

Two important phenomena are worth noting in this example. First, in computing $f_3(c)$ we had to resort to optimization over two decision variables, an act that runs counter to the grain of DP. Second, although the network possessed six activities, the number of stages of the DP was only five (i.e., $N=5$). This is due to the simultaneous optimization over two decision variables in computing $f_3(c)$. This phenomenon is generally true; N will be smaller than the number of activities whenever simultaneous optimization is undertaken.

§6.2 Conditional Optimization

The term is coined to reflect the fact that optimization is carried out under the condition that one (or more) of the decision variables is fixed at a specific value. In essence, this ploy transforms the optimization over two (or more) decision variables into two (or more) searches, each in one dimension only, which is the basic concept underlying DP. The ultimate effect is to transfer the multidimensionality of the problem from the decision variables to the state variables. Optimization is then carried out over the range of fixed values of the state variables, and then one chooses as the global optimum the optimum of the optima.

To illustrate, consider $f_3(c)$ of Eq. (2.51). Suppose we treat x_4 as a state variable. (A similar result is obtained if x_5 is used as the state variable rather than x_4.) Then for each value of x_4, say $\hat{x}_4$, the minimization operation is a straightforward one-dimensional search over x_5, in which $g_4(\hat{x}_4)$ is a constant. We denote such minimand by $\hat{f}_3(c,\hat{x}_4)$; it is given by

$$\hat{f}_3(c,\hat{x}_4) = \min_{0 \leqslant x_5 \leqslant c-\hat{x}_4} \left[\max(g_4(\hat{x}_4), g_5(x_5)) + f_2(c-\hat{x}_4-x_5) \right]$$

From this point onward the two-dimensional state vector must be carried through to yield

$$\hat{f}_5(c,\hat{x}_4) = \min_{x_l} \left[g_l(x_l) + \hat{f}_4(\hat{c}-x_l) \right]; \hat{c} = c-\hat{x}_4$$

Finally, the desired optimum is achieved from

$$f_5(c) = \min_{x_4 \leqslant c} \hat{f}_5(c,x_4)$$

For large networks that are not decomposable but threaten to result in a state vector of high dimensionality, the sufficient condition of Theorem 2.3 may aid in the selection of the state variables. Normally, such considerations would result in fewer state variables. In particular, the variables in use as state variables are precisely those that render decomposable a set of precedence-related pairs of activities that otherwise fail to meet the sufficient condition. Of course, we are interested in the smallest number of such variables. This seems to be an interesting set-covering problem whose optimal solution is not presently known, to the best of our knowledge. However, an answer is possible through the construction of the path–activity matrix, and the successive conditioning on variables that violate the condition of Theorem 2.3. The approach is best illustrated by an example. Consider the network previously given as an example of the A-on-N representation (Fig. 1.6) and reproduced as Fig. 2.28 with new activity numbers.

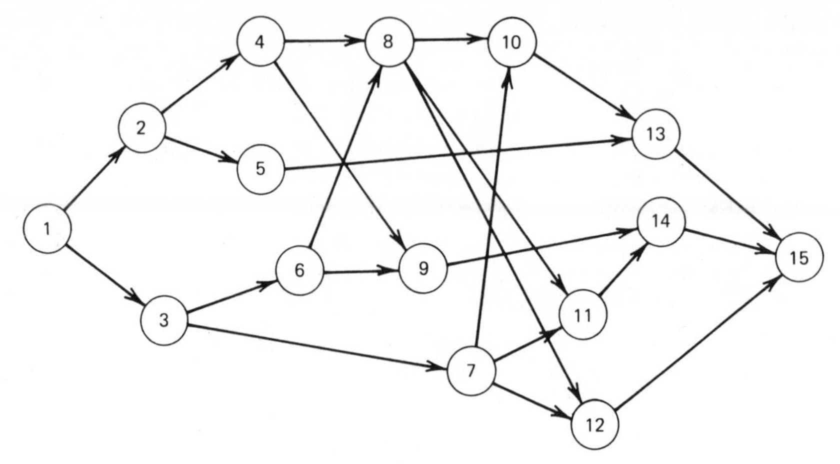

Figure 2.28.

Enumerating the paths in the network (there are 12 of them), we have:

$$(1,2,4,8,10,13,15);\ (1,2,4,8,11,14,15);\ (1,2,4,8,12,15);$$

$$(1,2,4,9,14,15);\ (1,2,5,13,15);\ (1,3,6,8,10,13,15);$$

$$(1,3,6,8,11,14,15);\ (1,3,6,8,12,15);\ (1,3,6,9,14,15);$$

$$(1,3,7,10,13,15);\ (1,3,7,11,14,15);\ (1,3,7,12,15);$$

from which we conclude that activities *1* and *15* compose the set σ, since they are common to all paths.

It is easy to see that the following pairs of activities violate the sufficiency condition of Theorem 2.3

$(4,8);(4,9)$ $\quad$ $(7,10);(8,10)$

$(4,8);(6,8)$ $\quad$ $(7,11);(8,11)$

$(4,9);(6,9)$ $\quad$ $(7,12);(8,12)$

$(6,8);(6,9)$ $\quad$ $(5,13);(10,13)$

$(7,10);(7,11);(7,12)$ $\quad$ $(9,14);(11,14)$

$(8,10);(8,11);(8,12)$

The question of the activities on which to condition the optimization is answered as follows. We construct the path–node matrix as shown in Fig.

2.29, in which rows represent paths, columns represent nodes, and a 1 is inserted when a node lies on a path.

Paths π_k	Activities (nodes) 1	2	3	4	5	6	7	8	9	10	11	12	13	14	15
1	1	1		1				1		1			1		1
2	1	1		1				1			1			1	1
3	1	1		1				1				1			1
4	1	1		1					1					1	1
5	1	1			1									1	1
6	1		1			1		1		1			1		1
7	1		1			1		1			1			1	1
8	1		1			1		1				1			1
9	1		1			1			1					1	1
10	1		1				1			1			1		1
11	1		1				1				1			1	1
12	1		1				1					1			1
	12	5	7	4	1	4	3	6	2	3	3	3	4	5	12

Figure 2.29. The paths–activities matrix.

It is easy to see that, proceeding from left to right in the matrix, one concludes that $2 \prec 4,5$ and $P_4, P_5 \subset P_2$; also $3 \prec 6,7$ and $P_6, P_7 \subset P_3$,[†] but every P_i for $i = 8, 9, \ldots, 14$ is not a subset of any other P_j. Consequently, we must condition on the seven variables $x_8, x_9, \ldots, x_{14}$. Evidently, the writing of the DP recursive expressions for this optimization problem is an onerous task, and the solution of the resulting function is well beyond the reach of practicality. This example serves to illustrate the computational difficulties encountered in this type of optimization problems.

§6.3 Additional Hints

In addition to capitalizing on structural properties of the network, and utilizing the approach of conditional optimization, the following are three almost self-evident hints:

1. *Activities of constant durations*: these may be completely ignored in the decomposition analysis, since their presence does not affect the decomposition process.

[†]Recall that P_i was defined as the set of paths to which activity i belongs.

2. *Dominance considerations*: an activity (or a path) may be "dominated" by other activities (or paths), in the sense that there will always exist at least one path, regardless of the allocation of resources, whose duration is longer than any of the paths on which the activity appears (or of the path itself). Then the activity (or set of activities on the path) will receive only the minimum investment, and its duration is taken as constant.
3. *Precedence modification*: eliminate (add) a precedence relation, reducing (increasing) the number of paths, in order to bring the network closer to satisfying the sufficient condition. This step is fraught with danger, since in any network of appreciable size it would be difficult to discern such precedence relations easily. Furthermore, an optimal allocation resulting from such action need not be optimal for the original network, depending on whether the removed (added) path is critical; if the path is critical, the allocation is not optimal.

§6.4 The Search Procedure

The DP formulation presented above is seen to lead to optimization of functions of the form

$$f(c)=\min_{x_1 \leqslant c}\left[\max\{g_I(x_I);g_2(c-x_I)\}\right] \tag{2.52}$$

in which the functions $g_i(\cdot)$ are monotone nonincreasing and discontinuous.

Our objective now is to exploit the very special structure of the cost-duration function to derive $f(c+\Delta)$ not ab initio, but from the knowledge of $f(c)$. This would result in a stepwise iterative procedure that yields the optimum in one pass; thus eliminating the need for any search procedure for each level of allocation.

Let x_I^c denote the value of the decision variable that yields the minimum in Eq. (2.52); in other words

$$\max\{g_I(x_I^c),g_2(c-x_I^c)\}\leqslant\max\{g_I(x_I);\ g_2(c-x_I)\}\ \text{for all}\ x_I\leqslant c$$

Theorem 2.4:

$$x_I^{c+\Delta}=\begin{cases} x_I^c \text{ if } g_2(c-x_I^c)>g_I(x_I^c) \\ x_I^c+\Delta \text{ otherwise}\end{cases} \tag{2.53}$$

Proof. Under the hypothesis that $g_2(c-x_1^c)>g_1(x_1^c)$, we have that if the additional resource increment (equal to Δ) is allocated to activity *1*, then

$$g_2(c-x_1^c) \geqslant g_1(x_1^c) \geqslant g_1(x_1^c+\Delta)$$

resulting in

$$\max[g_1(x_1^c+\Delta), g_2(c-x_1^c)] = g_2(c-x_1^c)$$

On the other hand, allocating the extra resource to activity *2* yields

$$\max[g_1(x_1^c); g_2(c+\Delta-x_1^c)] \leqslant \max[g_1(x_1^c); g_2(c-x_1^c)]$$

by the monotonicity of the g_2 function and the monotonicity of the "max" operator; in other words

$$\leqslant g_2(c-x_1^c) = \max[g_1(x_1^c+\Delta), g_2(c-x_1^c)]$$

Hence, it is optimal to allocate the additional resource to activity *2*.

For the case $g_2(c-x_1^c) \leqslant g_1(x_1^c)$, the proof is similar to that preceding but with activity *1* replacing *2*. This completes the proof of Theorem 2.4.

Theorem 2.5: $f(c) \geqslant f(c+n\Delta)$

Proof. The assertion is evident since the additional resource can only decrease the objective function, by the monotone nonincreasing property of the function g_i.

Thus, starting with $x_1^0=0$, an optimal value of x_1^m is determined by comparing the values of g_1 and g_2 using the optimal value of $x_1^0, x_1^\Delta, x_1^{2\Delta}, \ldots, x_1^{m-\Delta}$. Hence, the search procedure for these types of function is eliminated. This approach can be used for the entire sequence of optimization problems in the DP formalism since all $f(c)$ are also nonincreasing, that is, $f(c) \geqslant f(c+\Delta)$.

For an application of all these ideas, see Problem 23 at the end of the present chapter.

APPENDIX B. Elements of the Theory of Flow Networks†

Consider a network $G=[N,A]$ of directed or undirected arcs, in which each arc a possesses a *capacity* c_a, $0<c_a \leqslant +\infty$, which may be thought of intuitively as representing the maximal amount of some commodity flow-

†The development relies heavily on the book of Ford and Fulkerson (1962).

ing in arc a from one end node of the arc to the other end node. If the arc is assumed undirected, the flow can originate from either node and terminate at the other node. In particular, if arc a joins nodes i and j, a undirected, we let $c_a = c_{ij} = c_{ji}$, or, in other words, the capacity of the arc is an upper bound on the flow in either direction. Let f_{ij} denote the flow (i.e., the quantity flowing) in the arc (ij) in the direction $i \to j$. Clearly, $0 \leqslant f_{ij} \leqslant c_{ij}$. Similarly, for notational brevity, if node i connects with a subset of nodes Y, we let $f(i, Y)$ denote the total flow from i to all the nodes in Y and $f(Y, i)$ to denote the total flow into i from all the nodes in Y. We denote the set of nodes that "immediately precede" node i by $\mathfrak{B}(i)$, and the set of nodes that "immediately follow" it by $\mathfrak{A}(i)$. Let N denote the set of all nodes, (i.e., $N \triangleq \{1, 2, \ldots, n\}$) and suppose that a node $s \in N$ is chosen as the *source* and a node $t \in N$ as the *terminal* (or *sink*).

Consider a partition of the set N into two sets X and X^c s.t. $s \in X$ and $t \in X^c$. We define a *cutset between X and X^c*, denoted by $\mathcal{C}(X, X^c)$, as the set of arcs such that for each arc one end node is in X and the other is in X^c;

$$\mathcal{C}(X, X^c) \triangleq \{(ij) : i \in X \quad \text{and}\, j \in X^c\} \tag{B.1}$$

Clearly, $\mathcal{C}(X, X^c)$ separates s and t, in the sense that a commodity flowing from s into t must pass through one (or more) of the arcs of $\mathcal{C}(X, X^c)$. The term "cutset" stems from the fact that if one imagines all the arcs of $\mathcal{C}(X, X^c)$ actually "ruptured," it would be impossible to transmit any commodity from s to t; hence, the connection between s and t is "cut."

We define the capacity of the cutset $\mathcal{C}(X, X^c)$ as the sum of the capacities of its component arcs *in the direction X to X^c* and to denote such capacity by $c(X, X^c)$; in other words,

$$c(X, X^c) \triangleq \sum_{\substack{(ij) \in \mathcal{C}(X, X^c) \\ (i \to j)}} c_{ij}$$

By a *chain from s to t* we mean a sequence of nodes and arcs of the form

$$s \triangleq i_1; (i_1 i_2); i_2; (i_2 i_3); \ \ldots, (i_{r-1} i_r); i_r \triangleq t$$

We immediately have

Lemma B.1: Any chain from s to t must contain at least one arc in any cutset $\mathcal{C}(X, X^c)$

Proof. By definition, $s \in X$ and $t \in X^c$. Then there must exist some $i_k \in X$ and $i_{k+1} \in X^c$ so that $(i_k i_{k+1}) \in \mathcal{C}(X, X_c)$, for otherwise $i_{k+1} \in X$, $i_{k+2} \in X$, and so on, which implies that $t \in X$, a contradiction.

By the *total flow from s into t* we mean the sum of flows along all chains from s to t in which flow emanates from s and terminates at t.

Lemma B.2: Let v be the total flow from s into t. If $\mathcal{C}(X, X^c)$ is a cutset then

$$v = f(X, X^c) - f(X^c, X) \leqslant c(X, X^c)$$

Proof. The equality $v = f(X, X^c) - f(X^c, X)$ follows from the flow conservation constraints:

$$\sum_{j \in \mathcal{A}(i)} f_{ij} - \sum_{j \in \mathcal{B}(i)} f_{ji} = \begin{cases} v, & \text{for} \quad i = s \\ 0, & \quad\quad i \neq s, t \\ -v, & \quad\quad i = t \end{cases}$$

which is rephrased as

$$\begin{aligned} &f(s, N) - f(N, s) = v \\ &f(i, N) - f(N, i) = 0, \qquad i \neq s, t \\ &f(t, N) - f(N, t) = -v \end{aligned} \tag{B.2}$$

Summing Eq. (B.2) over all $i \in X$ (recall that $s \in X$), we obtain:

$$f(X, N) - f(N, X) = v$$

But $N = X \cup X^c$; therefore,

$$f(X; X \cup X^c) - f(X \cup X^c; X) = v$$

that is,

$$f(X, X) + f(X, X^c) - \left[f(X, X) + f(X^c, X) \right] = v$$

which yields the desired equality. Finally, from the definition of a cutset, we have that

$$f(X, X^c) \leqslant c(X, X^c)$$

while, by definition of f, $f(X^c,X) \geqslant 0$; hence, a fortiori,

$$v = f(X,X^c) - f(X^c,X) \leqslant c(X,X^c)$$

Next, we have a most fundamental result in flow networks. Its proof is constructive and provides the basis for the important labeling procedure.

Theorem B.1: (The maximum flow–minimum cut theorem). For any network, the maximum flow from s to t is equal to the value of the capacity of the minimal cutset; in other words,

$$v^* = \min_{\mathcal{C}(X,X^c)} \{c(X,X^c)\} \forall \mathcal{C} \text{ values}$$

Proof. In view of Lemma B.2 all we need to show is the existence of a cutset, say $\mathcal{C}(X^0,X^{0c})$, for which equality holds throughout. We construct such a cutset by letting v^* be the maximum flow from s to t. Such a flow exists, although it can be ∞. Define the set X^0 recursively as follows:

1. $s \in X^0$
2. if $i \in X^0$ and $f_{ij} < c_{ij}$, put $j \in X^0$
3. if $i \in X^0$ and $f_{ji} > 0$, put $j \in X^0$

We assert that $t \in X^{0c}$. The proof is by contradiction. Suppose it is not; then there exists a path $s = i_1, i_2, \ldots, i_r = t$, so that all its forward arcs have their flows $f < c$, and all their reverse arcs have their $f > 0$. Let δ_1 denote the minimal difference $(c-f)$ over all forward arcs, and let δ_2 denote the minimum of all f values along the reverse arcs. Let $\delta = \min(\delta_1, \delta_2) > 0$. Now construct a new flow from s to t by increasing the flow in all forward arcs by δ (so that the new flow in each arc is still $\leqslant c$) and reduce the flow in all reverse arcs by δ (so that the new flow is still $\geqslant 0$). This achieves a new flow from s to t equal to $v^* + \delta > v^*$, which contradicts the maximality of v^*. Therefore, $t \in X^{0c}$.

Now consider any arc $(i,j) \in \mathcal{C}(X^0,X^{0c})$. The following equality must be valid:

$$f_{ij} = c_{ij}$$

otherwise j would have been in X^0 by (2), and $f_{ji} = 0$, for otherwise j would have been in X^0 by (3). Therefore $f(X^0,X^{0c}) = c(X^0,X^{0c})$ and $f(X^{0c},X^0) = 0$; thus $v^* = f(X^0,X^{0c}) - f(X^{0c},X^0) = c(X^0,X^{0c}) \leqslant c(X,X^c) \forall \mathcal{C}(X,X^c)$. Hence, (X^0,X^{0c}) is a minimum cutset.

Definition: *A flow-augmenting path* from s to t is a path whose forward arcs have flow $0 \leqslant f < c$ and whose reverse arcs have flow $f > 0$.

The reason for the name is obvious, by an argument similar to that used in the maximum flow–minimum cut theorem, the flow from s to t can be augmented by a quantity $\delta > 0$.

Corollary. The flow v^* is a maximum flow iff there is no flow-augmenting path.

Proof. Of course, if v^* is maximum, then there can be no further increase in flow and hence no flow-augmenting path.

Conversely, if there is no flow-augmenting path, then define the sets X and X^c recursively as was done in Theorem B.1, starting with $s \in X$. Then $t \in X^c$, and the capacity of the cutset $\mathcal{C}(X, X^c)$ is exactly equal to v^*; hence, it is maximal.

The notion of a flow-augmenting path is fundamental and is the basic concept behind the labeling procedure. In essence, this latter is a systematic way for finding flow-augmenting paths. When none can be found we are assured by the corollary that we have the maximal flow. We discuss the labeling procedure next.

Despite the simplicity of the procedure and its intuitive appeal, its verbal description is rather awkward and, unfortunately, lengthy.

Definition: We define the *residual capacity* r_{ij} *in arc* (ij) *in the direction* $i \rightarrow j$ as given by

$$r_{ij} \triangleq \begin{cases} c_{ij} & \text{if } f_{ji} = 0 \\ c_{ij} + f_{ji} & \text{if } f_{ji} > 0. \end{cases} \tag{B.3}$$

Notice that a flow $f_{ij} > 0$ causes r_{ji} to be larger than c_{ji} [which is assumed equal to c_{ij} if arc (ij) is undirected], in particular

$$r_{ji} = c_{ji} + f_{ij} > c_{ji}$$

The interpretation of an $r_{ji} > c_{ji}$ is as follows: first it is possible to send a flow $f_{ji} = f_{ij}$ from j to i. Clearly, this new flow cancels the original flow from $i \rightarrow j$. Now, one can still admit a flow from $j \rightarrow i$ up to the capacity c_{ji}. The sum of the two segments of flow is precisely r_{ji} calculated above.

To start the procedure, assume any reasonable flow $\{f_{ij}\} \geqslant 0$. There is no question of feasibility since $f_{ij} = 0$ for all arcs $(ij) \in A$ is certainly feasible. But one can usually do better than that. The remainder of the procedure can be divided into two subroutines applied consecutively until the maximal flow is achieved: (a) identification of a flow-augmenting path and (b) carrying out the process of increasing the flow from s to t.

The search for a flow augmenting path proceeds in the following fashion. Node s is labeled with $(\infty, 0)$, where the ∞ indicates infinite availability at s and the second label, 0, identifies the source of such flow which, in the case of node s is some fictitious node 0. The general rule in labeling other nodes is that if node j connects with node i, j is not labeled and i is labeled, then node j can be labeled with $(q_j; i)$, where

$$q_j = \min(q_i; r_{ij}) \quad \text{if } r_{ij} > 0 \tag{B.4}$$

If $r_{ij} = 0$, node j is not labeled at all.

In labeling nodes, it is advisable to proceed in a systematic fashion such as at stage k, always start with the node bearing the smallest (or largest) number; label all nodes that connect with it that can be labeled, following the general rule stated above, then proceed to the next higher (or lower) numbered node and repeat the process until:

1. All labeled nodes at stage k are exhausted and some new nodes have been labeled. Then move to stage $k+1$, represented by the nodes that have just been labeled from stage k. Repeat this step until condition 2 or 3 results.
2. All labeled nodes at stage k are exhausted, but no new nodes have been labeled. This is the *nonbreakthrough* condition.
3. The terminal t is labeled. This is the *breakthrough* condition.

Consider the breakthrough condition first. When it is realized, it signifies the existence of a flow-augmenting path from s to t. Obviously, the flow can be increased by the amount q_t, as indicated by the label of node t. Furthermore, the second label of t indicates the source of such flow. A simple process of tracing the labels backward from t identifies the path that leads to such extra flow.

It is at this juncture that the second subroutine, which carries out the process of increasing the flow from s to t by exactly the amount q_t, is brought into play. In particular, once the path is identified, start at s and increase the flow in each forward arc by q_t and decrease the flow in each reverse arc by q_t, all the time adjusting the residual capacities along all the arcs of the path.

Under nonbreakthrough condition, we know that no flow-augmenting path exists; hence, by the corollary, the maximum flow has been achieved and is equal to the sum of flow in all arcs incident on either s or t.

One final remark. The labeling procedure leads, in a finite number of steps, not only to the value of the maximum v^*, but also to the identification of the minimal cutset between s and t, which we shall denote by $\mathcal{C}_{\min}(s,t)$. In particular, at the nonbreakthrough condition the nodes of the network are partitioned into two subsets:

X: the set of *labeled* nodes, and

X^c: the set of nodes that are *not labeled*.

Notice that, by construction, $s \in X$ and $t \in X^c$. Let $\mathcal{C}(X,X^c)$ denote the set of arcs that join nodes in X to nodes in X^c. We assert that this is $\mathcal{C}_{\min}(s,t)$. It has the following two characteristics: (a) $r_{ij}=0$ for all *forward* arcs, that is, for all arcs directed from $i \to j$, $i \in X$ and $j \in X^c$ and (b) $f_{ji}=0$ for all *reverse* arcs, that is, for all arcs directed from $j \to i$, $i \in X$, and $j \in X^c$.

APPENDIX C. The Primal-Dual Algorithm for Linear Programs†

This approach was developed by Dantzig, Ford, and Fulkerson (1956) in response to the Phase I–II approach of the revised simplex method. In particular, at the termination of Phase I, a feasible primal basis would be at hand (if the LP is feasible), but the effort expended (which may be considerable) does not guarantee that we are any closer to the optimum. The primal–dual algorithm corrects this deficiency.

The procedure is interesting and extremely efficient, especially when applied to specially structured problems (e.g., as in networks). It begins with a feasible solution to the dual. Such a solution is *always* readily available, as we demonstrate below. By duality and complementarity considerations, this dual solution "releases" certain vectors in the primal as candidates for inclusion in the primal basis. In particular, the "released" primal vectors are the ones whose levels may be increased to >0 without violating any complementarity conditions. Given this set of "released" vectors, we work on the primal until the optimality criterion is satisfied. Since we are dealing only with a subset of the total (primal) vectors, satisfying the optimality criterion over this subset falls short of satisfying the optimality criterion for *all* vectors. Thus, we have the *optimum for a restricted primal* problem. Still, this "restricted optimum" results in a new solution of the dual which, in turn, "releases" one or more additional primal vectors as candidates for entry into the (primal) basis. Complementary slackness is maintained at each step so that when a feasible solution to the primal is found it is also optimal. Hence, in the primal, we are working on feasibility (by the successive additions to the subset of "released" vectors) and optimality *at the same time*.

†The development relies heavily on the book of Hadley (1964), pp. 257–261.

Assume the primal problem is

$$\text{maximize } z = \mathbf{CX}$$

$$\text{s.t.} \quad A\mathbf{X} = \mathbf{B}; \quad \mathbf{X} \geqslant \mathbf{0} \tag{C.1}$$

A simple modification to this primal yields an immediate solution to the dual. To the constraints of (C.1) add the equality

$$\sum_j x_j + x_0 = b_0$$

where b_0 is sufficiently large so that the constraint imposes no limitation whatsoever on the original x_j values. Let the "value" of x_0 be zero in the objective function. Then the "modified primal" may be stated as

$$\text{maximize } z = \mathbf{CX}$$

$$\text{s.t.} \quad \begin{bmatrix} 1 & \mathbf{1} \\ \mathbf{0} & \mathbf{A} \end{bmatrix} \begin{bmatrix} x_0 \\ \mathbf{X} \end{bmatrix} = \begin{bmatrix} b_0 \\ \mathbf{B} \end{bmatrix}; \quad [x_0, \mathbf{X}] \geqslant \mathbf{0}^{\dagger} \tag{C.2}$$

The dual of (C.2), called the *modified dual*, is

$$\text{minimize } g = b_0 w_0 + \mathbf{B}'\mathbf{W}$$

$$\text{s.t.} \quad \begin{bmatrix} 1 & \mathbf{0} \\ \mathbf{1} & \mathbf{A}' \end{bmatrix} \begin{bmatrix} w_0 \\ \mathbf{W} \end{bmatrix} \geqslant \begin{bmatrix} \mathbf{0} \\ \mathbf{C}' \end{bmatrix} \tag{C.3}$$

The components of $\mathbf{W}$ are unrestricted in sign, although w_0 is constrained to be $\geqslant 0$. However, we can immediately obtain a solution to this modified dual problem as follows; simply put

$$w_0 = \max_j c_j, \quad \mathbf{W} = \mathbf{0}; \quad \text{where } c_0 = 0 \tag{C.4}$$

The inequalities in (C.3) can be changed to equalities by subtracting surplus variables $(w_{s0}, \mathbf{W}_s) \geqslant 0$, and the modified dual can be written as

$$\begin{bmatrix} 1 & \mathbf{0} \\ \mathbf{1} & A' \end{bmatrix} \begin{bmatrix} w_0 \\ \mathbf{W} \end{bmatrix} - \begin{bmatrix} 1 & \mathbf{0} \\ \mathbf{0} & I_n \end{bmatrix} \begin{bmatrix} w_{s0} \\ \mathbf{W}_s \end{bmatrix} = \begin{bmatrix} 0 \\ \mathbf{C}' \end{bmatrix} \tag{C.5}$$

If $w_0 = 0$, any solution to the modified dual of (C.3) will also be a solution

†Numbers in boldface are vectors of the appropriate dimensions.

of the dual of the original LP of (C.1). Furthermore, the complementary slackness conditions for the modified problem may be stated as

$$x_0 w_{s0} + \mathbf{X}'\mathbf{W}_s = 0 \tag{C.6}$$

Consequently, if we achieve a feasible primal vector $(x_0, \mathbf{X}) \geqslant 0$ and (C.6) holds, assuming that $(w_{s0}, \mathbf{W}_s) \geqslant 0$ and $w_0 = 0$, then indeed $\mathbf{X}$ is an optimal solution to the primal LP of (C.1). On the other hand, if we have a solution to the modified dual with $w_0 \neq 0$, and a feasible solution to the modified primal in which (C.6) is satisfied (this implies $x_0 = 0$), then

$$z = g = b_0 w_0 + \mathbf{B}'\mathbf{W}$$

Since b_0 is an arbitrarily large number, z may be made arbitrarily large; hence, the LP of (C.2) has an unbounded solution.

Suppose now we add artificial variables to the modified primal of (C.2) as if we were to make the usual Phase I calculations. We assign prices of $\gamma = -1$ to the artificial variables and $=0$ to the original variables:

$$\text{maximize } h = -x_{a0} - \mathbf{1}\mathbf{X}_a$$

$$\begin{bmatrix} 1 & \mathbf{1} \\ \mathbf{0} & A \end{bmatrix} \begin{bmatrix} x_0 \\ \mathbf{X} \end{bmatrix} + \begin{bmatrix} 1 & \mathbf{0} \\ \mathbf{0} & I_n \end{bmatrix} \begin{bmatrix} x_{a0} \\ \mathbf{X}_0 \end{bmatrix} = \begin{bmatrix} b_0 \\ \mathbf{B} \end{bmatrix} \tag{C.7}$$

$$(x_0, \mathbf{X}) \geqslant 0, \quad (x_{a0}, \mathbf{X}_a) \geqslant 0$$

We denote the $(m+1)$-component original activity vectors by a_j; $j = 1, 2, \ldots, n$, and a basis matrix for the constraints of (C.7) by $\boldsymbol{\beta}$. The vector containing the prices γ_j of the variables in the basis are denoted by γ_β (these prices are either 0 or -1).

We are now ready to start the primal-dual iterations. At the start, only artificial variables are in the primal basis.

Consider the modified dual of (C.7): an initial solution to this problem is given by (C.4), and for this solution

$$w_{sj} = w_0 - c_j; \quad j = 1, \ldots, n \tag{C.8}$$

It follows from the manner of selecting w_0 that at least one component of the vector $(w_{s0}, \mathbf{W}_s)$ will vanish, that is, $w_{sk} = 0$ for some k. Denote the set of w_{sj} that vanish by S.

The central concept of the primal–dual algorithm is to proceed in such a way that the complementary slackness condition of (C.6) is always satisfied. Then when $h=0$ and we have found a feasible solution to the primal, it will be optimal, and the maximum value of z is the g for the corresponding solution to the dual. This completely eliminates the need for Phase II.

Corresponding to the set S (of dual surplus variables of value zero) is a set of vectors P that are eligible to enter the extended primal basis. Since S is nonempty, P is nonempty. Furthermore, if a vector $\mathbf{a}_j \in P$ is introduced into the basis at a level $x_j > 0$, we have $x_j w_{sj} = 0$ by virtue of $w_{sj} = 0$. On the other hand, $w_{sj} > 0$ for any $w_{sj} \notin S$; but its corresponding primal variable $x_j = 0$ since it is not in the basis. Consequently, the complementary slackness condition of (C.6) is satisfied.

Now, the ordinary simplex method (or, better still, the revised simplex method) is applied to the extended primal, with only the vectors in P eligible for entry into the basis. This is usually referred to as the *restricted primal* problem. At the termination of this phase, let $\boldsymbol{\beta}_1$ be the extended primal basis matrix, and let γ_{β_1} be the corresponding vector of prices. Then it must be true that (letting $\boldsymbol{\Pi} = (\pi_0, \pi_1, \ldots, \pi_n) = \gamma_{\beta_1} \boldsymbol{\beta}_1^{-1}$):

$$\boldsymbol{\Pi}\mathbf{a}_j \geqslant 0 \quad \text{for all } \mathbf{a}_j \in P; \tag{C.9}$$

(recall that the price of $\mathbf{a}_j$ is 0), by the regular optimality criterion of the simplex algorithm for the restricted primal. Furthermore, if $\mathbf{a}_j$ is in $\boldsymbol{\beta}_1$, then $\boldsymbol{\Pi}\mathbf{a}_j = 0$.

We now come to the central concept in the primal–dual algorithm, namely, how to expand the restricted primal (by including more original vectors in P) while maintaining the complementary slackness conditions in case the current primal solution is not primal feasible.

It stands to reason that the way to generate more "eligible" vectors is to drive dual surplus variables to zero. The trick is to *accomplish this while still retaining the dual surplus variables corresponding to the basis vectors at their zero value*! This is achieved utilizing the fact of (C.9).

The original vectors ($\mathbf{a}_j$) are divided into three groups: (a) those in the current basis, $\boldsymbol{\beta}_1$ (and, consequently, also elements of the set P), (b) those in P but not in the current basis, and (c) those not in P.

Consider group (c). By construction, their corresponding $w_{sj} > 0$. Consequently, these w_{sj} values must be reduced if we are to augment the set P. Let the new values be denoted by $\hat{w}_{sj}$, given by

$$\hat{w}_{sj} = w_{sj} + k\boldsymbol{\Pi}\mathbf{a}_j, \quad j \notin P, \quad \boldsymbol{\Pi}\mathbf{a}_j < 0, \quad k \geqslant 0$$

(The eventuality that no $\boldsymbol{\Pi}\mathbf{a}_j < 0$ is discussed below.) We wish to choose k

such that $\hat{w}_{sj}$ remains $\geqslant 0$ for all $j \notin P$. Clearly, k must be $\leqslant K$, given by

$$K = \min_{\substack{j \\ \mathbf{\Pi a}_j < 0}} \frac{w_{sj}}{-\mathbf{\Pi a}_j} > 0; \quad j \notin P \tag{C.10}$$

Since we are interested in driving at least one $\hat{w}_{sj}$ to 0, we put $k = K$ in the above expression, yielding

$$\hat{w}_{sj} = w_{sj} + K\mathbf{\Pi a}_j; \quad j \notin P; \quad \mathbf{\Pi a}_j < 0 \tag{C.11}$$

The corresponding values of the dual variables are given by

$$(\hat{w}_0, \hat{\mathbf{W}}') = (w_0, \mathbf{W}') + K\mathbf{\Pi} \tag{C.12}$$

It is easy to demonstrate that $(\hat{w}_0, \hat{\mathbf{W}}')$ is, indeed, a solution to the modified dual:

$$\text{for } \mathbf{a}_j \in P\text{: } (\hat{w}_0, \hat{\mathbf{W}}')\mathbf{a}_j = (w_0, \mathbf{W}')\mathbf{a}_j + K\mathbf{\Pi a}_j \geqslant c_j$$

since $\mathbf{\Pi a}_j = 0$ for $\mathbf{a}_j \in \boldsymbol{\beta}_1$ and $\mathbf{\Pi a}_j \geqslant 0$ for $\mathbf{a}_j \notin \boldsymbol{\beta}_1$; and $(w_0, \mathbf{W}')\mathbf{a}_j \geqslant c_j$ by the starting algorithm.

$$\begin{aligned} \text{for } \mathbf{a}_j \notin P\text{: } (\hat{w}_0, \hat{\mathbf{W}}')\mathbf{a}_j &= (w_0, \mathbf{W}')\mathbf{a}_j + K\mathbf{\Pi a}_j + c_j - c_j \\ &= \left[(w_0, \mathbf{W}')\mathbf{a}_j - c_j\right] + K\mathbf{\Pi a}_j + c_j \\ &= w_{sj} + K\mathbf{\Pi a}_j + c_j \geqslant c_j \end{aligned}$$

because $w_{sj}/(-\mathbf{\Pi a}_j) \geqslant K$ for all $\mathbf{a}_j \notin P$, by (C.10). This establishes $(\hat{w}_0, \hat{\mathbf{W}}')$ as a feasible solution to the modified dual.

It remains to prove that $\hat{w}_{sj} = 0$ for $\mathbf{a}_j \in \boldsymbol{\beta}_1$ and that $\hat{w}_{sj} \geqslant 0$ for $\mathbf{a}_j \in P$ but $\mathbf{a}_j \notin \boldsymbol{\beta}_1$. For $\mathbf{a}_j \in P$ we have

$$\begin{aligned} \hat{w}_{sj} &= (\hat{w}_0, \hat{\mathbf{W}}')\mathbf{a}_j - c_j \\ &= (w_0, \mathbf{W}')\mathbf{a}_j + \mathbf{\Pi} K\mathbf{a}_j - c_j \begin{cases} = 0 & \text{for } \mathbf{a}_j \in \boldsymbol{\beta}_1 \\ \geqslant 0 & \text{for } \mathbf{a}_j \notin \boldsymbol{\beta}_1 \end{cases} \end{aligned}$$

The first equality follows from $w_{sj} = (w_0, \mathbf{W}')\mathbf{a}_j - c_j = 0$ by the construction of P and $K\mathbf{\Pi a}_j = 0$ by virtue of $\mathbf{a}_j$ being in $\boldsymbol{\beta}_1$. The second inequality follows from $w_{sj} = 0$ by the construction of P and $K\mathbf{\Pi a}_j \geqslant 0$ by virtue of the simplex optimality criterion.

The above argument leads to the following conclusions:

1. At least one new vector currently not in P is now added to P. Its $\hat{w}_{sj}=0$, hence it is eligible for introduction into the basis.
2. All basic (original) variables still retain their $\hat{w}_{sj}=0$.
3. All nonbasic variables still retain their $\hat{w}_{sj}\geqslant 0$.

We conclude that the complementary slackness conditions of (C.6) are satisfied. Furthermore, the set S is expanded by at least one element, and the set P is expanded by at least one element.

The regular simplex procedure may now be applied to the expanded P, and the cycle repeats. Iteration halts when one of the following conditions occurs:

1. The restricted primal is feasible. This occurs when all the artificial vectors are driven out of the basis. If $w_0=w_{s0}=0$, then the desired optimum is at hand. If $w_0>0$, then the primal has unbounded solution.
2. $K=\infty$. This occurs when all $\mathbf{\Pi a}_j\geqslant 0;\ j\notin P$. Then

$$\begin{aligned}\hat{g}&=(\hat{w}_0,\hat{\mathbf{W}}')(b_0,\mathbf{B})\\&=(w_0,\mathbf{W}')(b_0,\mathbf{B})+K\mathbf{\Pi}(b_0,\mathbf{B})\\&=g+K\mathbf{\Pi}(b_0,\mathbf{B})=g+Kh.\end{aligned}$$

Since $h<0$, $\hat{g}<g$; in other words, the iterations reduce the dual objective function monotonically. However, $K=\infty$ yields $\hat{g}=-\infty$, which implies that the primal is infeasible.

The considerable advantage of the primal–dual algorithm over other algorithms is manifested in network problems, where an extremely efficient method for optimizing the primal can be used. In the general case, the simplex (or its revised version) is suggested because it appears to be the most efficient one available.

The specialization of the primal–dual algorithm to the project cost-duration calculations runs as follows. The computations start with a primal *infeasible* solution (total project duration is $\overline{T}$, while the desired duration is $-t_1+t_n=T<\overline{T}$). Here, the roles of the primal and dual are reversed: we drive some *primal* slacks to zero (to form the set S) in order to generate the "eligible" set of vectors (the set P) of flows that are eligible for introduction into the *dual* basis (i.e., eligible for increase to positive levels). Thus, an $s_{ij}^{(1)}=0$ permits the corresponding flow $f_{ij}^{(1)}$ to be increased to positive level. The range of $f_{ij}^{(1)}$ is known:

$$0\leqslant f_{ij}^{(1)}\leqslant a_{ij}$$

Optimizing the restricted dual is tantamount to maximizing the flow in the designated subnetwork. This maximization is easily accomplished by the labeling procedure. Its termination is signaled by the nonbreakthrough condition. Now, we must return to the primal and secure a few (one or more) additional slacks that are at zero level *while retaining the complementarity conditions between the flows and their corresponding slacks*. This is accomplished through the *node-change subroutine*. Essentially, this subroutine retains $s_{ij}^{(1)}=0$ for those arcs whose flows, $f_{ij}^{(1)}$, are $<a_{ij}$ while reduces some $s_{ij}^{(1)}$ or $s_{ij}^{(2)}$ to 0, by reducing the duration of those arcs that are saturated ($f_{ij}^{(1)}=a_{ij}$). Recall that this latter set of arcs forms the cutset. Having enlarged the set S (of primal slacks at zero level) by at least one element, we return to the dual and optimize *its* value (through maximization of flow), and so on. Iterations stop when the total duration of the project reaches T for the first time, since we would then have a primal feasible as well.

Notice that the special characteristics of the complementarity conditions impose requirements somewhat different from the general primal–dual algorithm. For instance, a flow $f_{ij}^{(1)}$ may be at a positive level ($f_{ij}^{(1)}=a_{ij}$) while its corresponding primal slack has a nonzero value ($s_{ij}^{(1)}<0$), since the complementarity conditions require that $f_{ij}^{(2)}$ may be increased to a positive level only if $s_{ij}^{(2)}=0$ and $f_{ij}^{(1)}=a_{ij}$. But the condition $s_{ij}^{(2)}=t_i-t_j+l_{ij}=0$ implies that $s_{ij}^{(1)}=t_i-t_j+u_{ij}>0$ by virtue of the fact that $u_{ij}>l_{ij}$. All this points out the need for care in interpreting and applying complementarity conditions.

PROBLEMS

1. Consider the following project network and the associated upper and lower limits on its arc durations, u and l, together with the marginal cost of reduction, a:

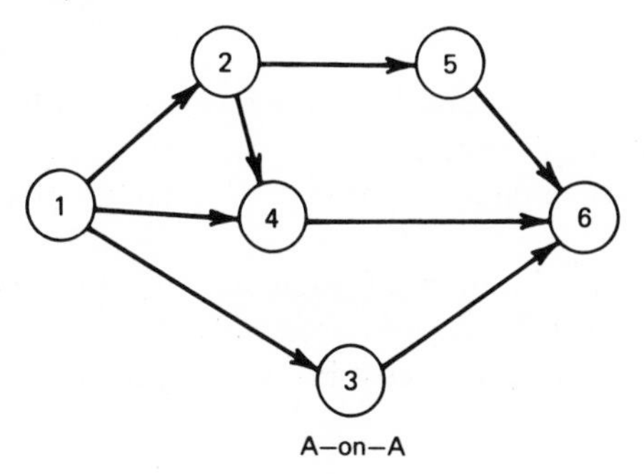

Arc	l_{ij}	u_{ij}	a_{ij}
(1,2)	4	6	8
(1,3)	4	8	9
(1,4)	3	5	3
(2,4)	3	3	∞
(2,5)	3	5	4
(3,6)	8	12	20
(4,6)	5	8	5
(5,6)	6	6	∞

You are asked to develop the complete optimal cost function as the duration of the project is shortened from its "normal" duration to its "crash" duration.

Do the above by both "dual" and "primal" approaches of Sections 2 and 3 of this chapter.

2. Determine the optimal cost-duration function for the following project with the given arc parameters, assuming that the cost of shortening individual activities is linear with coefficient a_{ij}.

Activity (ij)	Lower Limit l_{ij}	Upper Limit u_{ij}	Cost Coefficient a_{ij}
(1,2)	2	10	2
(1,3)	5	7	5
(1,4)	3	9	1
(2,3)	1	5	5
(2,4)	4	10	4
(2,5)	4	8	2
(3,5)	6	9	9
(4,5)	3	6	8

The limits, l_{ij} and u_{ij}, are the lower and upper limits on the duration of the activity; respectively.

(You may use either the primal or dual approach to solve this problem.)

a. Determine the duration of each activity and the corresponding minimum cost if the desired project completion time is 17.
b. What is the incremental cost in reducing the project duration from 17 to 13?

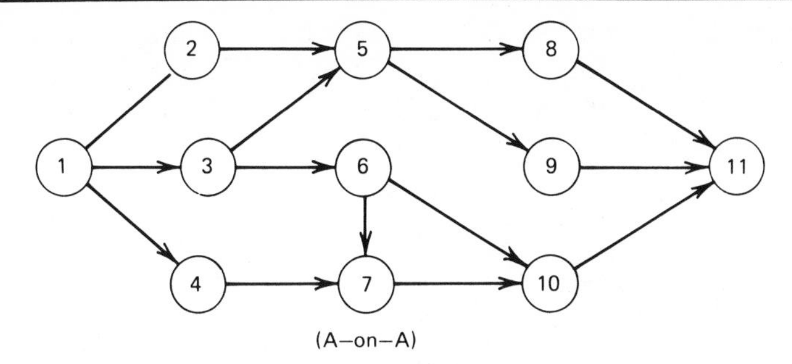

(A–on–A)

3. In the above network, some of the activities can be performed in shorter or longer durations, depending on the desired completion time of the project and the total level of expenditure desired. The following table gives the pertinent data.

Activity (ij)	l_{ij}	u_{ij}	a_{ij}
(1,2)	6	12	4
(1,3)	1	6	2
(1,4)	2	5	13
(2,5)	5	5	∞
(3,5)	1	1	∞
(3,6)	2	5	4
(4,7)	6	12	2
(5,8)	4	7	20
(5,9)	1	4	12
(6,7)	3	3	∞
(6,10)	3	4	9
(7,10)	6	11	2
(8,11)	3	5	10
(9,11)	5	10	11
(10,11)	2	7	4

a. Determine the CP assuming all activities are at their minimum duration l.
b. Determine the CP assuming all activities are at their maximum duration u.
c. Determine the project duration-cost (T–C) curve following the procedure of Sections 2 or 3.
d. At each "breakpoint" of the T–C curve determine the subgraph of CPs.

4. Determine the complete T–C curve for the following AN, in which the notation on each arc is that used in Section 2. In particular, u is the longest duration of the activity, l is its shortest duration, and a is the marginal cost.

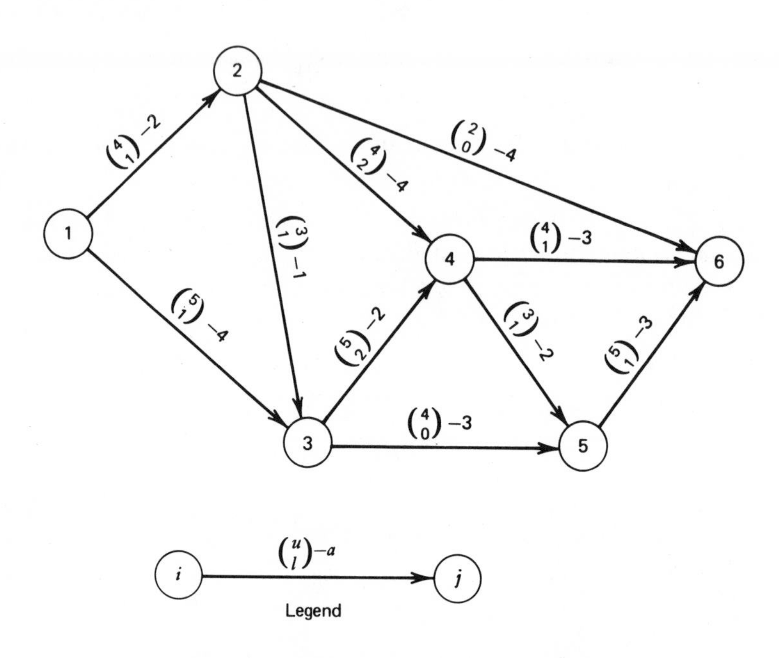

5. Determine the optimal cost-duration function for the following project (the notation on each arc represents its longest duration, its shortest duration, and its marginal cost, assuming linear cost functions):

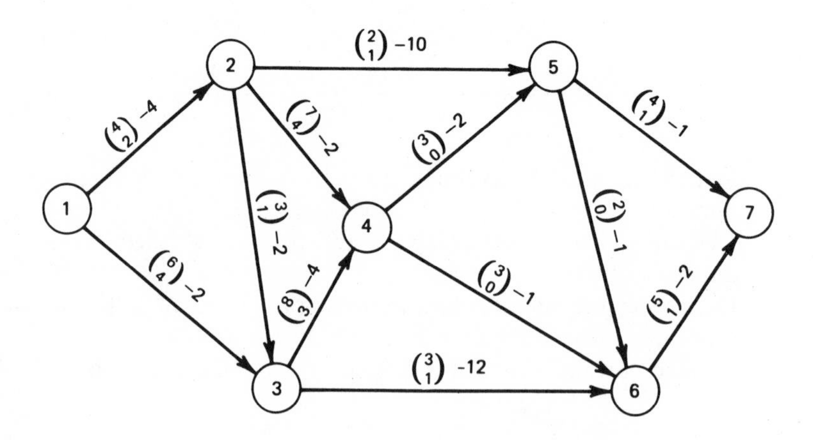

6. The following data pertain to a project of 11 activities and seven events.

Activity	l_{ij}	u_{ij}	a_{ij}
(1,2)	7	10	5
(1,3)	6	8	7
(1,4)	6	13	6
(2,4)	6	6	∞
(2,7)	20	28	9
(3,4)	3	5	8
(3,6)	15	23	11
(4,5)	3	8	3
(5,6)	4	9	8
(5,7)	6	10	10
(6,7)	5	7	6

Determine, using the "primal" method of Section 3, the optimal duration of each activity to yield a project completion time of $T=31$ (the symbols l, u, and a are those used in Problem 2).

7. Consider a given AN as specified by its graph $G=(N,A)$, where N is the set of nodes (events) and A is the set of arcs (activities) in A-on-A mode. Suppose that there exists a subset of events (the so-called key events) that are contracted to be realized by given target dates—say $d_{j_1}, d_{j_2}, \ldots, d_{j_R}$ in the case of $R \geqslant 1$ key events. A penalty is incurred if any of these events is realized after its target date d_{j_r}; $r=1,2,\ldots,R$. Assume that the penalty is linear in the amount of delay, with coefficient $p_{j_r} > 0$.

Assuming that the activities possess linear cost functions as specified in Fig. 2.1 between upper and lower limits, u and l, you are asked to construct a model for minimizing the total cost of the project, which now involves the costs of shortening the activities *and* the costs of violating the target dates of the key events.

Apply your model to the project of Problem 5 with the following two key events:

$$\text{node } 4; \quad d_4 = 12; \quad p_4 = 10$$
$$\text{node } 7; \quad d_7 = 18; \quad p_7 = 20$$

8. Devise a procedure by which the optimal "flows" can be constructed at any iteration of the procedure of Section 2 given the last node time (assuming that the first node is realized at time 0).

This procedure should enable one to step into the procedure "in the middle."

9. What is the relationship, if any, between the variables $\{t_j\}$ defined in the LP of Eqs. (2.1)–(2.2) and the node numbers $\{w_j\}$ defined for the dual LP model of Eqs. (1.8).
10. Consider the primal LP of Eqs. (2.3)–(2.4) and its corresponding dual LP of (2.5)–(2.7). How many constraints does each formulation have? Based on the number of constraints, which of the two LPs would be easiest to solve? When are the number of constraints in both formulations equal?
11. The rationale for seeking the minimum-cost duration of activities has been implicitly assumed to be the need to satisfy an externally imposed completion time for the entire project.

    However, in many instances no such "deadline" is imposed, and the project may be completed anywhere between times $\underline{T}$ and $\overline{T}$ which are imposed "from the outside." In such instances the rationale for determining the minimum-cost durations of the activities, and hence the optimum time–cost curve, must be because of knowledge of the utility of early completion. In many instances this can be measured in direct monetary cost of late completion. (Such would be the case of the construction of a distribution center for a major wholesaler, a new plant for a manufacturer, or a major repair facility for an airline company.) But in many other cases it is not immediately obvious how such a utility (or disutility) can be measured, for instance, if the project is related to the rendering of service or enhancing the general welfare of a community (e.g., building a new hospital or expanding an existing high school).

    You are asked to hypothesize a "utility function" of the early completion of a sanitorium of 300 beds and to discuss the approach you would use for establishing the optimal target "deadline."
12. Consider a project composed of a large number of activities (or tasks, jobs, etc.) related by precedence constraints of the form $i \prec j$, meaning that activity i precedes activity j. Suppose that a completion time, or deadline, is established for each activity (usually by the standard "latest realization time" calculations of the CPM or variants thereof).

    Suppose that the activities can be performed one at a time (due, perhaps, to limited availability of a scarce resource). For each activity i there is a choice between a long processing time b_i at low cost β_i or a short processing time a_i at high cost α_i; $\beta_i < \alpha_i$ and $b_i > a_i$.

    It can be shown [see Lawler and Moore (1969)] that it is optimal to consider the activities for processing in the order established by the following procedure: let the activities be numbered such that $i \prec j \Rightarrow i < j$. Let d_i be the given deadline for activity i. Determine the *gener-*

alized deadline $\hat{d}_i$ according to:

$$\hat{d}_i = \min_k (d_k | i \prec k)$$

breaking ties in favor of the smaller-numbered task. In other words, the generalized deadline of activity i is given by the earliest deadline of its immediate successors. Then the activities are ordered according to nondecreasing $\hat{d}$.

Construct a DP model to determine the minimum-cost plan by which all the tasks are processed so that all deadlines are met, assuming such a schedule is feasible.

13. Assume that all marginal costs of shortening an activity duration are constant, although they may vary from activity to activity.

    Based on the logic of the procedures of Sections 2 and 3, devise a heuristic procedure that accomplishes the reduction in project duration to a specified completion time.

    Apply your heuristic procedure to the project of Problem 6 and compare the results. Discuss briefly.

14. In the discussion of Step 3(b) of the "primal" method of Section 3 (p. 85), explain why the "state of flow (i)" was ignored.

15. Consider the network of Fig. 2.16 with the given cost functions of Eq. (2.27) and Table 2.4.

    Determine the optimal activity durations and node times if it is desired to complete the project at $T = 28$.

16. Suppose that the activities of the project of Problem 4 have strictly convex costs, rather than the linear costs hypothesized then, and that the cost functions possess the same quadratic form:

$$\text{cost} = b + (a - x)^2; \, l \leqslant x \leqslant a$$

The parameters of the problem are as follows:

Activity	b	a	l
(1,2)	3	4	1
(1,3)	4	5	1
(2,3)	2	3	0
(2,4)	1	4	2
(2,6)	4	9	5
(3,4)	1	5	2
(3,5)	1	4	0
(4,5)	3	3	1
(4,6)	1	4	1
(5,6)	1	5	1

a. Determine the project duration assuming each activity has duration equal to the value of a.
b. Determine the least cost duration of the activities to complete the project at $T=15$ and at $T=8$.
c. Is there any difference in the approach used for these two values of T?

17. What modification, if any, would you expect the bounds on the activity duration, as given in Eqs. (2.30)–(2.31), to introduce in the statement of Theorem 2.2?

    Give a heuristic reasoning for your conclusion, and illustrate by a small numerical example (or examples).

18. Could the iterative, B&B procedure of Section 5 be adapted to *convex* functions? If your answer is "yes," state the procedure and illustrate its use on a small numerical example. If your answer is "no," give your rationale.

19. The discussion of the iterative procedure of Section 5 indicated that other heuristics may be used in the selection of the B&B node chosen for branching, as well as in the manner of branching itself.

    You are asked to enumerate as many heuristics as possible that may be used in all steps of the procedure, including multiple activity division and so on. Discuss the rationale, advantages and disadvantages of each proposed heuristic.

20. Consider the project network of Problem 2, and suppose the costs are concave functions of the activity durations as follows:

Activity	l	u	Cost Function
(1,2)	2	10	$c=70-(y-2)^2$
(1,3)	5	7	$c=20-(y-5)^2$
(1,4)	3	9	$c=40-(y-3)^2$
(2,3)	1	5	$c=11.2-1.2y;\ 1\leqslant y\leqslant 3$
			$13.6-2y\ ;\ 3\leqslant y\leqslant 5$
(2,4)	4	10	$c=36-1.5y\ ;\ 4\leqslant y\leqslant 6$
			$37.8-1.8y;\ 6\leqslant y\leqslant 10$
(2,5)	4	8	$c=35+y-0.5y^2$
(3,5)	6	9	$c=50+2y-0.8y^2$
(4,5)	3	6	$c=10-(y-3)^2$

a. Determine the least-cost project duration, and the value of that cost.
b. Determine the optimal duration of the activities to complete the project by $T=16$. What is the corresponding project cost?
c. Compare with the results of Problem 2 and comment briefly.

21. In the study of DANs in Sections 2 and 3 we encountered the "design" problem of determining the optimal cost of the project for any specified duration, after assuming a linear cost-duration "trade-off" for each activity (or conversely, determining the optimal duration of each activity for a given total cost of the project).

    Formulate a meaningful problem of "project cost- duration trade-off" for the case of PAN (PERT) and discuss its solution. (You may still assume a linear cost-duration relation for individual activities, but notice that the duration of each activity is itself a random variable.)

22. Write the DP functional equation for the determination of the optimal activity durations under total investment availability c for the project of Fig. 2.28 in the text, following the conditioning procedure outlined on p. 117.

23. Determine the optimal allocation of the available funds $c = 100$ to the eight activities of the project depicted in the figure. The cost-duration functions, $g_i(\cdot)$, are given in the table that follows

$g_i(x_i)$

x_i	Activity 1	2	3	4	5	6	7	8
0	15	14	31	10	18	40	65	23
5	12	12	30	9	10	38	40	20
10	8	10	22	8	10	25	40	17
15	7	8	20	7	10	25	30	13
20	—	8	16	7	5	20	30	13
25	—	8	—	6	—	18	20	—
30	—	6	—	—	—	16	20	—
35	—	—	—	—	—	12	10	—

(*Hint*: Condition on activities *2* and *4*.)

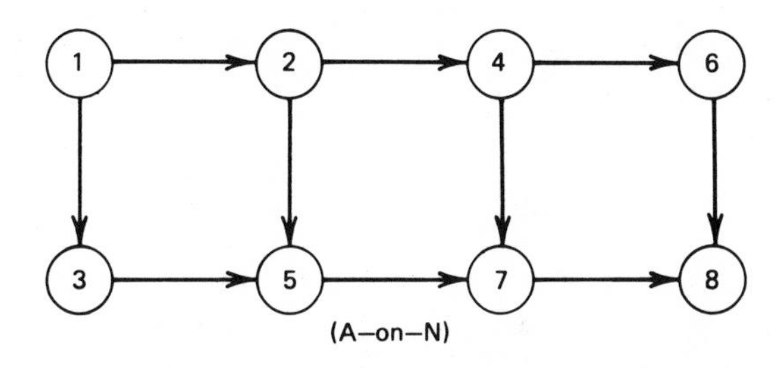

(A–on–N)

24. Determine the minimum number of activities to be "conditioned on" in order to satisfy the sufficient condition of Theorem 2.3 for the

following networks (A-on-N representation). Then state the DP functional equations for all steps of recursion.

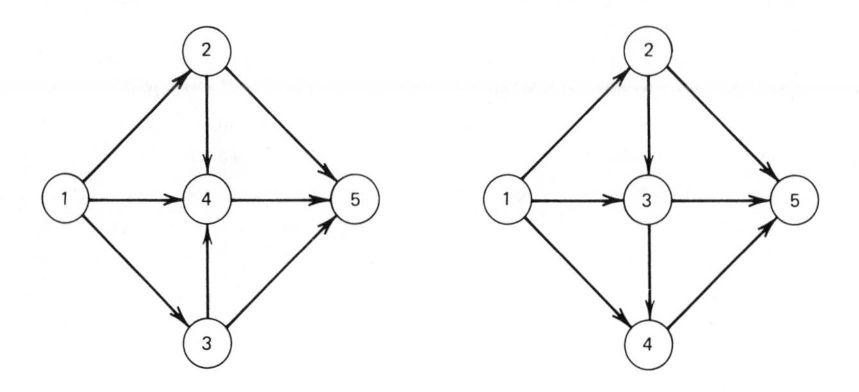

25. You are asked to determine the optimal duration (i.e., minimal-cost) of each activity in the following network, using the methods of Section 6 when the total additional funds available are 180.

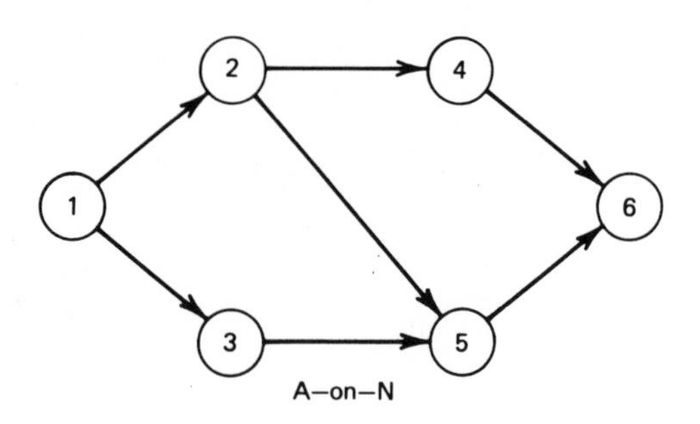

A–on–N

Activity	l_{ij}	u_{ij}
1	2	5
2	3	8
3	6	10
4	2	4
5	3	5
6	1	4

The cost function of all activities is a discontinuous nonincreasing function of the duration derived from the following function:

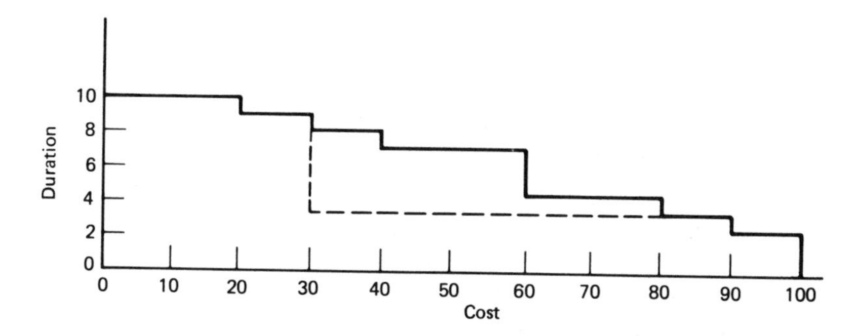

This summary cost-duration function is truncated on both axes (i.e., the origin is shifted) to correspond to the individual activity lower and upper limits on duration. For instance, activity *2* with $l_2 = 3$ and $u_2 = 8$ would have the following cost-duration function:

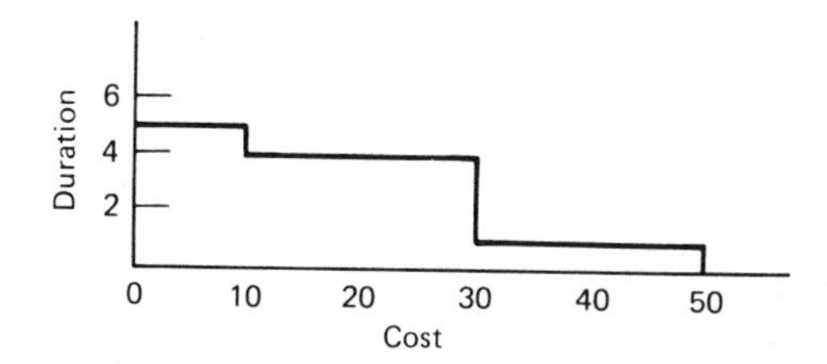

You may compute the valuations of $f_i(c)$ using the results of Theorem 2.4.

REFERENCES

Alpert and Orkand, 1962.
Barnes and Gillespie, 1972.
Berman, 1964.
Clark, 1961.
Clingen, 1964.
Dantzig, 1963.
Dantzig, Ford and Fulkerson, 1956.
Dijkstra, 1959.
Elmaghraby, 1968.
Falk and Horowitz, 1972.
Falk and Soland, 1969.
Feiler, 1974.
Ford and Fulkerson, 1962.
Fulkerson, 1961.
———, 1964.
Goyal, 1975.
Hadley, 1964.
Hartley and Hocking, 1963.
Jewell, 1965.

———, 1971.
Kapur, 1973.
Kuhn and Tucker, 1951.
Lamberson and Hocking, 1970.
Lawler and Moore, 1969.
Minty, 1957.
Moder and Phillips, 1964.
Prager, 1963.
Rech and Barton (1970).
Robinson, 1975.
Scherer, 1966.
———, 1966.
Shackelton, Jr., 1973.
Siemens, 1971.
Simonnard, 1966.
Shier, 1974.
———, 1974.
Sobczak, 1962.
Soistman, 1962.
———, 1965.
———, 1966.
Special Projects Office, 1958.
Steinberg, 1963.
Sumray, 1965.
Thomas, 1969.
Wiest and Levy, 1969.

3

PROJECT PLANNING UNDER LIMITED RESOURCES

§1. THE CONTEXT

Basically, ANs are management tools intended to answer questions of analysis and synthesis relative to projects involving a multitude of activities and drawing on a variety of resources. The determination of the CP, the second CP, and so on, up to the kth CP; and determination of the event slack or activity floats and of both the earliest and latest start and completion times are all examples of *analysis of temporal relationships* among the entities of the network. They do not involve any consideration of the resources required to perform these activities, which includes the inevitable interaction among activities that utilize the same resources, because the availability of these resources was assumed unlimited.

Yet, the problems of project planning and control under resource constraints are inescapable. If the resources available are constrained to certain limits it may be that the scheduling of the available resources to meet these constraints (and also the project completion time) is a major problem. Or, with sufficient resources available, it may be desirable to have them committed at as constant a rate as possible, in which case "leveling" or smoothing of resource usage is the allocation problem.

As the reader well knows, it is almost always possible to translate "upperbound" constraints into *cost* terms and replace the dictum, "use no more than k units" by, "if you use up to k units it costs you $\phi_1(x)>0$ per unit, but beyond that it costs you $\phi_2(x)>\phi_1(x)$ per unit," where ϕ is a nondecreasing function of the usage x. To prohibit the usage of more than k units, one simply puts $\phi_2(x)=\infty$ for $x>k$. This is a standard device and is of great practical value and intuitive appeal. In its application, no upper bound on the usage of any resource appears per se. In its place there appears a time/cost trade-off relationship in which the resources' availability is absorbed and the higher consumption of the resources is reflected either in higher cost (i.e., higher consumption of money) or in higher cost *and shorter duration* of the activities.

Our discussion in Chapter 2 is an illustration of such absorption. Throughout that chapter, resources were assumed available in unlimited quantities but, of course, at a price. A higher expenditure on any activity presumably meant the acquisition of more resources (in an optimal, or near optimal, combination) which made it possible to shorten the duration of the activity. As we have seen, different functional relationships between cost (or, equivalently, the resources made available) and activity duration give rise to different models of analysis.

Without the assumption of unlimited availability of resources, one runs immediately into the conglomerate of problems that we have grouped under the title "the problems of scarce resources." In brief, these problems arise naturally in any project in one of two forms:

1. It may be that management has fixed amounts of each resource that it either cannot or does not desire to exceed, and wishes to evaluate the impact of such scarcity in the resources on performance (e.g., as measured by the total duration of the project or the maximum delay in certain activities).
2. It may be that management has a free hand, more or less, in acquiring (or securing) any amounts of the necessary resources, and the query arises regarding what the optimal amounts are to be acquired to achieve a given target date. Naturally, optimality will have to be defined relative to a well-specified criterion.

The thing to remember is that the very existence of limited resources plays havoc with some of the very basic concepts in purely temporal considerations. For instance, in the case where resources are not limiting, the concept of *float* is simple and unambiguous. Thus, there is a single total float value associated with activity (ij) and is defined by [see Eq. (2.6a)] $S_{ij}^1 = t_j(L) - t_i(E) - y_{ij}$. This simplicity was the immediate result of $t_i(E)$ and $t_j(L)$ being uniquely determined for each node of the network. Such is not necessarily the case when resources are limited, $t_i(E)$, as well as $t_j(L)$, may not be so easily determined but are rather ambiguous.

Consider, for example, the simple project shown in Fig. 3.1, in which all activities require the same resource and are of duration one. The quantity of the resource consumed by each activity is shown next to each activity (Fig. 3.1 adopts the A-on-N representation). It is assumed that the project should take no more than 4 time units, and that only 10 units of the resource are available. Notice that if all activities are started at their earliest start times, activities *2* and *4* will have no "float" because neither can be delayed at all due to the resource constraint. On the other hand, if either of these two activities "takes a jump over the hump" and is started in period 3, then it will have a "float" of one unit! More interestingly, the other activity may not be started in periods 2 or 3, but only in period 4. The interpretation of this activity's "float" certainly requires some analysis. Compare all this complication with the simple case of unlimited resources: activities *2* and *4* would have 3 units of float each! (This confusion, or rather indefiniteness, in determining the float of an activity should add weight to our previous warning about the need for utter caution in discussing the concept of float; see p. 22.)

For another example, it must be evident from the above discussion that the concept of a CP as previously defined assuming unlimited availability of resources loses its meaning when the resources are limited. Now, some

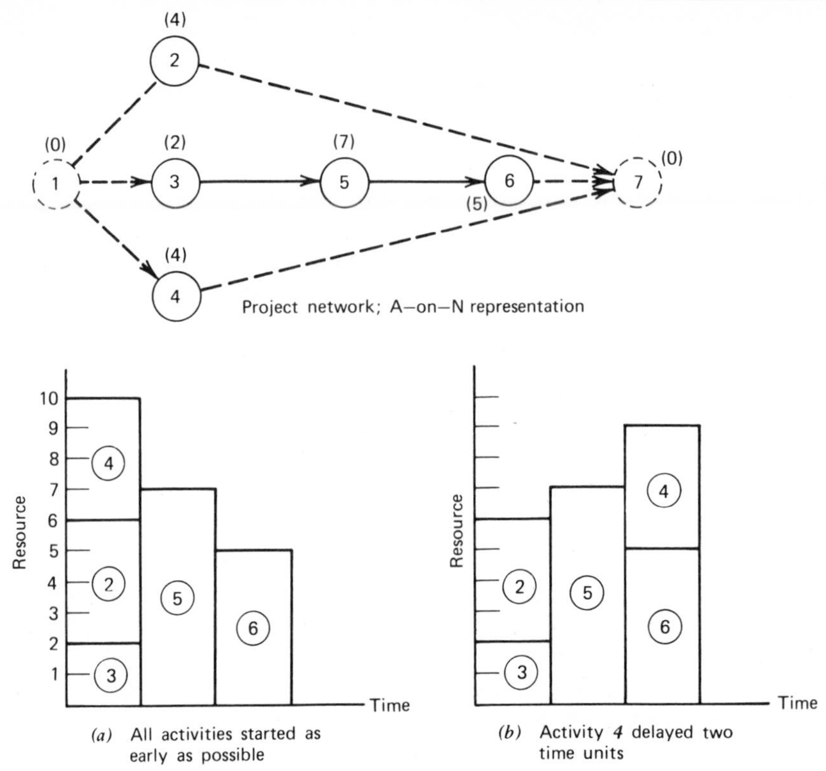

Figure 3.1. Illustration of complexity of concept of activity float under constrained resources.

activities on the so-called CP may have to be delayed because of insufficient resources. This would occur if the integrity of the CP is maintained, if at all feasible, at the expense of prolonging the project unnecessarily. Considering the CP as not an inviolate entity leads to shorter project-completion time!

An example to illustrate this point is shown in Fig. 3.2*a*. The CP is *1, 3, 5, 6,* and the resource availability is exceeded. Notice that there exists no feasible schedule of the activities that respects the limitation on total resource availability of 6 units and maintains the integrity of the CP (since activity *1* must precede activity *6*)! On the other hand, if we adhere to maintaining the integrity of the CP *as long as possible,* we are forced to insert activity *4* before activity *6* (thus breaking the continuity of the CP), which is then followed by activity *2*, as shown in Fig. 3.2*b*. The total duration of the project under this schedule is 11 time units. Finally, consider the schedule shown in Fig. 3.2*c*, in which the CP is "split" after

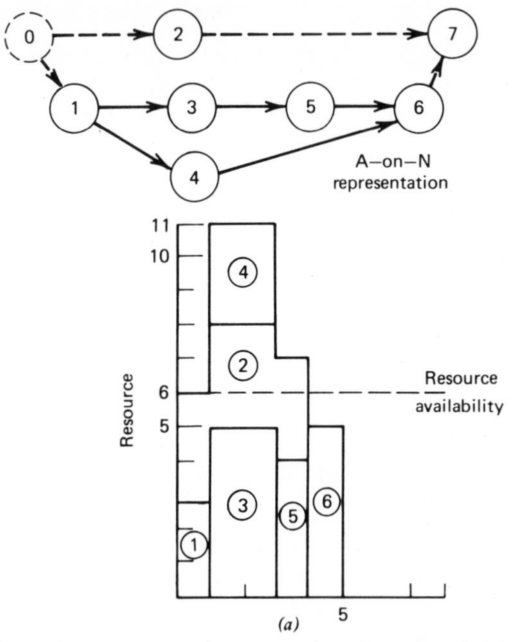

(a) Skyline of resource requirements assuming all activities start early.[a]

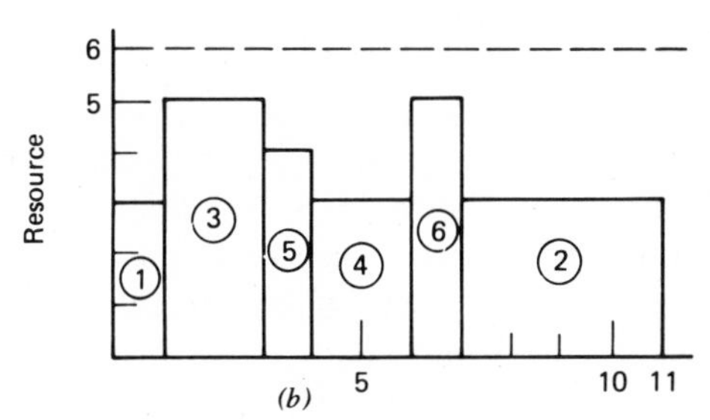

(b) Skyline of resource requirements under latest interruption of CP.

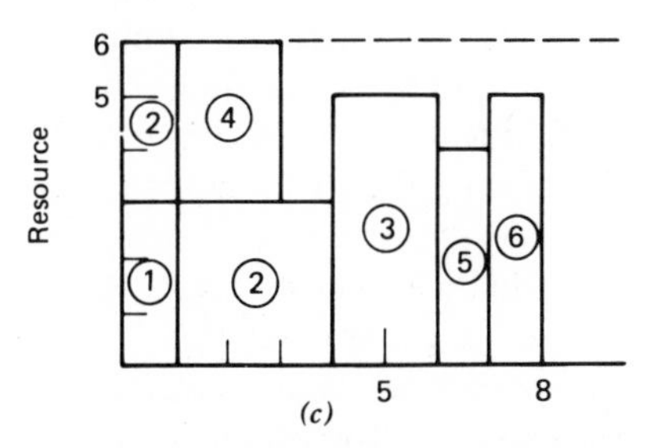

(c) Skyline of resource requirements to minimize project duration.

Figure 3.2

[a]For definition of "skyline," see p. 152.

activity *1*; the total duration of the project is only 8 time units! Hence there emerges the concept of a *critical sequence* (CS) of activities in a project, which is dependent on both the duration of the activities and the resource constraints; furthermore, it is a function of a given feasible schedule. The CS need not form a connected path in the network, though, similar to a CP, activities on the CS have zero "float," where the latter is obviously defined differently.

Activity	1	2	3	4	5	6
Duration	1	4	2	2	1	1
Resource requirement	3	3	5	3	4	5
	Resource availability = 6.					

Before proceeding any further, we should clarify the type of resources with which we are concerned here. Generally speaking, these are: (a) labor, engineering, and management skills, (b) machinery, (c) equipment and productive units, (d) facilities such as stores and receiving docks, and (e) capital in its different forms. Despite the radical difference in the nature of these resources, it is evident that *for the purposes of our present analysis* they can be treated in the same manner. For example, suppose that management knows that because of its other financial commitments it will not be able to borrow more than $1.5 million over the next year to meet its cash requirements. Clearly, this figure of $1.5 million can be treated as a limiting capacity on borrowing for cash needs in the very same way that the productive capacity of a piece of equipment over the next year is treated. Henceforth, we need not identify the various resources beyond a label to distinguish one from the other.

From the very beginning of development of PERT/CPM, suggestions have been made for extending the general utility of network planning methods by including variables other than time. Indeed, the initial CPM paper formulated an approach to the time/cost trade-off problem and suggested methods of application, see Kelley (1963). Another more ambitious approach to generalizing the PERT approach was offered in an early letter by Freeman (1960a) where he suggested the desirability of including cost and technical performance data in the PERT model. In a subsequent paper (1960b) he outlined a general scheme for accomplishing this idea. Unfortunately, even with very restrictive assumptions, his method encounters severe obstacles to solution for practical sized problems and for that reason the concept has not been widely pursued. An equally ambitious, and likewise dormant, concept has been suggested by Dogrusoz (1961).

The real-life problems of resource considerations in ANs have thus far defied rigorous mathematical treatment and the achievement of optimal solutions, except in some very special cases to be detailed below. In the majority of cases, one has to be content with "approximate" solutions and "heuristic" or "common sense" approaches. Some of these approaches are reviewed below, but first let us investigate the complete and rather ghastly problem as it presents itself in the day-to-day operations.

§1.1 Some Difficulties in Resolving the Resource Problem

Two reasons stand out as the causes for the current uncertainty and lack of a definitive solution to these problems. These reasons can be summarized under two general headings: (a) the difficulty of stating the problem and (b) the difficulty of formulating a mathematical model even if a complete and clear statement were possible.

Difficulties in the Statement of the Problem.

First of all, the same comments made previously concerning the difficulties encountered in estimating the *duration* of an activity (see p. 26) can be repeated almost verbatim relative to estimating the projected *consumption* of the various resources, especially if the activity in question is a gross activity, in the sense that it was not subdivided into its elementary components.

This is true for several good reasons. Oftentimes, only a subset of the resources are required for the full duration of the activity; the other resources are needed only for a fraction of that duration. The analyst must then choose between subdividing the activity according to the combination of resources required at any point of time, which may lead to a prohibitively complicated network, or leaving the activity as an entity, a course of action that must lead to a gross exaggeration of the total requirements, especially of those resources needed intermittently!

It is generally recognized, although very rarely explicitly stated, that the time estimates of an activity are based on subjective knowledge of the availability of resources. The functional relationships between these two variables may not be known, which in no sense vitiates their existence. It immediately follows that subdividing the activity in correspondence to its resource requirements at various points of time may change the time estimates of the subdivisions, which in turn may affect the resource requirements! A vicious cycle results which may disrupt any possibility of obtaining meaningful results.

Another major roadblock to a clear and precise statement of the problem is the fact that an activity can usually be *started*, and possibly *maintained* for a long time, with fewer resources than are ideally needed. In other words, the statement of resource requirements is inherently subject to a "band of indifference," in which duration is not affected, and another wider "band of feasibility" at the price of prolonged duration of the activity. If the functional relationship between resource availability and the duration were known, analysis similar to the duration-cost analysis could be conducted. In general, this functional relationship is not defined, as was stated above. Moreover, activities that require more than one resource would require a completely different treatment from those activities requiring only one resource.

The minimum requirement of a resource to start an activity is called *threshold.* If the various resources are totally independent, the threshold of an activity which requires a combination of resources (e.g., an assortment of skills) would be the threshold of its most limiting resource. Unfortunately, the possibility of resource substitution (e.g., labor of one skill substituting for labor of another skill for a period of time) plays havoc with the concept of threshold for complex activities.

Leaving, for a moment, the problems centered around the activity, we find that difficulties are encountered on a completely different plane. In particular, the availabilities of the various resources are set arbitrarily. A more meaningful statement would be a cost-availability level functional relationship, but this is difficult to obtain. The result is the existence of a threshold on the high side of the specified capacity, which we call *margin*, over and above the threshold (on the lower side of the requirements of an activity) mentioned in the previous paragraph. This extra margin of availability of resources is often present in the form of overtime capacity, possible temporary acquisitions, or shifting in the resource labeling (e.g., from one skill to another). Consequently, if at any point of time the stated availability of one or several resources is exceeded, the analyst must recognize the ever-present possibility of relaxing the constraint(s). The limits to such relaxation are, necessarily, dependent on the set of resources in question, and cannot be stated in advance.

Finally, there is an implicit and incorrect assumption that an activity is an indivisible entity. This may be true from certain points of view, (e.g., precedence relationships) but need not be true from the point of view of resources. It is conceivable, and by no means rare, that an activity can be carried out intermittently so that its demand on the resources is also intermittent. The question, of course, is the statement of this fact within the framework of activity networks, especially if an activity *may* be (but *need not* be) split only at a finite number of points. The combinatorial character of the various possibilities that present themselves in the case of several activities demanding the same resources at the same time can easily be seen.

Difficulties in the Formulation of the Mathematical Model.

Unfortunately, it seems that even if the problem were correctly and precisely stated, it would be very difficult indeed to give a reasonably comprehensive mathematical model that would incorporate the various variables in a *solvable* formulation. This is perhaps due to four factors acting individually or in unison to limit the power of analytical models.

First, there is the interdependence among the activities due to the sharing of the same resources, hopefully in an efficient manner, the dependence of resource consumption on the manner in which the activity is subdivided, and the dependence on the limits of availability of the various resources. All three dependencies are basically nonlinear in character. Thus, even if the functional relationships describing all three types of dependencies were determined—a formidable task in its own right—there would still remain the task of combining these individual relations into a meaningful whole! More significantly, resources are often substitutable with others, but *the duration of the activity is dependent on the resource used!* Here is a problem of a radically different nature: the very duration of the activity cannot be determined a priori but is a consequence of the resource used. But the decision concerning the resource (say the use of high-skilled rather than low-skilled labor) is usually made after one knows that the shortening (or prolongation) of the activity is profitable (or harmless) relative to another criterion. Here is evidently a case in which the duration of the activity and the specification of the resource (and its consumption) must be determined jointly, leading to an even more complex mathematical formulation. In Section 4 of the present chapter we propose such a model. The reader can verify its complexity (and computing difficulty) despite the drastically simplifying assumptions made in its construction.

Second, it is often difficult to ascertain the objectives of management, particularly when these objectives are poorly formulated and far from crystallized in management's own mind. For example, the objective may be any one of the following three:

1. Minimize total project cost assuming unlimited availability of the various resources *at a price.* The price may or may not be linear with the quantity ordered. But clearly the trade-off here is between small capacities with a long project duration and large capacities with short duration.
2. Minimize the duration of the project under limited capacity levels.

3. "Level" the resources' consumption while achieving a specified target completion date.

Management is frequently desirous of "having its cake and eating it too," in the sense that it specifies the minimization of all three variables—total duration of project, capacity levels, and variation about the average usage—and remain within a limited budget.

Third, the freedom of subdividing some activities, combined with the freedom of scheduling an activity within an interval equal to its duration plus the activity's float time, give rise to a combinatorial problem of formidable magnitude. In the heuristic approaches currently proposed for the problem of resource allocation, examples of which are discussed below, some arbitrary ranking rule is usually adopted to break the ties among contending activities and thus get rid of the combinatorial problem.

Fourth, and finally, the notions of "threshold" and "margin" are necessarily subjective in nature, and their incorporation into a model is difficult if not impossible. For example, if more than one activity compete for the same scarce resources, should we start more than one activity at their threshold, or should we devote the resources to fewer activities? How would this affect the duration of the project in both cases? As another example, suppose an activity that started with a partial crew is stopped—how much of this activity is accomplished?

The above discussion is not intended to discourage the reader from undertaking active research and investigation to solve the problem of optimal resource allocation. On the contrary, it is intended to encourage such work by identifying difficulties and pointing out possible pitfalls. Needless to say, any attack on the problem must start with simplifying assumptions. The qualitative discussion presented above should assist in the choice of the necessary assumptions to render the problem analytically tractable.

§1.2 The "Skylines" of Resource Requirements

Let activity $u \equiv (ij)$ be of duration y_u and require $c_u(r,\tau)$ units of resource r, $r = 1,2,\ldots,R$, in time period τ. Let the start time of activity u be s_u and its finish time be f_u. Clearly, since u is synonymous with (ij), $s_u \geqslant t_i(E)$, $f_u \leqslant t_j(L)$, and $f_u - s_u = y_u \geqslant 0$; where $t_i(E)$ and $t_j(L)$ are, as always, the earliest realization time of node i and the latest realization time of node j, respectively, in the A-on-A representation mode. Thus, to each activity $u \equiv (ij)$ corresponds a matrix of consumption of resources, $C_u = [c_u(r,\tau)]; r = 1,2,\ldots,R; \tau = 1,2,\ldots,T$ where $T = t_n(E)$, the earliest completion time of the project. Clearly, the $R \times T$ matrix will have zeros everywhere except

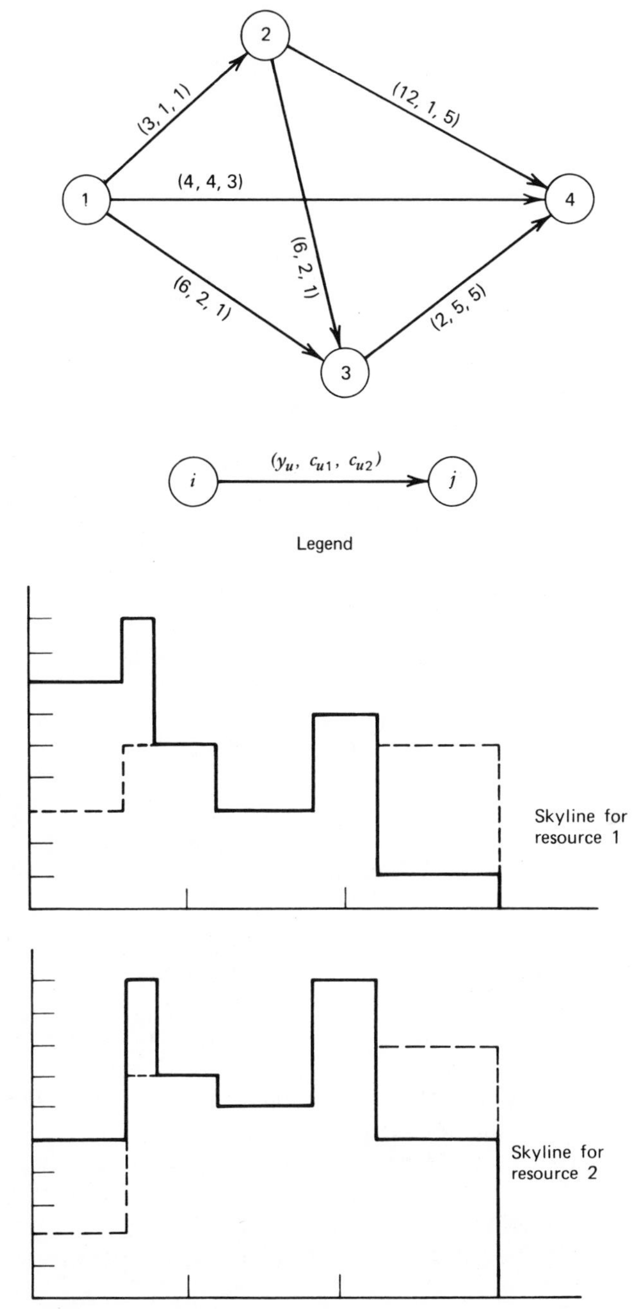

Figure 3.3. Example of the matrices C_u and $\mathcal{C}$ and of skylines.

Activity a	Matrix C_u														
(1,2)	1	1	1												
	1	1	1												
(1,3)	2	2	2	2	2	2									
	1	1	1	1	1	1									
(1,4)	4	4	4	4											
	3	3	3	3											
(2,3)				2	2	2	2	2	2						
				1	1	1	1	1	1						
(2,4)				1	1	1	1	1	1	1	1	1	1	1	1
				5	5	5	5	5	5	5	5	5	5	5	5
(3,4)										5	5				
										5	5				
Matrix of total consumption $\mathcal{C}=$	7	7	7	9	5	5	3	3	3	6	6	1	1	1	1
	5	5	5	10	7	7	6	6	6	10	10	5	5	5	5

Figure 3.3. (*Continued*)

possibly for the entries in the $R \times y_u$ submatrix occupying columns $s_u, s_u + 1, \ldots, f_u$.

It follows that to every feasible schedule of activities (where feasibility is relative to the precedence constraints) in which the start and completion times of each activity are determined, there exists a unique matrix of total consumption of resources $\mathcal{C}=[C(r,\tau)]$, where $C(r,\tau)=\Sigma_u C_u(r,\tau)$; or, more succinctly, $\mathcal{C}=\Sigma_u C_u$.

Consider the rth row of the matrix $\mathcal{C}$; the row vector $(C_{r_1}, C_{r_2}, \ldots, C_{r_T})$ represents the total consumption of resource r in periods $1,2,\ldots,T$. If the values (C_{r_τ}) are plotted against the periods $\tau = 1,2,\ldots,T$ the result is a step function termed the *skyline* of the consumption of resource r. Clearly, there are as many skylines as there are resources. Figures 3.1 and 3.2 contained examples of such skylines assuming only one scarce resource. Figure 3.3 further illustrates the above notions relative to a simple activity network of six activities and only two resources. The skylines (solid lines) are drawn assuming that $s_u = t_i(E)$ for all u, that is, assuming that all activities are started as early as possible. Simple as it is, the example illustrates in a forceful fashion the dilemma of the scheduler. If activity $(1,4)$ is delayed as late as possible, since this seems to "even out" the consumption of resource 1, the consumption of resource 2 is made even more uneven (see dotted lines in Fig. 3.3)! Furthermore, it is obvious that if management has limited availability of either (or both) resources, the prolongation of the project may be inevitable. Finally, there are as many skylines as there are ways of rearranging the starting times of the activities, which is a very large number, indeed.

§2. THE HEURISTIC APPROACH

Heuristic reasoning has been defined as "reasoning not regarded as final and strict, but as provisional and plausible only, whose purpose is to discover the solution of the present problem" [Polya (1957)]. Heuristic reasoning is pragmatic and opportunistic. It claims no "best properties" for a solution (e.g., optimality), but evaluates a solution on its operational utility as an improvement over previously available solutions. A specific heuristic is justified because it is plausible, because it narrows the domain of search for optimal solutions, and because experimentation proves that it "works," that is, it improves on previous solutions.

The heuristic approach stands in sharp contrast to the analytical (or algorithmic) approach. While the former is inductive, pragmatic, open-ended, and with no claim to optimality or even attainment of a desired degree of accuracy, the latter is deductive and inalterable and achieves optimality or approaches it with a known degree of accuracy.

Finally, while it is easy to see the distinction between the analytical (or algorithmic) and heuristic approaches, it is not equally easy to grasp the difference between the heuristic approach and the simulation approach. On the surface, the two appear to be the same since both approaches involve trial and repeated experimentation, and both may use the Monte Carlo technique in the conduct of individual experiments.

Briefly, the difference lies in the absence of *internal* decision rules capable of modifying the answer (or some other decision rule) in the case of simulation, and the presence of such rules in heuristic problem solving. Stated differently, in simulation alternatives must be supplied "from the outside," but in heuristic problem solving, the alternatives are generated *internally* following a set of rules called *the heuristics*.

Heuristic procedures for the resolution of the scarce resources problem of ANs abound, and there is little doubt that their number increases as more and more investigators are attracted to the field.† Unfortunately, several among them are proprietory programs on which no detailed information concerning internal logical structure is available. Furthermore, very

†For a review and critique of the various approaches, see Bennington and McGinnis (1973), Davis (1973) and Woodworth and Dane (1974).

little computing results have been published, a fact that renders comparative analysis well nigh impossible. And in any event, there is a good reason for not making such comparative analysis even if computing data were available,† namely, that heuristic procedures are usually "tailor made" to fit a particular set of conditions. Consequently, it is understandable that a procedure performs well under such a set of conditions, but poorly under another set of conditions; and no general statement can be made on the "goodness" of a procedure relative to others.

We trust that the reader will be able to construct his own heuristic procedure(s), once sufficient insight is gained into the manner in which such procedures are constructed. Basically a heuristic or a combination of heuristics are used to resolve the conflict arising from the scarcity of the resources. In other words, all heuristics are approaches to allocate the resources to the activities. Indeed, if the resources are in ample supply, there would be no need for any heuristic.

The following ideas have been used in one procedure or another:

1. *The serial approach*, which derives its name from the fact that activities are considered sequentially, starting at the originating node and proceeding in a stepwise fashion toward the last node. The activities are ranked in some order—for instance, in order of ascending arrowhead (or tail). They are then scheduled in that order and also according to the availability of resources. An activity whose turn it is to be scheduled but that requires unavailable resources is passed over for other activities that can be scheduled with available resources, always respecting the precedence constraints. The delay in an activity may or may not cause a delay in the termination time of the project, depending on the available float of the activity. Despite the apparent simplicity of the procedure, its application involves problems of its own. For example, if the activities are ranked according to the number of the arrowhead, two disturbing phenomena occur:
   a. Since the numbering of the nodes is not unique, different numbering would lead to different ranking which, in turn, leads to different resource utilizations and project durations.
   b. The ranking is not unique when more than one arc impinge on the same node. It is then necessary to have a secondary ranking. This may be based on the float time of the activities (whichever of the four floats), the number of their originating nodes, the subjective feeling of the "importance" of the activity, or some random rule. It is conceivable that an activity can be delayed due to scarcity of the

†This is not to deny the need for testing and evaluation of a newly proposed heuristic procedure—see Section 2.3 for a more detailed discussion of this question.

necessary resources beyond a limit which is considered tolerable. Under such conditions it may be deemed necessary to acquire more of the necessary resources rather than delay the activity. Our discussion presented above concerning the "margin" on capacity limitations is pertinent in this respect. What is interesting is that now we are obliged to define what is meant by "acceptable" delay in each activity.

2. *The parallel approach*, where we consider one time interval and all the activities that can be scheduled during that interval. The activities are ranked in some fashion (see below) and scheduled according to resource availability; otherwise, the activities are delayed. Then the next time interval is considered, and so forth. The fashion in which the ranking of competing activities is accomplished may be according to one of the following four rules:
   a. *Greatest remaining resource demand*, which schedules first those activities with the greatest amount of work remaining to be completed. This rule characteristically produces schedules that result in the least amount of resource idle time.
   b. *Least total float*, which schedules first those activities possessing the least total float. The program would continuously update the float present in an activity, so that any activity sufficiently postponed would eventually become critical. This rule has been found to be superior to others in reducing the total weighted delays in the completion of projects.

      It has been reported by Moder (1975) that equivalent results have been achieved without requiring the updating of the activity float through the use of the "latest start time" rule.
   c. *Shortest imminent activity*, which schedules first those activities requiring the least time to complete. This rule mimics the well-known "shortest processing time" rule in job shop scheduling. It has the property of minimizing the delay beyond the (resource free) CP completion time, compared to other rules.
   d. *Greatest resource demand*, which schedules first those activities requiring the greatest quantity of resources from the outset. This rule appears to be valuable when one desires to satisfy a combination of objectives, rather that satisfy one objective in preference to others.

A particular program that utilized this parallel approach was based essentially on three heuristics [Moder (1975)]:

1. Allocate resources serially in time (and in parallel in the activities): start with the first period and schedule all activities possible, then do the same for the second period, and so on.

2. When several activities compete for the same resources, give preference to the activity with minimum remaining total float. In case of ties, choose at random among tied activities.
3. Reschedule noncritical activities, if possible, in order to free resources for scheduling critical activities.

These heuristics are incorporated in the program as shown in the flow diagram of Fig. 3.4. This approach has the advantage of eliminating the dependence on the numbering of the nodes, unless, of course, such numbers are again used for secondary ranking. To give the reader an idea about how these, and other heuristics are "put together" in a cohesive procedure, we explain in some detail the construction and operation of the SPAR-1 Model due to Wiest (1967).

§2.1 An Example of a Heuristic Procedure: The SPAR-1 Model (Scheduling Program for Allocation of Resources)

The model focuses on available resources, which it serially allocates, period by period, to jobs listed in order of their early starting times. Activities are scheduled, starting with the first period, by selecting from the list of those currently available, and ordered according to their total float (based on technological constraints only and normal resource assignments). The most critical jobs have the highest probability of being scheduled first,† and as many jobs are scheduled as available resources permit. If an available job fails to be scheduled in that period, an attempt is made to schedule it the next period. Eventually all jobs so postponed become critical and move to the top of the priority list of available jobs.

The basic program described above is modified by a number of additional scheduling heuristics or subroutines generally designed to increase the use of available resources and/or to decrease the length of the schedule. In the brief description of these heuristics that follows we refer to resources as men, to the amount of resources applied to a job as the crew size, to a group of homogeneous resources as a shop, and to the scheduling

†Jobs in the list are scanned sequentially. Each job has a probability p (where p is an input parameter) of being considered for scheduling. If a job is passed by, the next one is considered with probability p, and so on. All other jobs in the list are scanned before a by-passed job is reconsidered. Scanning is repeated until all jobs have been considered. A similar procedure is discussed in the work by Levy, Thompson, and Wiest (1962). Probabilistic elements in the program provide some randomness of job assignments and the likely production of different schedules on repeated applications of the model to a project. Thus, a number of different schedules can be generated (depending on the size of the project and computer time available), and the best one can be selected.

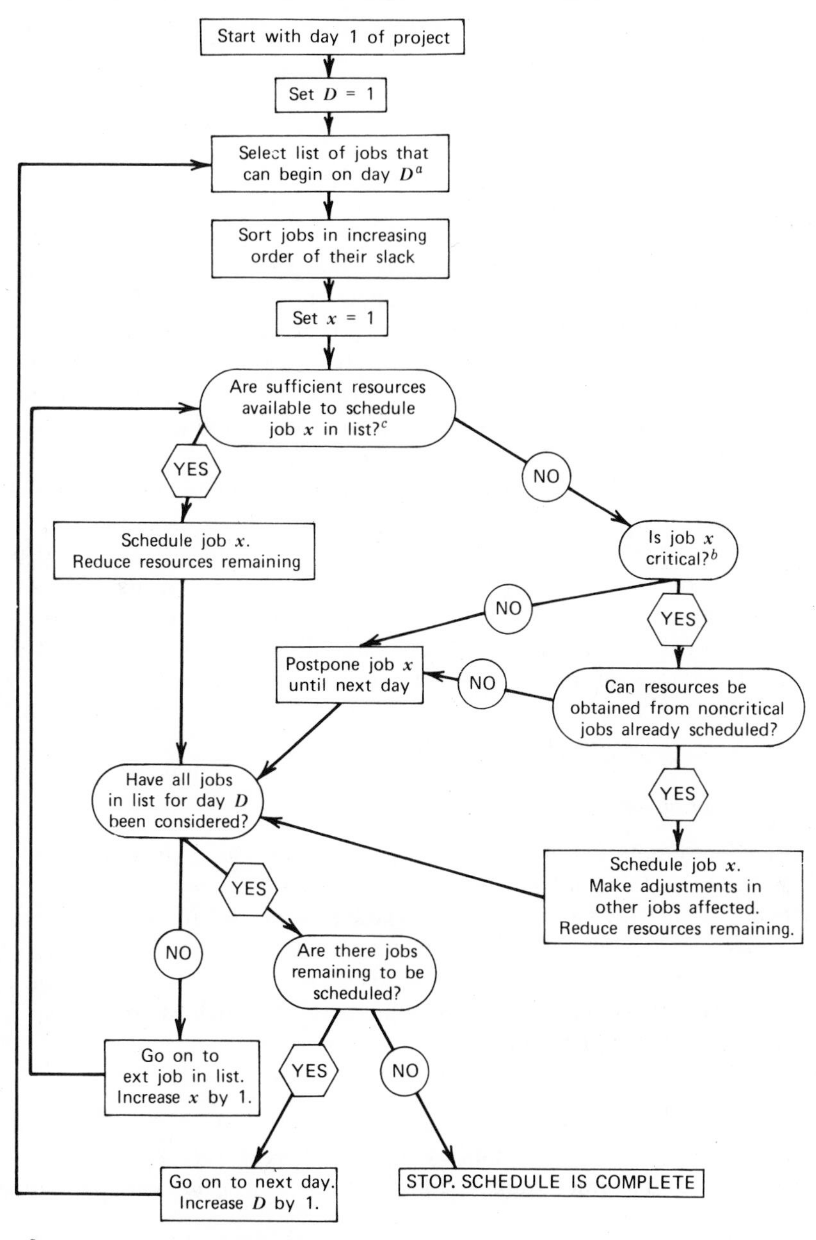

[a] Day D is the day under consideration.
[b] A job is *critical* if it has no slack.
[c] Job x is the job under consideration.

Figure 3.4.

period as a day. (Any other units, of course, could be used in place of these.)

Crew Size:

With each job is associated a *normal* crew size, or the number of men or other resources normally assigned to the job, a *maximum* crew size, or the maximum number of men required for crashing the job, and a *minimum* crew size, or the smallest number of men that can be assigned to the job. Normally these three crew sizes will differ from each other, although in some cases two or all three of them may be equal when jobs cannot be stretched out or "crashed." The rules for crew size selection are as follows: If a job to be scheduled is critical,† it is placed on a priority list and given special treatment. If sufficient men are available, the job is scheduled at its maximum crew size,‡ that is, it is "crashed." If insufficient men are available to do so, or even to schedule the job at normal crew size, then an attempt is made to obtain the required men by means of "borrow and reschedule" routines described below. If all efforts fail, however, and the job cannot be scheduled even at minimum crew size, then its early start date is delayed one period and an attempt is made to schedule the job on the following day. Noncritical jobs are scheduled at normal crew size if the required number of men are available, but if resources available are insufficient for scheduling even the minimum crew size, then the job is delayed for consideration until the next day.

Augment Critical Jobs:

Repeated attempts are made to speed up critical jobs (including those which become critical after being scheduled) that have crew sizes less than their maximum crew size. Before any new jobs are scheduled on a given day, jobs previously scheduled and still active are examined. If any of these is critical and has a crew size assigned less than its maximum, and if resources are available, its crew size is increased as much as possible up to the maximum (see † footnote).

Multiresource Jobs:

Sometimes a job requires a number of different resources, such as men of different skills, machines, money, and so on, each of which may be limited

†A job is considered critical if its slack does not exceed some value k where k is an input parameter.

‡Actually, a "critical" crew size is chosen—that is, one sufficient to increase the job's slack to $k+1$ (if possible), with the maximum crew size as an upper bound.

in quantity. For such multiresource jobs, separate jobs are created for each resource and the jobs are constrained to start on the same day with the same level of resource assignment—that is, normal, minimum, or maximum crew size (or some intermediate level).

Borrow from Active Jobs:

If resources available are insufficient for scheduling some critical job j, then the model enters into a procedure for searching currently active jobs to see if sufficient men might be borrowed from them for scheduling job j on that day. Men are borrowed from a job only when the resultant stretching of the job will not delay the entire project. If enough resources cannot be borrowed, then j is delayed for one day.

Reschedule Active Jobs:

Sometimes a critical job j could be scheduled if other jobs previously scheduled that utilize the same resources had been postponed to start at a later time. The model scans the list of currently active jobs and picks out those that could be postponed without delaying the project due date. If sufficient men could be obtained in this way and/or from the borrow routine described earlier, then job j is scheduled and the necessary adjustments are made in previous assignments.

The reschedule routine has much the same effect as a predictive feature. Instead of attempting to look ahead to future needs of critical jobs (which is difficult in the limited resource case since jobs are not always scheduled at their early start), the model schedules all jobs possible as it moves along from day to day, "repenting," so to speak, of previous scheduling errors if jobs are encountered with more critical need of resources than those to which the resources were assigned at earlier dates.

Add on Unused Resources:

After as many jobs as possible are scheduled on a given day, there still may be unused resources in some of the shops for that day. The model compiles a list of active jobs to which these resources might be assigned and arranges them in ascending order of their total slack. Proceeding down the list, the model increases the crew size of these jobs until either the unused resources or the list of jobs is exhausted. The increment is temporary; jobs supplemented in this way return to their assigned crew size the next day, unless unused resources are available then also.

After going through the above scheduling routines each day, the model then records the results in a manpower-loading table that notes the number of men assigned to each job and in each shop, and it updates the critical path data for unscheduled jobs. (Note that several of the above routines may alter a job's criticality; e.g., when the maximum crew size is assigned to a critical job the shortening of the job may result in its gaining slack and perhaps some other job or jobs becoming critical.) The model thus proceeds from day to day until all jobs have been completed.

Evaluation of Schedules:

A schedule should be evaluated in terms of its effects on schedule-related costs. Relevant costs are a function of the particular project being scheduled and its setting, but in general they include: (a) resource costs (in particular, costs of unused resources and premium or overtime resources which are affected by the schedule), (b) overhead costs (directly related to the length of the schedule), and (c) costs of changing resource levels (including hiring and layoff costs, etc.).

Two particular cost functions reflecting these items have been used in the model so far.

$$\text{(i)} \quad \text{Total cost} = cz + \sum_{s=1}^{m} q_s^* w_s z$$

where c is the daily cost (e.g., overhead expenses and/or due-date penalties, charged on a perdium basis); z is the length of the schedule; q_s^* is the maximum number of men required in shop s; w_s is the average daily wage in shop s; and m is the number of shops. The implicit assumption is that shop crew are maintained at all times and are paid whether active or idle.

Should the circumstances of a specific project justify it, a nonlinear cost function of z could be substituted for the linear one above,

$$\text{(ii)} \quad \text{Total cost} = cz + \sum_{s=1}^{m} a_s^* w_s z + \sum_{s=1}^{m} \sum_{d=1}^{z} (q_{sd} - a_s^*) v w_s \delta^{\dagger}$$

where c, z, and w are as defined above; q_{sd} is the number of men in shop s used in day d; a_s^* is the "optimum" crew size of shop s; $\delta = 0$ if $(q_{sd} - a_s^*) \leqslant 0$ and $=1$ otherwise; and the parameter $v(\geqslant 1)$ is the ratio between premium and regular pay rates. The parameter a_s^* is determined by some incremental analysis. For instance, let $n(a_s)$ denote the number of days out of the duration z that manpower requirements in shop s exceed a_s. Then,

†Writing δ in this manner is shorthand notation for $\delta(q_{sd} - a_s^*)$. The argument of δ is to be understood from the context of the equation.

evidently, $n(a_s)=\Sigma_{d=1}^{z}\delta(q_{sd}-a_s)$. Let $Q_s(a_s)=\Sigma_{d=1}^{z}(q_{sd}-a_s)\delta$ denote the total man-days of overtime worked in the span of time z in shop s. Then a_s^* is required to satisfy the inequality

$$vw_s n(a_s^*)Q_s(a_s^*)-vw_s n(a_s^*+1)Q_s(a_s^*+1)\leqslant w_s z^{\dagger}$$

The ℓ.s. represents the savings in overtime cost due to the addition of one man, and the r.s. is the cost of paying the man the regular wage for z days.

Various additional criterion functions for evaluating schedules have been suggested by Burgess and Killebrew (1962), Carruthers (1968), and others. Since schedule-related costs are likely to be influenced by the particular circumstances surrounding a given project, it will often be expedient for cost functions to be designed on an ad hoc basis.

Search Routines:

A given schedule is based on a preassigned set of shop limits that may either be constant over the scheduling period or may vary each day. It is possible that in some cases, the cost of increasing shop resource limits above this initial set of levels would be more than offset by the resulting decrease in overhead charge and/or due-date penalty, and the reverse situation might also be true. Hence, a search procedure—either manual or by means of the computer—should be employed to seek some optimum combination of shop-resource levels and resulting completion date.

Two rather simple computer approaches have been explored. The first search routine starts with minimal resource levels or those just sufficient to ensure that all jobs can be scheduled, serially if not in parallel. After a schedule has been generated and its costs calculated, the resource levels of all shops associated with critical jobs are increased by some increment, a new schedule is generated, and its costs are calculated. The process is repeated as long as some improvement in schedule cost is noted. The second search method starts at the other extreme. Ample resource levels are assumed; that is, levels high enough to permit starting all jobs at their early start. Then resource levels are incrementally decreased—first, all shops at once and later, one shop at a time, until no further improvements in scheduled costs are noted.

A computer program that incorporates the characteristics and features of the general model described above has been written and tested. Called SPAR-1, it is displayed in Figure 3.5 in simple flow-diagram form. The basic program is able to handle single or multiple projects, fixed or

†Note that we determine a_s^* differently from Wiest.

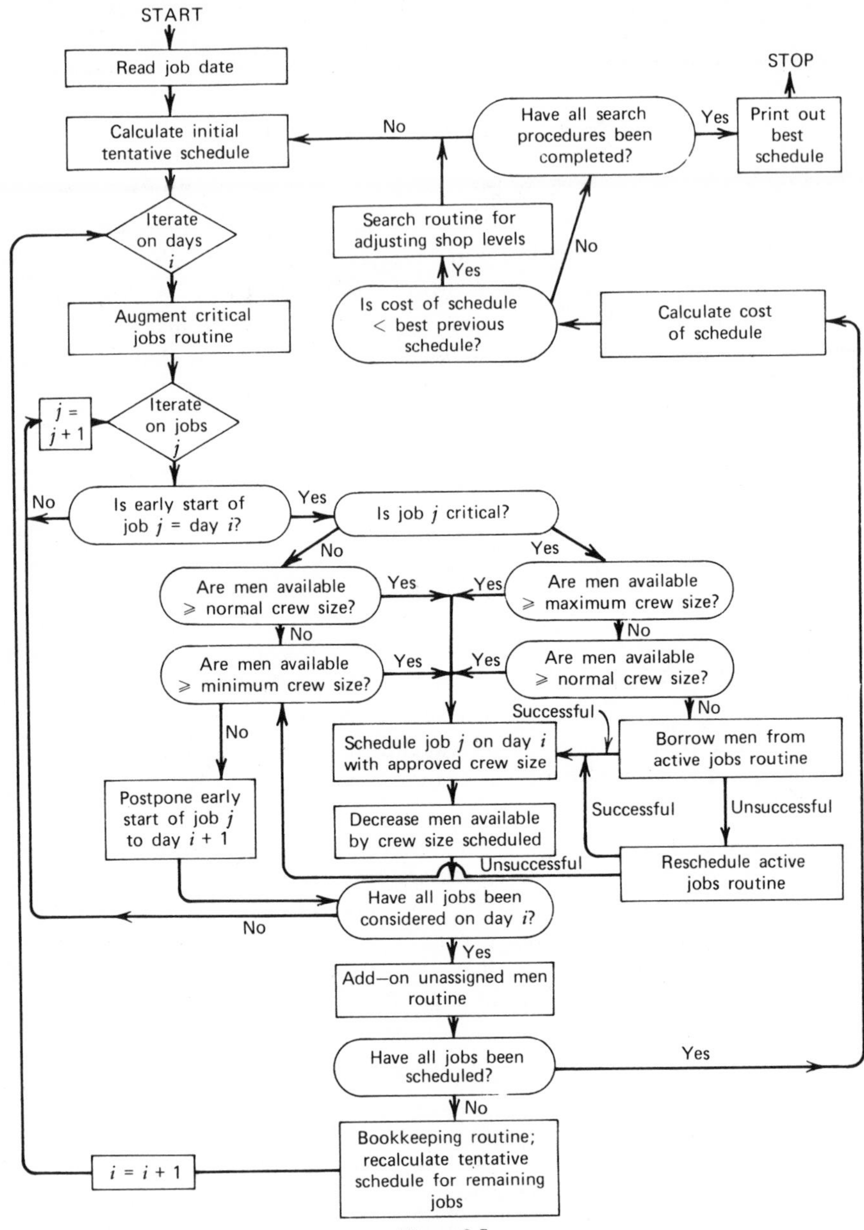
START
Read job date
Calculate initial tentative schedule
Iterate on days i
Augment critical jobs routine
Iterate on jobs j
j = j + 1
Is early start of job j = day i?
No
Yes
Is job j critical?
No
Yes
Are men available ⩾ normal crew size?
Are men available ⩾ maximum crew size?
Are men available ⩾ minimum crew size?
Are men available ⩾ normal crew size?
Yes
No
Successful
Schedule job j on day i with approved crew size
Borrow men from active jobs routine
Postpone early start of job j to day i + 1
Decrease men available by crew size scheduled
Successful
Unsuccessful
Reschedule active jobs routine
Unsuccessful
Have all jobs been considered on day i?
Add–on unassigned men routine
Have all jobs been scheduled?
Bookkeeping routine; recalculate tentative schedule for remaining jobs
i = i + 1
Calculate cost of schedule
Is cost of schedule < best previous schedule?
Search routine for adjusting shop levels
Have all search procedures been completed?
Print out best schedule
STOP

Figure 3.5.

variable crew sizes, constant or variable shop limits (over the scheduling period), and various functionals and search routines as described above. By means of input data flags, jobs may be marked individually as uninterruptible (once started) and crew sizes as unchangeable (once assigned), if so desired. These and other options are useful in fitting the model to a particular project situation. SPAR-1 is written in FORTRAN. On a 32K machine, the model may be dimensioned to handle a project with 1200 single-resource jobs, 500 nodes and 12 shops over a time span of 300 days.† Multiple projects may be scheduled within the same total jobs constraint.

§2.2 The Objective of Analysis

Minimizing the total project duration under known availabilities of resources appears to be the most natural objective of analysis. In fact, it is the most widely used objective among the majority of heuristic (and other) procedures. It implicitly assumes a high penalty for delaying the completion of the project. We wish to emphasize, at the risk of appearing redundant, that this is neither the only objective sought by management nor is it the most desirable in many cases. And it is up to the analyst to glean the more pertinent objective.

For instance, the above objective is sometimes overshadowed by considerations of how closely the skyline follows resource availability, in which case being on or below availability level is not enough, and some measure of *deviation* from such availabilities is needed (e.g., minimization of the sum of absolute deviations, maximum absolute deviation, or sum of squares of deviations).

On the other hand, we have mentioned that management is frequently interested in determining how much of the resources it should *acquire*, in which case the trade-off is between securing more of the resources versus prolonging the duration of the project. This would necessitate an objective function that combines the total cost of acquiring these resources and the cost of delaying the project or parts thereof. The SPAR procedure discussed above is an excellent example of such considerations. Typical of heuristic procedures, such cost considerations are introduced in an after-the-fact fashion, with the attitude of "try this and see how it fares." Evidently, this is an unsatisfactory state of affairs, and the analytical

†Most of the experience with the model has been obtained on a Control Data G-20 (Carnegie Institute of Technology) and an IBM 7094 (Western Data Processing Center, UCLA). Some trade-offs in the above dimensions are possible. A variation of the model with more extensive use of tapes increases the job limitation to 1500, with practically no limit on shops or days.

models presented in the following sections constitute the beginnings of a concerted attempt on the part of several researchers in this field to resolve some of the resulting issues.

§2.3 Testing and Evaluating Heuristic Procedures

There is little doubt that heuristic models, by their very nature, pose their own peculiar problems in their evaluation since they often preclude a guarantee of finding an optimum solution (which is not to say that optimality is never reached). Quality of the solution, however, is only one among several criteria to be considered, which also include computational efficiency (relative to time, storage requirements, background of operators, etc.) and the degree to which the model reflects "real world" complexities.

The criteria of such evaluation are somewhat fuzzier than could be desired, because the objectives are not necessarily well-defined and, in any event, depend on the situational requirements. In some instances, meeting a fixed due date may be of paramount importance, while in others respecting resource constraints may have higher value.

Evaluation of computational efficiency also lacks absolute standards because of differences in the computers used as well as in the programers' skill. Of course, some comparison may be made on the basis of the count of "elementary operations," but even that may be of dubious validity when such count cannot be made with any accuracy.

In general, the following three approaches have been proposed:

1. *Laboratory tests*. Schedule projects of sufficiently small scale for hand solution (or explicit determination of the optimum by use of computers in larger projects) and compare the model's output with the known optimum schedules. Simple as this approach may seem, it has at least two major drawbacks. The first is that "small enough" projects amenable to explicit determination of the optimum need not be representative of the real-life situations—for example, such projects may not exhibit sufficient parallelism of CPs or sufficient interference in the demand for resources. The second is that "the deck may be loaded" for, or against, a particular heuristic approach—that is, projects may be fabricated that produce good or bad results, as desired, with any particular set of heuristics.
2. *Field tests*. Apply the model to a variety of real-world scheduling problems to study its behavior when subject to various scheduling environments and constraints. Typically, the major objection to this approach is that there are so few such real-world applications in which a *conscious* attempt has been made to optimize relative to some objec-

tive function. Consequently, any comparison between the output of the heuristic model and "what happened in practice" seems unfair. Furthermore, "practice" involves the day-to-day conduct of business, with the myriad of everyday problems that can never be replicated in the model. To compare the actual performance under those conditions to the sterilized and germ-free atmosphere of the heuristic model reinforces the sense of unfairness. The only alternative is to run the model *in parallel* with the actual conduct of the project as well as with all competing models. Unfortunately, such an experiment may take too long and cost too much to be worthwhile. But even if it were conducted, the end result is rather fragile; we would have one sample point, and few people would risk any generalizations based on such meager data! Of course, it is possible to compare several heuristic models on the same set of real-life data, without the need for parallel operation with the actual conduct of such projects. We refer to this model as *modified field tests*. Unfortunately, one can only perform few such comparisons because of the dirth of such real-life data; and the results of comparison are always subject to criticism relative to the degree of confidence one can invest in them.

3. *Comparative large-scale simulated test*. Compare schedules generated by several models on projects that are constructed artificially. If such projects are large in size, and if a large number of them are constructed, almost all the objections raised above can be obviated, and a valid "test" may be conducted. True as this may be, the reader is alerted to the size, the time, and cost of the experiment to be conducted, if meaningful statistical conclusions are to be drawn.

In summary, then, it is clear that the issue of testing and evaluation is a knotty one, with no easy resolution. One has the choice between "rough-and-dirty" comparisons that lead to low confidence level in the conclusions drawn and statistically reliable tests that require a vast expenditure of time and effort.

There is no question about where the answer currently rests: it is in the former approach. The SPAR-1 model, for example, was tested under laboratory-test conditions as well as under modified field tests. The conclusions are, as is almost always the case, tentative and cautious; "Additional development and testing of the model—on a variety of project types and with a number of new scheduling heuristics—is underway. A major modification of the model (SPAR-2) based on concepts of "critical sequence" and "conditional slack" is being programmed.

On the whole, it seems fair to say that these models are fulfilling an immediate need and are also laying a basis for further progress in heuristic programing. While they demonstrate in a specific way that heuristic programing, aided by the computational power of a digital computer, helps to solve complex scheduling problems in large project management, in a broader view they contribute to our understanding of the use of this approach on a more general class of large combinatorial problems frequently met by managers and researchers, many of which have not previously yielded to analytic solutions" [Wiest (1967), p. B376].

§3. A PROLOGUE TO ANALYTICAL MODELS

The formalization of scheduling procedures under scarce resources—which we have termed the "analytical models"—occupy the remainder of this chapter. It is a recent and welcome development because the problem is of sufficient importance to warrant serious and more rigorous analysis and also because some benchmark had to be established for the ever-proliferating "approximate" procedures.

The models reflect different assumptions and hypothesized conditions in their attempt to represent varying degrees of realism concerning the contextual constraints of the problem and the objectives of its analysis. Generally speaking, the models may be classified as follows:

1. Models that view the project as a *process* that changes its state each epoch. Such change is brought about by the consumption of resources, and yields completed (or partially completed) activities. In such models, attention is naturally focused on the concept of *state* as well as on the *state-transformation functions* that govern the evolution of the process. Furthermore, one is naturally led to DP models, such as those proposed by Petrovic (1968) and Lofts (1974). Such models exhibit a great deal of ingenuity in their construction but unfortunately lack operational utility. This is mainly due to the large dimensionality of the state vector. In fact, little has been reported in the literature on their computing performance. Consequently, we do not dwell here on their detailed description; the interested reader may consult the references cited above.
2. Models that view the project as a collection of *entities* (the so-called variables) and *relations* among these entities (the so-called constraints) and attempt to determine the magnitude of these entities, which optimize a given objective function. The outcome of such a view is a mathematical programing model of the garden variety, usually in integer variables.

The solution of these models is still a stumbling block except for small-scale problems. However, the intensive research conducted today in integer and mixed programing promises the resolution of this difficulty in the not-too-distant future.

These models occupy Section 4, in which several representations are offered. In addition to serving the immediate purpose of modeling the scheduling problem, they are instructive as specific examples of the "kit of tools" of integer program representations. The reader is well advised to dwell on these models to savor their manner of modeling rather complex relations.

3. Models that sift through all the possible forms that a project may assume, searching for an optimal (or a feasible) combination without necessarily exhibiting the mathematical formalism of the model.

   These models typically utilize some implicit enumeration approach and stand out as offering the only bright prospect, at the present time, of operational utility. One can obtain answers to medium-scale problems, while large-scale problems are consistently studied to reduce them to the realm of computing feasibility.

   An interesting aspect of one type of these models (discussed in Section 5) to which we wish to draw the reader's attention is that it achieves a link between ANs and models of certain types of assembly-line problems. Prima facie, the relation between the latter types of problem and project planning under scarce resources may seem at best artificial and tenuous. But close scrutiny reveals that one problem is very much akin to the other, and the relationship is illuminating to both fields of study.

One final remark: in the sequel we have restricted ourselves to *linear* models because they are simpler and constitute the fundamental base of understanding of the problem and because nonlinear models are well beyond the grasp of current computing schemes. The researcher in the field is advised to consult the references on nonlinear models cited in the bibliography [Chapman (1970), Clark (1961), Dogrusoz (1961), Kapur (1973)].

§4. INTEGER LINEAR PROGRAMING MODELS

The very nature of the precedence relation between an activity and its successors brings into play the "either-or" nature of the problem: either the activity is completed, hence its successors may start, or it is not completed, hence its successors cannot start. This, in turn leads to integer programing models, in particular, 0, 1 ILP models.

However, this is not the sole circumstance in which we find that we must resort to the definition of such 0, 1 variables. As we see below, we need such variables to identify "counters," to indicate resource ceilings, to denote the start (or completion) of activities, to distinguish among resources, and so forth.

This section is devoted to the discussion of the various circumstances that lead to such ILP models. Some of these representations are straightforward, while others are ingenious and signify a great deal of economy in the number of structural variables and/or constraining relations.

Our discussion proceeds as follows. First, in Section 4.1 we present isolated representations of some common phenomena usually encountered in the context of project planning. Then, in Section 4.2 we synthesize a complete model for a particular design situation. The exercises at the end of the chapter complement our (necessarily brief) treatment, and give the reader a chance to practice his own modeling capabilities.

§4.1 Some Representations

Suppose that the activities may be interrupted any number of times during their progress, but if an activity is interrupted and later on resumed at a different intensity of application of the resources, a cost is incurred that is a function of the *difference* between the two intensities. This situation is typical of the so-called smoothing problems, which characteristically discourage *variations* in the level of resource application and, therefore, penalize such variation.

To represent such variation from period to period, let y_{ij} denote the fraction of activity i accomplished in period j and α_{ir} denote the total quantity of resource r required to complete activity i (measured in units of resource periods, say man days). Then the change in the level of use of resource r from period $j-1$ to period j is given by $\sum_{i \in A} \alpha_{ir}(y_{ij} - y_{i,j-1})$, a quantity that may be either positive or negative. Since a different cost may be incurred in either case, it is important to identify each eventuality separately. This is accomplished by the equality

$$\sum_{i \in A} \alpha_{ir}(y_{ij} - y_{i,j-1}) = v_{rj} - w_{rj};\ v_{rj}, w_{rj} \geqslant 0 \text{ for all } r \text{ and } j \tag{3.1}$$

Now v_{rj} represents the increase in resource utilization in period j over period $j-1$, while w_{rj} represents the decrease in resource utilization. In an *optimal* program it is evident that we cannot have both v_{rj} and w_{rj} at positive levels simultaneously; at most one of them is positive and the other is zero (why?)

To develop the precedence constraints in this case, let $\mathfrak{B}(i)$ denote the subset of activities that immediately precede i, and hence must be completed before ($\mathfrak{B}$ denotes "before") activity i is started. The precedence relationship may be translated as follows: $y_{ij}=0$ unless all the activities in $\mathfrak{B}(i)$ are completed; that is, unless $\Sigma_{k=1}^{j-1} y_{vk}=1$ for each $v \in \mathfrak{B}(i)$. Let $\delta_{ij}=0,1$ variable. Then we can write a system of two constraints which, *in unison*, guarantee the desired dichotomy:

$$y_{ij} \leqslant \delta_{ij};$$

$$\sum_{v \in \mathfrak{B}(i)} \sum_{k=1}^{j-1} y_{vk} \geqslant \delta_{ij} |\mathfrak{B}(i)| \tag{3.2}$$

where $|\mathfrak{B}(i)|$ is the number of activities in the set $\mathfrak{B}(i)$. The rationale for this set of two constraints is as follows. If all activities in the set $\mathfrak{B}(i)$ are not completed by day j (i.e., at least one of them is still in progress) then the sum $\Sigma_{k=1}^{j-1} y_{\cdot k}$ is <1, and the second constraint can be satisfied only if $\delta_{ij}=0$. This, in turn, forces y_{ij} in the first constraint to remain at 0, which implies no action on activity i. On the other hand, if all activities preceding i were completed by day j, then δ_{ij} may assume the value 1, and the first constraint now permits a $y_{ij}>0$. Whether y_{ij} will be positive is governed by the other constraints and the optimization procedure.

Suppose that a penalty p per period is incurred for a delay in the completion of the project beyond a fixed date, D. (The value of D may be the duration of the CP ignoring resource constraints or a contractual delivery date.) Once more we are faced with an either-or dichotomy; the penalty p will be incurred in each period $j>D$ *if* the project is still in progress in that period; *otherwise*, no penalty is incurred. Let $\rho_j=0,1$ variable, which serves as a "counter" for the periods in which the project is in progress beyond D, and consider the following constraint:

$$\sum_{k=j}^{H} \sum_{i \in A} y_{ik} \leqslant \rho_j |A|; \quad j=D+1, D+2, \ldots, H \tag{3.3}$$

where $|A|$ is the number of activities in the project and H is some known upper bound on the duration of the project. Alternatively, one may constrain ρ_j by the following constraint

$$\sum_{k=1}^{j} \sum_{i \in A} y_{ik} \geqslant (1-\rho_j)|A|; \quad j=D+1, D+2, \ldots, H \tag{3.3a}$$

In either formulation if some activities are still in progress in period $j > D$, then ρ_j must $=1$ for the constraint to be satisfied. This is so because the ℓ.s. of (3.3), or (3.3a), will be strictly positive and less than $|A|$.

The above three representations (3.1)–(3.3) were predicated on the definition of the structural variables $\{y_{ij}\}$ which, in turn, were devised based on the need to measure the variation in resource consumption from period to period. If such measure were not needed—that is, if we are not concerned with *smoothing* considerations—then one has more freedom in defining the structural variables.

For instance, suppose that we are interested in minimizing the total project duration subject to limited resource availabilities (a subject treated more extensively, but from a different point of view, in the following section). How can one achieve the representation of precedence and project completion?

To this end, let

$x_{ij} = 0, 1$ denote the start time of activity i, $x_{ij} = 1$ if i is started in period j, and $x_{ij} = 0$ otherwise;

t_i: the duration of activity i

$\mathcal{Q}(i)$: the set of activities immediately succeeding i

Then we may represent the precedence constraints as follows:

$$-\sum_j jx_{ij} + \sum_j jx_{vj} \geqslant t_i; \quad \text{for all } v \in \mathcal{Q}(i), \text{ and all } i \in A. \tag{3.4}$$

Note that since x is a 0, 1 variable, each sum on the ℓ.s. of the inequality will "pick up" exactly one period, say j_1 in the first sum and j_2 in the second. The constraint simply states that $j_2 - j_1 \geqslant t_i$, which is the desired restraining relation since activity v can be initiated only after i is completed.

The range of the index of summation of both sums in (3.4) deserves some attention. One can always let j run over the maximum duration, H. But this is wasteful since we know that the great majority of the x_{ij} and x_{vj} variables are zero. Therefore, it behooves one to determine the smallest possible range of each index. This is accomplished by capitalizing on the precedence constraints themselves. We leave such determination as an exercise to the reader (see Problem 7 at the end of the present chapter). For our immediate purposes we assume that it is known, and let a_i denote the time of earliest availability of activity i and $\bar{a}_i$ its latest start time; hence, its latest completion time is $\bar{a}_i + t_i - 1$. Then the range of summation in (3.4) extends from $j = a_i$ to $j = \min(\bar{a}_i, \bar{a}_v - t_i + 1), v \in \mathcal{Q}(i)$.

The representation of the period of completion of the project is simply stated as

$$z \geqslant \sum_j jx_{ij} + t_i; \quad \text{all } i \in L \tag{3.5}$$

where L is the set of last activities and z denotes the time of completion. Evidently the objective is to minimize z.

The representation of the constraints on the availabilities of the resources under this definition of the structural variables $\{x_{ij}\}$ is interesting. Clearly, if $x_{ij}=1$, then activity i will "occupy" positions $j, j+1, \ldots, j+t_i-1$. Let $\mathbf{V}_{ij}$ be a vector of dimension H with 0 values everywhere except positions at $j, j+1, \ldots, j+t_i-1$, where it has 1 values. Suppose that resource r is available in quantity m_j^r in period j, and let the vector $\mathbf{M}^r$ denote the availability of resource r over the planning horizon of H periods. Finally, suppose that activity i requires r_i units of resource r throughout its duration. Then we must have

$$\sum_{i \in A} \sum_j x_{ij} r_i \mathbf{V}_{ij} \leqslant \mathbf{M}^r \tag{3.6}$$

A further generalization of the scarce resources problem is not to assume a priori knowledge of the availabilities of the scarce resources, m_j^r, but to inquire about their optimal availabilities. This is easily accomplished in the above representation by considering them as structural variables in (3.6) rather than as given constants.

§4.2 A Cost-minimization Model with Resource-duration Interaction

What we mean by "resource-duration interaction" is simply the following. It often happens that two or more resources are interchangeable and that *the duration of the activity depends on which resource is utilized.* Then we speak of the dependence of the activity *duration* on the *resources* or the "interaction" between the two. This is the case, for example, when there are two classes of workers, high- and low-skilled. If the high-skilled labor is utilized, the activity will be of shorter duration than if the low-skilled labor is utilized. As another example, there may be two or more types of machinery capable of performing the activity (say different types of trucks) with varying capacities and efficiencies, which result in different activity durations. Normally, the resource yielding the shorter duration is also the more expensive. One would then be interested in balancing the costs of resource utilization against the economies accruing from shorter project duration. But sometimes, as in the numerical example treated below, attention is not focused on the costs of the resources but on their utilization, in which case one assumes a limited availability of each resource and it is then desired to allocate them to the activities in such a way as to satisfy some other objective, such as to minimize the total project duration.

The reader will recognize these eventualities to be quite common in real life. Unfortunately, the problem of resource-duration interaction has scarcely been treated in the literature. And yet, we see below that a slight variation of the representations discussed above handles it with equal ease.

To fix ideas, suppose there are two classes of resources: (a) those of which the activity durations are *independent*, which we denote by the set I and (b) those that are interchangeable and relate to the durations of the activities in the sense described above, which we denote by the set J. (If there are $q>1$ such groups of interchangeable resources, we denote them by the sets $J_1, J_2, \ldots, J_q$.)

The novel and significant element introduced is that now *we do not know a priori the durations of the activities*, since these depend on the resource in the set J applied to each activity. In other words, the duration of the activity is inexorably entwined with the resource utilized in the set J. The reader should satisfy himself that he can no longer write neither the resource constraints (3.6)—since the vectors $\mathbf{V}_{ij}$ are not predetermined as in the previous model—nor the precedence constraints of Eqs. (3.4), since the r.s. of that constraint set is not a given constant but a deduced result!

To extricate ourselves from this seeming dilemma, let t_i^r denote the duration of activity i if it is allocated to resource $r \in J$.† We assume that the activity would then consume r_i units of the resource per period of its duration. Define the variables x_{ij}^r to denote the fraction of job i started in period j under the assumption that activity i is allocated to resource $r \in J$. As before, x_{ij}^r has meaning only if it is equal to either 0 or 1: it is equal to 1 only when i is allocated to a particular resource $r \in J$ and is started at a particular time j; $a_i \leqslant j \leqslant \bar{a}_i$. Determination of the bounds a_i and $\bar{a}_i$ becomes somewhat more involved, but is still elementary. One approach is to assume that *every* activity in the project is performed in the shortest possible time. Then a_i and $\bar{a}_i$ are the earliest and latest start times of i under such regime of durations. Actually, one can often improve on these crude a limits by considering the *longest* possible durations of the activities, with considerable saving in the number of variables. But we postpone such calculation to our discussion of the numerical example that follows, where it is easier to explain the underlying procedure (see p. 178 and Problem 7).

†An activity that does not require, ab initio, any resource $r \in J$ is assigned arbitrarily to any resource in the set at its fixed duration and zero consumption of that resource.

There are four kinds of constraints:

1. Each activity must be started in some period within the limits specified by its a_i and $\bar{a}_i$. Evidently, one need not write a constraint for each activity; it is sufficient to thus restrict only the last activities; that is, the set of activities with no successors, denoted by L (see Fig. 3.6). The precedence relations should compel the previous activities to be started also, by the transitivity of these relations. Thus, we have

$$\sum_{j=a_i}^{\bar{a}_i} \sum_{r \in J} x_{ij}^r = 1 \quad \text{for } i \in L \tag{3.7}$$

2. The precedence constraints must be respected [this is the analog of Eq. (3.4)]:

$$\sum_j \sum_{r,r' \in J} \left[(j + t_i^r) x_{ij}^r - j x_{vj}^{r'} \right] \leqslant 0 \tag{3.8}$$

for all $v \in \mathscr{Q}(i)$, all i; and j running from $j = a_i$ to $j = \min(\bar{a}_i, \bar{a}_v - t_i + 1)$

3. Do not exceed the resource availabilities. Here we must deal with the two classes of resources separately. First, for the resource set J, we define the vector $\mathbf{V}_{ij}^r$ to be the H-dimensional vector of both 0 and 1 values, where the 1 values appear in positions $j, j+1, \ldots, j + t_i^r - 1$. Then we have

$$\sum_{i=1}^{n} \sum_j x_{ij}^r r_i \mathbf{V}_{ij}^r \leqslant \mathbf{M}^r \quad \text{for each } r \in J \tag{3.9}$$

where $\mathbf{M}^r$ is the column vector of availabilities $(m_1^r, m_2^r, \ldots, m_H^r)$ of resource $r \in J$. Notice that $x_{ij}^r = 1$ for only *one* combination of i, j, and r. This will add the vector $r_i \mathbf{V}_{ij}^r$ to the sum in the ℓ.s., as desired.

Next, for resource r in the set I we have the constraint

$$\sum_{i=1}^{n} \sum_j \left(\sum_{\rho \in J} x_{ij}^\rho \mathbf{V}_{ij}^\rho \right) r_i \leqslant \mathbf{M}^r \quad \text{for each } r \in I. \tag{3.10}$$

Again, one and only one combination of i, j and $\rho \in J$ will yield an $x_{ij}^\rho = 1$, which will immediately add the corresponding vector $r_i \mathbf{V}_{ij}^\rho$ to the sum, as desired. Note that here $a_i \leqslant j \leqslant \bar{a}_i$.

The difference between Eqs. (3.9) and (3.10) lies in the inner sum over the resources in the set J in the latter constraints. For now we do not wish to identify that particular $\rho \in J$ consumed by activity i; we are interested only in capturing its corresponding vector $\mathbf{V}_{ij}^{\rho}$—hence the summation over all $\rho \in J$ in Eq. (3.10).

Finally, we must specify the criterion function. For the problem to be meaningful we must balance the cost of utilization of the more expensive resources with the savings accruing from the early completion of the project. Therefore, let p denote the penalty per period for project prolongation beyond a given (shortest possible) time D, and let c_r denote the cost of utilizing resource r per period. (For logical consistency, c_r should be $> c_{r'}$, if $t_i^r < t_i^{r'}$; otherwise, management is incurring the double penalty of a resource that costs more *and* prolongs the activities!)

In order to compose the criterion function, we need a measure of lateness of the project. This leads to the fourth type of constraint.

4. Define the completion time of the project. Let z be an integer variable denoting the completion time of the project. Then it must be true that

$$\sum_{j=a_i}^{\bar{a}_i} j\left(\sum_{\rho \in J} x_{ij}^{\rho}\right) + \sum_{j=a_i}^{\bar{a}_i}\left(\sum_{\rho \in J} x_{ij}^{\rho} t_i^{\rho}\right) \leqslant z \tag{3.11}$$

This constraint is written for each activity i in the set L of last activities in the project (i.e., activities with no successors). The first term in the ℓ.s. gives the start of the activity, and the second term gives its duration, which is a function of the resource $\rho \in J$ to which it was assigned. In case of activity $i \in L$ requiring only resources in the set I, the second term of (3.11) is replaced by a constant (the activity duration).

With this last relation established we can easily write the criterion function as (**1** is a row vector of 1 values of dimensionality H):

$$\text{minimize } C = \sum_{r \in J} c_r \mathbf{1}\left(\sum_i \sum_j x_{ij}^r r_i \mathbf{V}_{ij}^r\right)$$

$$+ \sum_{r \in I} c_r \mathbf{1}\left(\sum_i \sum_j \sum_{\rho \in J} x_{ij}^{\rho} r_i \mathbf{V}_{ij}^{\rho}\right) + p(z - D) \tag{3.12}$$

The quantity in parentheses in the first and second terms is a column vector of dimensionality H, representing the total consumption of the resource under consideration over the planning horizon. The last term is the penalty for lateness. Notice that as long as $p > 0$, the value of z will be as small as is economically feasible, and that at least one of the inequalities (3.11) must be satisfied as *equality*.

Example 3.1. Consider the application of this model to the project of six activities shown in Fig. 3.6. There are two resources in the set I (the "independent" resources) of availabilities 6 and 8 units, respectively, and two resources in the set J (the "substitutable" resources) representing skilled labor (one person available) and unskilled labor (two persons available). It is assumed that any activity requires only one unit of labor, and the decision is whether to assign a skilled or an unskilled worker to each activity. It is desired to complete the project in the shortest possible time. The pertinent data on the activities and their needs of the resources are given in the first five columns of Table 3.1. Column 2, labeled s_i, gives the duration of the activity when the skilled labor is applied to it. Column 3, labeled u_i, gives the duration of the activity when the unskilled labor is

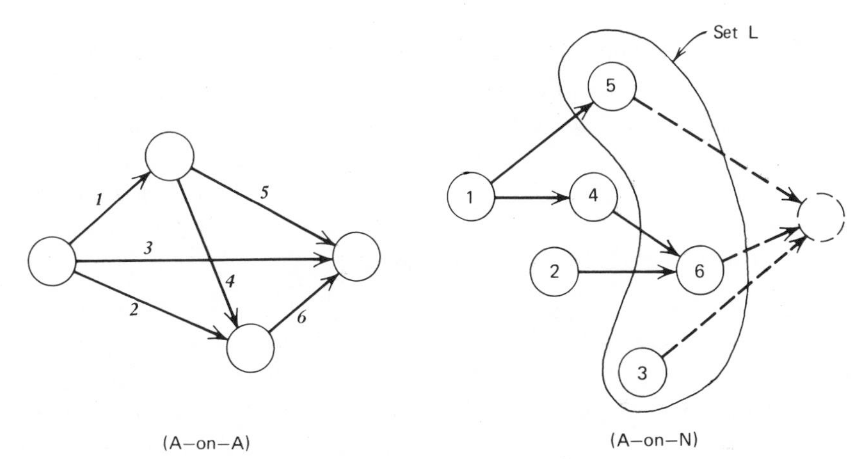

Figure 3.6. Example network.

Table 3.1. Data of numerical example.

Activity Number	Set J Skilled s_i	Unskilled u_i	Set I (1) r_i	(2) r_i	Bounds a_i	$\bar{a}_i$
1	2	3	2	1	1	3
2	1	3	3	2	1	9
3	3	4	1	4	1	8
4	5	7	1	3	3	5
5	4	6	2	2	3	7
6	1	4	3	4	8	10
Availabilities	1	2	6	8		

applied to it. Columns 4 and 5 give the needs for the "independent" resources, here numbered 1 and 2, respectively. Thus, reading across activity *4*, we find that it requires one unit of resource 1, three units of resource 2, and its duration would be 5 days if the skilled worker is scheduled, and 7 days if an unskilled worker is scheduled, and so on.

The last two columns of Table 3.1 give the earliest start times $\{a_i\}$ and the latest start times $\{\bar{a}_i\}$ of the activities. The derivation of these parameters is as follows. The parameter a_i is the earliest start time of activity i when all the activities have their *shortest* possible durations. Thus, it is evident that since activities *1*, *2*, and *3* have no predecessors, they can start in period 1; hence, $a_1 = 1 = a_2 = a_3$. Activity *4* is preceded by *1*; hence, it cannot start before *1* has been completed. The fastest possible duration of *1* is 2 days; hence, *4* can start, at the earliest, on day 3 (i.e., $a_4 = 3$) and so on. The derivation of the parameter $\bar{a}_i$ is more complicated, unfortunately. For now we have three values to consider:

1. We need an estimate of the latest completion time, $\bar{T}$, of the project. A very crude estimate is available directly from the data of the problem and is given by $\Sigma_i u_i$ ($=27$ in the example). This is the most pessimistic estimate, since it assumes that the activities will be done one at a time. Usually the application of any simple heuristic, or a set of heuristics, reveals a completion time much shorter than this value. In our example, a few minutes' reflection results in the feasible schedule of Fig. 3.7, which is only of duration 10 days. Hence, we take $\bar{T} = 10$. (It is advisable to obtain as small a value of $\bar{T}$ as is reasonably possible—without the expenditure of an inordinate amount of effort—because of the impact it has on the number of variables and the time of solution.) In anticipation of the final result, it is interesting to remark that the optimal project duration, z^*, obtained from the solution of the ILP detailed below, is equal to 9 days! The optimal schedule is shown in Fig. 3.8.
2. We must consider the *maximal* subset of activities that *may* precede the activity under consideration without violating the precedence constraints. The maximum duration of this set occurs when they are done in series, with each activity at its longest duration (i.e., by $\Sigma_{k \prec i} u_k$). Then, evidently, activity i should not be delayed beyond that point in time. Let $\alpha_{i1} = 1 + \Sigma_{k \prec i} u_k$; then obviously $\bar{a}_i \leqslant \alpha_{i1}$.
3. Finally, we must consider the *minimal* subset of activities that *must* succeed the activity under consideration (according to the precedence constraints). The minimal duration of this set is the duration of the CP among the activities in the set when each activity is at its shortest duration. Denote that value by v, and put $\alpha_{i2} = \bar{T} - (v + s_i - 1)$, where $\bar{T}$

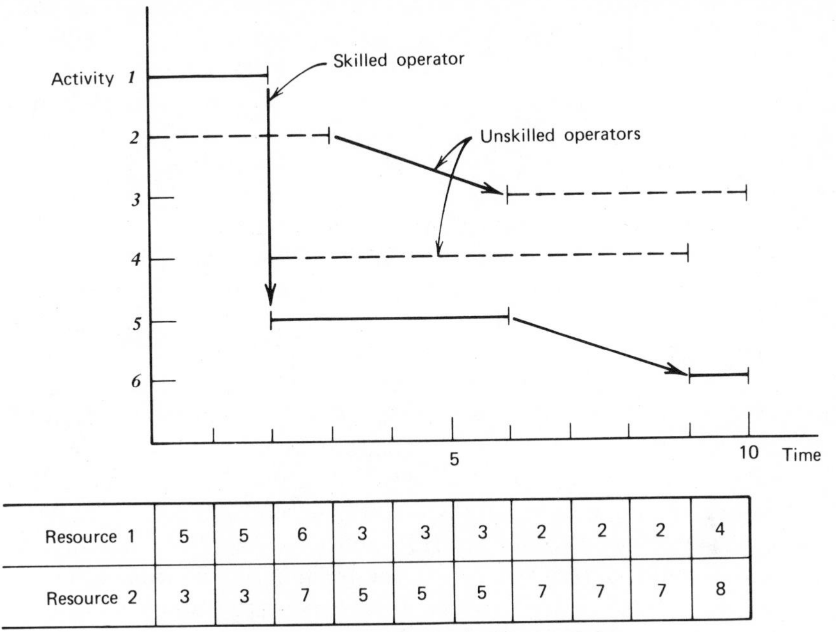

Resource 1	5	5	6	3	3	3	2	2	2	4
Resource 2	3	3	7	5	5	5	7	7	7	8

Figure 3.7. Schedule obtained by heuristics.

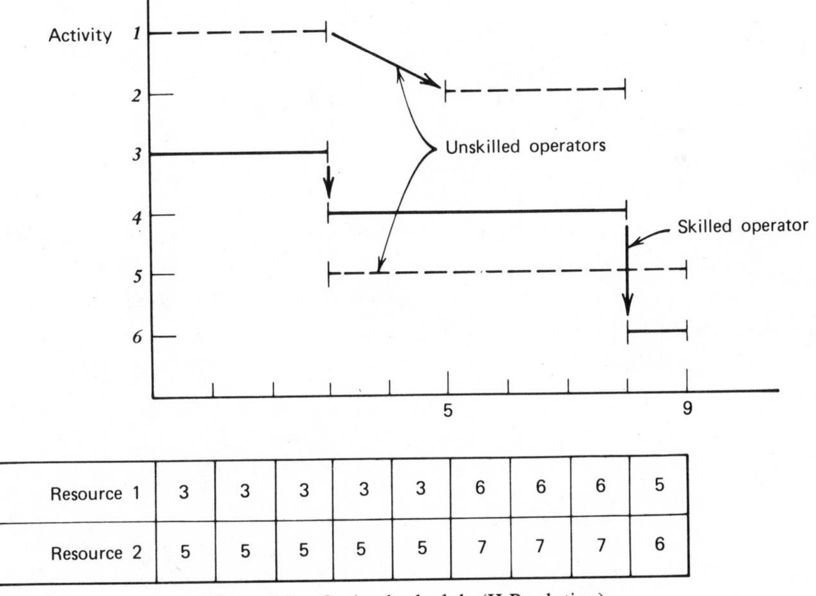

Resource 1	3	3	3	3	3	6	6	6	5
Resource 2	5	5	5	5	5	7	7	7	6

Figure 3.8. Optimal schedule (ILP solution).

is the value specified in (1) above. Evidently, it is impossible to delay the start of activity i beyond day α_{i2}; in other words, $\bar{a}_i \leqslant \alpha_{i2}$. Combining this result with that of (2) above, we conclude that $\bar{a}_i = \min(\alpha_{i1}, \alpha_{i2})$.

The reader will note that the need to go to all this trouble in defining, as precisely as possible, the values of the parameters $\{a_i\}$ and $\{\bar{a}_i\}$ stems from our desire to minimize the number of (integer) variables, which has a serious impact on the computing time.

To illustrate the determination of $\bar{a}_i$, consider activity *1*: it is evident that, at worst, it will be preceded by *2* and *3* (the only other activities which are not constrained to succeed *1*). Since these two activities may, at most, be of durations 3 and 4 days, respectively, activity *1* will start immediately thereafter. Hence, $\alpha_{1,1} = 8$.

On the other hand, *1* must be succeeded by *4* and *6* (=the CP in the succeeding set), whose shortest possible durations are 5 and 1, respectively. Recognizing that $\bar{T} = 10$ and $s_1 = 2$, we deduce the value $\alpha_{1,2} = 10 - (6 + 2 - 1) = 3$. Hence, $\bar{a}_1 = \min(8, 3) = 3$, as shown in Table 3.1. Similar analysis leads to the remaining values in the last column of Table 3.1.

We are now ready to state the ILP model. To simplify notation (and eliminate a superscript), let x_{ij} denote the fraction of activity i started on day j *on skilled labor* and let y_{ij} denote the same but *on unskilled labor*.

The constraints of (3.7) yield the following three equations:

$$\sum_{j=1}^{7} (x_{3j} + y_{3j}) + x_{3,8} = 1;$$

$$\sum_{j=3}^{5} (x_{5j} + y_{5j}) + x_{5,6} + x_{5,7} = 1; \qquad \sum_{j=8}^{10} x_{6j} = 1$$

The variables $y_{3,8}, \ldots, y_{6,9}, y_{6,10}$ are missing because it is impossible to terminate the project on day 10 if these variables are given the value 1, as can be seen from the "bar chart" of Fig. 3.9 [which is drawn on the basis of $\{a_i\}$ and $\{\bar{a}_i\}$ of Table 3.1]. Hence, we know à priori that these variables have the value 0, and we simply ignore them in all subsequent analysis (their corresponding dashed lines are also absent from Fig. 3.9).

The precedence constraints of (3.8) yield the following set of seven inequalities:

$$1 \prec 4: \sum_{j=1}^{3} \left[(j+2)x_{1j} + (j+3)y_{1j} \right] - \sum_{j=3}^{4} j(x_{4j} + y_{4j}) - 5x_{4,5} \leqslant 0$$

$$1 \prec 5: \sum_{j=1}^{3} \left[(j+2)x_{1j} + (j+3)y_{1j} \right] - \sum_{j=3}^{5} j(x_{5j} + y_{5j}) + \sum_{j=6}^{7} jx_{5j} \leqslant 0$$

$$2 \prec 6: \quad \sum_{j=1}^{8} \left[(j+1)x_{2j} + (j+3)y_{2j} \right] + 10x_{2,9} - \sum_{j=8}^{10} jx_{6j} \leq 0$$

$$4 \prec 6: \quad \sum_{j=3}^{4} \left[(j+5)x_{4j} + (j+7)y_{4j} \right] + 10x_{4,5} - \sum_{j=8}^{10} jx_{6j} \leq 0$$

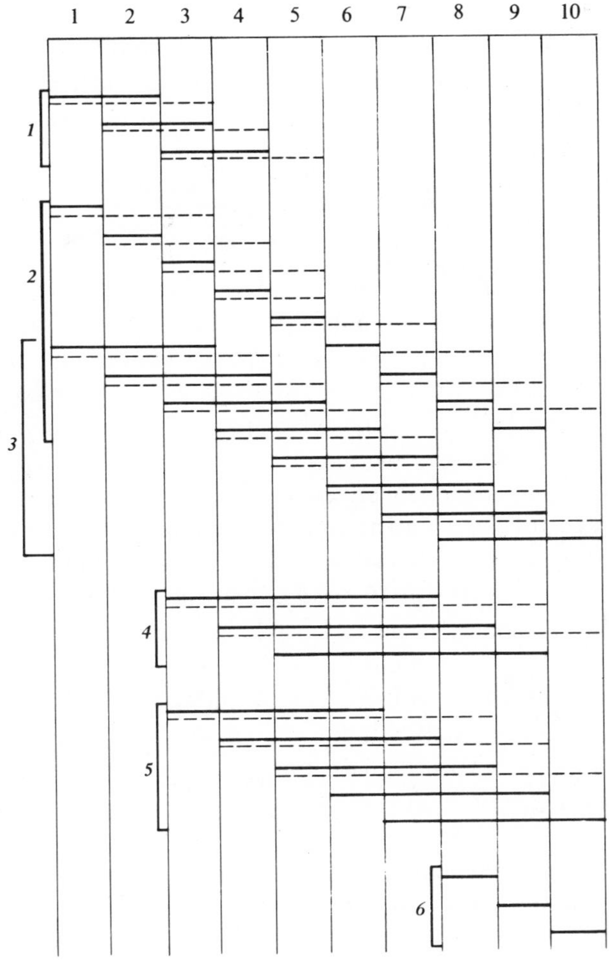

Code: Solid line ——— ⟹ skilled labor

Dashed line – – – – ⟹ unskilled labor

A "bar" represents the position of 1 values in the $\mathbf{V}_{ij}$ vector

Figure 3.9. The "Bar Chart" of Example 3.1.

For the set L (the set of last activities) we have

$$\sum_{j=1}^{7}\left[(j+3)x_{3j}+(j+4)y_{3j}\right]+11x_{3,8}-z\leqslant 0$$

$$\sum_{j=3}^{5}\left[(j+4)x_{5j}+(j+6)y_{5j}\right]+\sum_{j=6}^{7}(j+4)x_{5j}-z\leqslant 0$$

$$\sum_{j=8}^{10}(j+1)x_{6j}-z\leqslant 0$$

The resource-availability constraints of (3.9), for resources in the set J, are exemplified, for the sake of brevity, by the constraints on the unskilled labor. Consider any day, say $j=4$. The "active" variables are (scanning downward column 4 of Fig. 3.9 for the dashed lines):

$$y_{1,2}, y_{1,3}, y_{2,2}, y_{2,3}, y_{2,4}, y_{3,1}, y_{3,2}, y_{3,3}, y_{3,4}, y_{4,1}, y_{4,2}, y_{5,1}, y_{5,2}$$

which yield the constraint:

$$y_{1,2}+y_{1,3}+y_{2,2}+y_{2,3}+y_{2,4}+y_{3,1}+\cdots+y_{3,4}+y_{4,1}+y_{4,2}+y_{5,1}+y_{5,2}\leqslant 2$$

A similar equation is written for each day j; $j=1,2,\ldots,10$. Alternatively, one may read off the $\mathbf{V}^u_{ij}$ vectors corresponding to the y_{ij} variables immediately from Fig. 3.9. For instance, $\mathbf{V}^u_{1,1}=(1,1,1,0,0,0,0,0,0,0)$; while $\mathbf{V}^u_{5,4}=(0,0,0,1,1,1,1,1,1,0)$, and so on. Then one simply adds the column vectors $y_{ij}\mathbf{V}_{ij}$ to obtain the desired set of constraints (here the vector $\mathbf{M}^r$ is the vector of 2 values).

Finally, the resource availability constraints of (3.10), for resources in the set I, are exemplified by the constraints on resource 1. Again, these are constructed by reference to Fig. 3.9, either on a day-by-day basis or through the $\mathbf{V}$ vectors. Consider once more day 4; now we must take into account all "active" x *and* y variables [notice the inner summation in (3.10)]; we have (scanning downward column 4 of Fig. 3.9):

$$\begin{aligned}&2(y_{1,2}+x_{1,3}+y_{1,3})+3(y_{2,2}+y_{2,3}+x_{2,4}+y_{2,4})+1(y_{3,1}+x_{3,2}\\&+y_{3,2}+x_{3,3}+y_{3,3}+x_{3,4}+y_{3,4})+1(x_{4,3}+y_{4,3}+x_{4,4}+y_{4,4})\\&+2(x_{5,3}+y_{5,3}+x_{5,4}+y_{5,4})\leqslant 6\end{aligned}$$

The criterion function is to minimize z. This ILP was solved by a special purpose code for 0, 1 ILP, due to Dr. U. Suhl,[†] in 1.486 sec on the CDC CYBER 72 (including the problem setup). It yielded the schedule shown in Fig. 3.8 with $z^*=9$.

This solution possesses three rather interesting features that are worth noting. First, in one respect it is highly unintuitive since activity *1*, which is the activity with the most number of successors, is allocated to the *unskilled* labor; hence, its duration is the longest possible while activity *3*, which has neither predecessors nor successors is allocated to the *skilled* labor, hence its duration is the shortest possible! Second, it loads the skilled labor all the time while leaving the unskilled labor partially idle. This result conforms to expectations under the adopted criterion of minimizing the project completion time. Third, and finally, at no time is resource 2 fully utilized; its availability is eight units while the maximum usage is only seven units in days 6, 7, and 8; which indicates that its availability may be reduced with possible savings and with no adverse effect on the project completion time. In comparison, the heuristic solution of Fig. 3.7 reached the full capacity of both resources, and we venture to say that such performance will result from the majority of heuristic approaches.

§4.3 A Footnote: The Solution of ILPs

The solution of ILPs, whether of the 0, 1 garden variety discussed here or of the more general integer (or mixed) LPs, is currently an active field of research. The objective of this feverish activity is to enhance the power of the computing algorithms to enable one to cope with real-life problems. The interested reader is directed to the specialized literature on ILP for the details.

For our purposes, and restricting ourselves to the field of ANs, we briefly review a few basic ideas that may be of value in reducing the computational burden.

The most elementary idea, and one that seems to have been used by several researchers [e.g., McGinnis (1973)], before it was published in the open literature by Patterson and Huber (1974), is to foresake the search for the optimum of the ILP and, instead, test for the feasibility of an assumed completion time, z. Thus, a problem of optimization is replaced by a sequence of problems that test for feasibility.

[†]The Freie Universität Berlin, Institut für Unternehmungsführung, Fachbereich Wirtschaftswissenschaft, 1 Berlin 33, Garystrasse 21, the German Federal Republic.

Note that a lower bound on the value of z can be determined from the expression

$$\max\left\{CP;\ \max_{r}\sum_{i\in A}\frac{r_i t_i}{\overline{m}^r}\right\} \tag{3.13}$$

where $\overline{m}^r$ is the average availability of resource r per unit time. Let w be the smallest integer larger than or equal to the value given in expression (3.13). Then one can assert that the optimum value $z^* \geqslant w$.

The search procedure itself is patently simple. Start with $z = w$ and test for the feasibility of the ILP. If feasible, then it is, indeed, also optimal. Otherwise, put $z = w + 1$ and repeat, and so on. The first time feasibility is achieved, so is optimality. The saving in computing, which is quite considerable, stems from the fact that ILP codes consume the majority of the time seeking, and proving, optimality after having already achieved feasibility.

Of course, approaching the value z^* may also proceed from the "tail end," namely, from a known upper bound such as the duration of a feasible schedule. Patterson and Huber made no attempt to achieve the least upper bound; however, they used four heuristic rules† to schedule the activities and then picked the shortest duration as the upper bound, which we denote by W. They started the search with $z = W - 1$. If the ILP is not feasible, then clearly $z^* = W$. But if it is feasible, then put $z = W - 2$ and repeat, and so forth. If infeasibility is first detected at the value $z = W - k$, then $z^* = W - k + 1$. One may also attempt to optimize the search procedure itself when both lower and upper bounds are known, such as by using dichotomous search over the interval of uncertainty.

The computing experience of Patterson and Huber demonstrates conclusively the superiority of this approach over a frontal attack on the ILP with any existing, commerically available, computer code. Projects with up to 30 activities and three resources were used as test problems. The time to achieve the optimum varied between a fraction of a second and about 16 seconds. Furthermore, the procedure was either superior to, or competitive with others.‡

†Namely, the "shortest imminent operation," the "least total float," the "greatest resource demand," and the "greatest resource usage"; see Section 2 of this chapter.

‡Comparisons were made with the models of Brand, Meyer, and Shaffer (1964), Davis (1968) [see also Davis and Heidorn (1971)]; Moodie and Mandeville (1966), Martino (1965), and Shaffer, Ritter and Meyer (1965).

Other ideas under investigation by several researchers at the time of this writing include improving the bounds (both lower and upper) suggested by Patterson and Huber on the value of z^*, improving the a bounds (of p.178) of individual activities by conditioning them on the resources as well as on the precedence relations, and by replacing some of the constraints (say the resource constraints) by a "penalty function" in the objective, following the ideas of Fisher (1973) and Nepomiastchy (1974). Of course, one can always approach the ILP from the point of view of implicit enumeration, which is the subject of Section 6 of this chapter.

§5. THE MINIMUM PROJECT DURATION PROBLEM

In the absence of resource substitutability, and when interest is focused solely on the minimization of project duration, Davis (1968) constructed a model different from the 0, 1 models of Section 4. Our discussion of his model depends heavily on the above reference as well as on the lucid paper of Davis and Heidorn (1971).

Since the model depends on concepts and procedures developed in conjunction with the so-called assembly line balancing (ALB) problem, we devote Section 5.1 to introduce the ALB problem and to establish the link between it and our minimum-project-duration problem.[†] We then return in Section 5.2 to the application of these concepts to the problem of the minimum project duration subject to resource availability constraints.

§5.1 The Assembly Line Balancing (ALB) Problem

Very few engineering feats parallel the assembly line in its profound effect on production methods, philosophy, and impact on the general society in which it is embedded. Henry Ford (1863–1947) is usually credited with the creation of the "assembly line era," which has had a stupendous and profound effect not only on the development of the automobile (a piece of machinery which, itself, has had a phenomenal impact on the American way of life), but on the occupation, education, leisure, and the social fabric of Americans and other people.[‡]

[†]The connection between these two problems was first established by Wilson (1964).

[‡]A vast literature exists on "mass production" and the "mass-production society." The reaction to the assembly line varied from the visionary ridicule of the great movie actor Charlie Chaplin in his film "Modern Times," to the sober acceptance by the British philosopher Aldous Huxley in his *Brave New World,* where history starts from the year one of our Ford!!

Yet, the idea of the assembly line is patently simple; suppose that the end product can be decomposed into subassemblies, and subsubassemblies, and so on down the line until the level of commodities† is reached. Now, if this subdivision is reversed, we can define the operations, henceforth referred to as *tasks,* which are necessary to group the commodities into components, the components into subassemblies, and so on, until the end product is reached. This is the *assembly* operation; and since it is unidirectional (i.e., linear) it defines the assembly line process.

The tasks to be performed for such production are the focus of our attention. We assume that each task is well defined and that its duration is a given number that is independent of other tasks, of the person performing the task, and of the time at which the task is performed. Furthermore, we assume that all other necessary factors of production (e.g., material, machines, and manpower) are available.

It is clear that while some of the tasks can be performed at almost any point of the production line (e.g., fitting the nameplate on the cover of a motor), the majority of the tasks must be performed in a specific sequence relative to some other tasks (e.g., a hole has to be drilled before it is tapped; the rotor of a generator must be wound before it is inserted in the stator). This gives rise to technological *precedence relationships,* which are all too familiar from our studies on ANs. This is the first "link" between the ALB and ANs; both problems start from the definition of a set of tasks partially ordered by the binary relation "precede."

Such precedence relationships can be expressed as a matrix **M** in which an entry $m_{ij}=1$ indicates that task i immediately precedes task j, written, as usual, $i \prec j$, and $m_{ij}=0$ indicates the absence of such precedence relationship. As is well known, the precedence relationship is transitive; if $i \prec j$ and $j \prec k$, then $i \prec k$; nonreflexive: $i \nprec i$; and nonsymmetric: if $i \prec j$, then $j \nprec i$. Consequently, if the tasks are represented as nodes of a graph, and the *directed* arc (ij) from task i to task j exists iff $i \prec j$, then the resulting graph is directed and contains no cycles or self-loops. All this is very similar to ANs with the A-on-N representation; see Chapter 1.

Clearly, a unit of the product is "built up" a little bit at a time through the stages of the assembly line, and the unit does not exist as a clearly identifiable "end product" except after the last stage of production. Consequently, if q is the rate of production in units of product per unit time and c denotes the time between units completed, the so-called cycle time, in units of time per unit of output, then, $c = 1/q$.

†A commodity is defined as an input to the production system. It is sometimes referred to as "raw material," which need not be "raw" at all!

Problems in Line Balancing.

The general ALB problem is an extremely complex one whose solution has defied strict analytical treatment thus far. This is mainly because the assembly line has become the object of a number of conflicting criteria and a wide variety of constraining conditions. For instance:

1. Some tasks must be performed at the *same* work station for reasons of quality control, special job classifications, time consumed in getting into and out of work positions, or the availability of special tools and skills.
2. Some tasks require teamwork of two or more operators, and therefore must be performed together.
3. Some tasks are divisible; hence such tasks can be performed by two or more operators, with the concomitant reduction in the total time.
4. Some tasks cannot be grouped with other tasks, due to health, safety, technical, routing, or other reasons.
5. The dynamic nature of the product causes the design of the product, as well as the design of the production method itself, to change. These, in turn, may change the task times and upset any "balance" previously determined.
6. The product may not be unique; in many applications, the same line is used for a number of different styles or product designs, which are introduced in batches or even perhaps in what appears to be a random, mixed sequence.

Typically, we assume away almost all of these complications; after all, this is not a treatise on assembly line techniques, but a mere introduction to the subject to investigate its relation to the minimum-duration problem in ANs. Thus, we assume that the assembly line produces only one product, that task i has processing time t_i, that each workstation is composed only of one operator (or machine), and that the only constraints on the tasks are the precedence constraints.

Still, there are two possible questions that can be raised:

1. Given a cycle time c, a set of tasks $J=\{1,\ldots,i,\ldots,n\}$ to be performed according to a precedence matrix M; and given the processing time t_i of task i, it is desired to find the *minimum number of groupings* of these tasks into stations so that: (a) each station consumes no more time than the cycle time c and (b) the precedence relations are respected.
2. *Minimize the cycle time* c for a given number of stations subject to the precedence constraints and the condition that the total time at any station does not exceed c.

Henceforth, we concern ourselves only with formulation (1) of the problem.

A Network Model.[†]

A more formal statement of the ALB problem of (1) may run as follows. Given a finite set J (the set of tasks), a partial order $\prec$ defined on J (the precedence relationship), a positive real-valued function t defined on J (the processing time of each task), and a number $c>0$ (the cycle time), find a collection of subsets of J: $S_1, S_2, \ldots, S_R$ (the work stations) satisfying the following five conditions:

1. $\cup_{r \in R} S_r = J$ (the set of stations accounts for all the tasks);
2. $S_r \cap S_s = \phi$, $r \neq s$ (each task is in one and only one station);
3. $t(S_r) = \Sigma_{i \in S_r} t_i \leqslant c$; $r = 1, 2, \ldots, R$ (station time $\leqslant$ cycle time);
4. if $k \prec l$, and $k \in S_r$ and $l \in S_s$, then $r \leqslant s$ (maintain precedence relationship);
5. minimize $\Sigma_{r=1}^{R}[c - t(S_r)]$ (minimize total idle time).

It is easy to see (and equally easy to verify) that the minimization of the total idle time stated in 5 is equivalent to the minimization of the *number of stations R*. The mathematical model we are about to discuss constructs a very special kind of network, each arc of which represents a possible work station. Hence, minimum R is translated into a minimum number of arcs in that network, or, in other words, the problem is translated into a *shortest path problem.*

The network, which we refer to as the A-network, is constructed in the following fashion. Let A_s (for assignment), $s = 1, \ldots, S$, be a collection of tasks (i.e., a subset, $A_s \subseteq J$) so that:

1. $A_0 = \phi$, the empty set;
2. $A_S = J$, the set of all tasks:
3. if task $j \in A_s$, and $i \prec j$, then $i \in A_s$; in other words, each A_s is a collection of tasks that can be regarded as a "first station" set of tasks (without regard to the cycle time c). In essence, each assignment A_s is the union of all possible work stations which contain the tasks in A_s.

We define the processing time of assignment A_s to be the sum of the processing times of the tasks in A_s; in other words,

$$t(A_s) = \sum_{i \in A_s} t_i, \quad s = 1, \ldots, S$$

[†]Due to Gutjahr and Nemhauser (1964).

Let assignment A_s be represented by node s. An arc (s,u) is directed from node A_s to node A_u iff:

4. $A_s \subset A_u$ (A_s is a proper subset of A_u)
5. $t(A_u) - t(A_s) \leqslant c$ (the added tasks in A_u constitute a feasible station)

The very manner of construction of this A-network leads to the conclusion that the set of tasks given by $(A_u - A_s)$ obey the precedence constraints and consume no more than c units of time. Hence, they constitute a feasible work station. The idle time in this station is given by $c-[t(A_u)-t(A_s)] \geqslant 0$.

No (directed) arc enters node A_0 (the empty assignment), and no arc leaves node A_S (the complete set of tasks, J). Thus, the resulting A-network is a finite, directed, acyclic network. Any path from A_0 to A_S represents a sequence of feasible stations that account for *all* the tasks. Hence, the optimal allocation minimizes the *length* of the path from A_0 to A_S (which corresponds to the minimization of the total idle time, if each arc is given a "length" equal to the idle time) or, equivalently, minimizes the *number* of arcs in the path.

A slight reflection reveals that the problem of finding the path with the smallest number of arcs between nodes A_0 and A_S can be accomplished simultaneously while constructing the network itself. The construction proceeds in stages in the following manner: the empty assignment, A_0, is the first assignment. The tasks without any predecessors are placed in stage 1, and all possible subsets of this set of tasks are formed. Each such subset is an assignment. At the end of stage 1, the tasks are deleted from the list of tasks, leaving only the newly generated assignments.

By a task that is an "*immediate follower*" of an assignment A_s we mean a task that is an immediate successor of at least one task in A_s and is not preceded by any task not in A_s. For example, consider the 11 tasks given by Table 3.2 and Fig. 3.10. (Notice that Fig. 3.10 can be interpreted as an activity network.) Consider the assignment $A = \{1,2\}$: tasks *3* and *4* are "immediate followers" of A, but not task *11* because while it is true that *11*

Table 3.2 Data for Illustrative Problem

Task	1	2	3	4	5	6	7	8	9	10	11
Processing time	4	3	5	6	4	5	3	2	4	1	5
Immediate successors	3,4	4,11	5,6	8	9	7	—	9	10	11	—

Cycle time $c = 12$

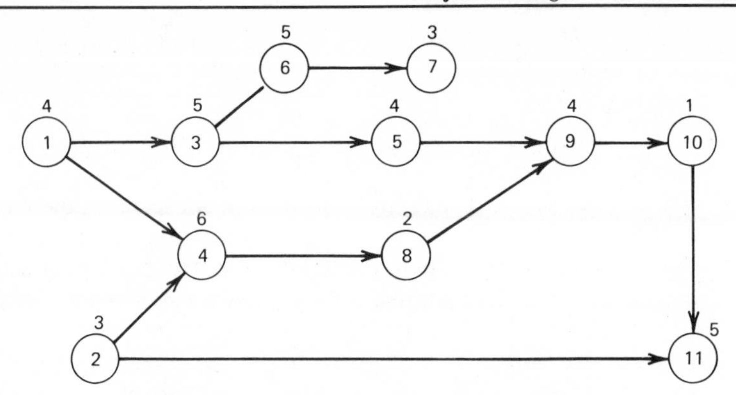

Figure 3.10. Precedence diagram of tasks in Table 3.2.

is the immediate successor of *2*, which is in A, task *11* must be preceded also by task *10*, which is *not* in A.

In general, for any assignment A_s in stage k we define the *undeleted immediate followers* of A_s, denoted by UIF(A_s), as the set of tasks immediately following A_s. For each subset θ of tasks in UIF(A_s), the union $\theta \cup A_s$ is formed, and constitutes a new assignment to be placed in stage $k+1$. When all assignments possible have been generated for each assignment A_s in stage k, we delete the elements in the union of UIF(A_s) for all A_s in stage k and proceed to stage $k+1$.

Now the shortest paths to the nodes of stage $k+1$ are determined as follows. For each assignment A_u in the newly generated stage $k+1$, determine the length of the shortest path(s) to it, denoted by $\mu(A_u)$, which is obtained from the expression

$$\mu(A_u) = \min[\ \mu(A_s) + 1\ ]; \mu(A_0) = 0$$

The minimand is taken over all subsets A_s in stage k, satisfying the two conditions

$$A_s \subset A_u$$

$$t(A_u) - t(A_s) \leqslant c$$

Iteration is continued until the final assignment $A_S \equiv J$ is generated. $\mu(A_S)$ is the "length" of the shortest path from A_0 to A_S, which is equivalent to the minimum number of stations sought. The workstations themselves and the tasks contained in each station are obtained by tracing backward the shortest path from $A_S(\equiv J)$ to $A_0(\equiv \phi)$.

The procedure, when applied to the set of tasks of Fig. 3.10, yields the 38 assignments of Table 3.3 and the assignment network of Fig. 3.11. For

Table 3.3. The Generation of Subsets $\{A_s\}$ for Fig. 3.10

Iteration	s	A_s	UIF(A_s)	$t(A_s)$	Length of Shortest Path
0	0	ϕ	1,2	0	0
1	1	1	3	4	1
	2	2	–	3	1
	3	1,2	3,4	7	1
2	4	1,3	5,6	9	1
	5	1,2,3	5,6	12	1
	6	1,2,4	8	13	2
	7	1,2,3,4	5,6,8	18	2
3	8	1,3,5	–	13	2
	9	1,3,6	7	14	2
	10	1,3,5,6	7	18	2
	11	1,2,3,5	–	16	2
	12	1,2,3,6	7	17	2
	13	1,2,3,5,6	7	21	2
	14	1,2,4,8	–	15	2
	15	1,2,3,4,5	–	22	2
	16	1,2,3,4,6	7	23	2
	17	1,2,3,4,8	–	20	2
	18	1,2,3,4,5,6	7	27	3
	19	1,2,3,4,5,8	9	24	2
	20	1,2,3,4,6,8	7	25	3
	21	1,2,3,4,5,6,8	7,9	29	3
4	22	1,3,6,7	–	17	2
	23	1,3,5,6,7	–	21	2
	24	1,2,3,6,7	–	20	2
	25	1,2,3,5,6,7	–	24	2
	26	1,2,3,4,6,7	–	26	3
4	27	1,2,3,4,5,6,7	–	30	3
	28	1,2,3,4,5,8,9	10	28	3
	29	1,2,3,4,6,7,8	–	28	3
	30	1,2,3,4,5,6,7,8	–	32	3
	31	1,2,3,4,5,6,8,9	10	33	3
	32	1,2,3,4,5,6,7,8,9	10	36	3
5	33	1,2,3,4,5,8,9,10	11	29	3
	34	1,2,3,4,5,6,8,9,10	11	34	3
	35	1,2,3,4,5,6,7,8,9,10	11	37	4
6	36	1,2,3,4,5,8,9,10,11		34	3
	37	1,2,3,4,5,6,8,9,10,11		39	4
	38	1,2,3,4,5,6,7,8,9,10,11		42	4

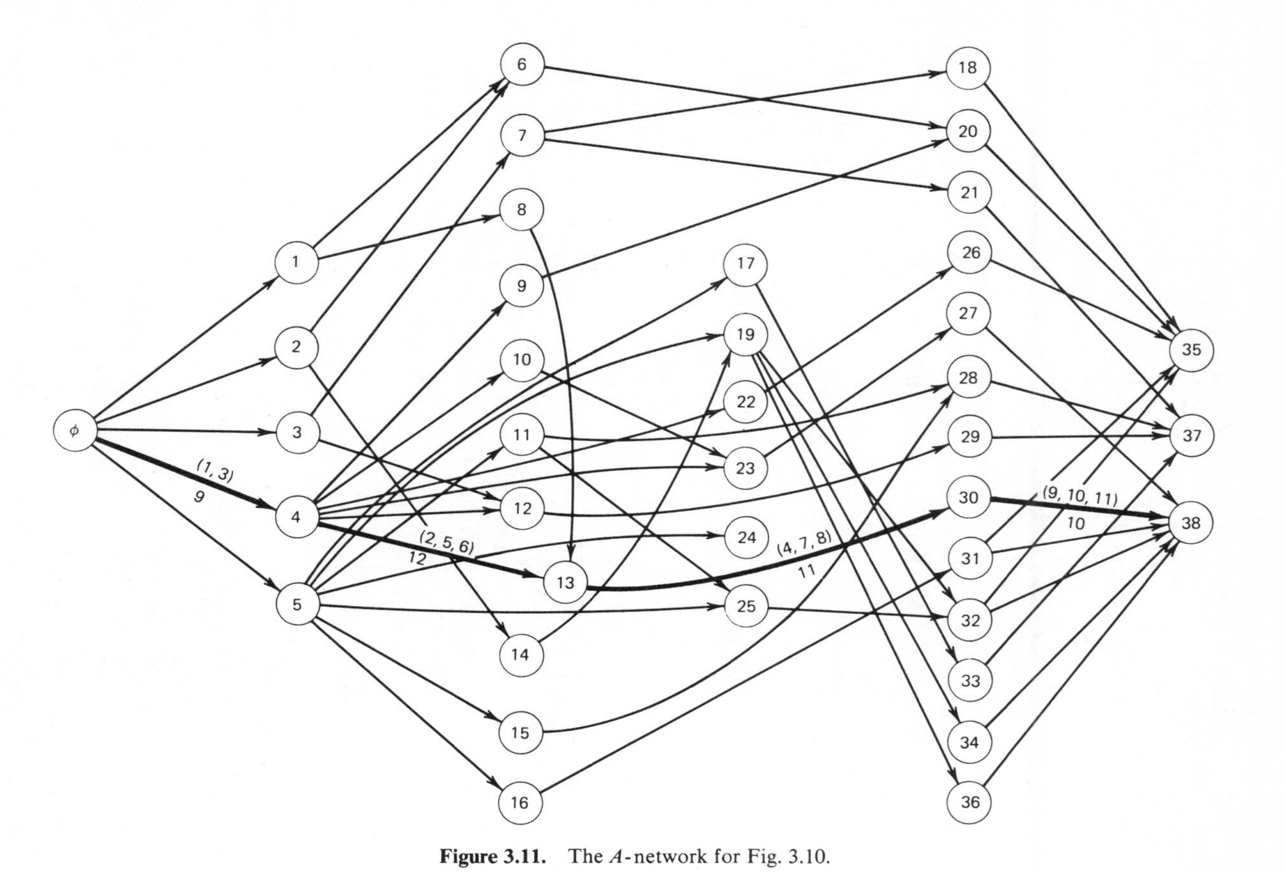

Figure 3.11. The *A*-network for Fig. 3.10.

instance, assignment $A_5=\{1,2,3\}$ has as UIF(A_5) the pair of tasks $\{5,6\}$. These give rise to assignments $A_{11}=\{1,2,3,5\}$, $A_{12}=\{1,2,3,6\}$, and $A_{13}=\{1,2,3,5,6\}$, which correspond to the subsets $\theta_1=\{5\}$, $\theta_2=\{6\}$, and $\theta_3=\{5,6\}$. After all the "descendent" assignments of iteration 2 are generated—that is, the descendents of A_4 to A_7—tasks *5*, *6*, and *8* are deleted and we move to iteration 3; and so on.

The shortest path to A_{38} is of length $\mu(A_{38})=4$, denoting the need for 4 stations. These are: $S_1=\{1,3\}$, $S_2=\{2,5,6\}$, $S_3=\{4,7,8\}$, and $S_4=\{9,10,11\}$. The reader is invited to verify that, indeed, $t(S_r)\leqslant 12$ for $r=1,2,3,4$.

§5.2 Relation to Activity Networks

The approach to the ALB problem; namely, to the grouping of tasks into the minimum number of stations, bears direct generalization to the problems of ANs under scarce resources when it is desired to determine the shortest project duration. This should come as no surprise: there are just too many elements in common for the two problems to be unrelated.

It is easy to see that the ALB problem is identical to the scarce-resources problem in ANs with the following structure: (a) there is only one scarce resource, (b) each activity is of duration one period; (c) t_i represents the quantity of the resource required by i in that period, and (d) c represents the resource availability. A "station" now corresponds to a period, and minimizing the number of stations is equivalent to minimizing the project duration. The crossover from ALB to ANs was accomplished by Black (1965), whose modification of the Gutjahr–Nemhauser approach runs as follows.

Consider the simple case of only *one* scarce resource. It is evident that an activity exists, so to speak, in *two* dimensions: the first is *time*, similar to a task in the ALB, and the second is *resource*, which is a new element, indeed. Naturally, the reduction of an AN to the point where it is amenable to treatment by the same approach as the ALB must involve the transition from two dimensions to only one. This is accomplished by "breaking up" each activity of duration t_i into t_i segments, each of duration one unit of time. In this manner the availability of the resource is identical to the availability of the cycle duration c, and the consumption of a resource is similar to the occupancy in a station of a portion of the cycle time. Obviously, the case of more than one resource is handled in a similar fashion by vector rather than scalar occupancy.

Let (ij) be a subscript denoting the jth segment of activity i, which from now on will be "a task." Clearly, a "task" is an activity of unit duration.

Thus, the first segment of activity i is $(i1)$, while the last segment of the same activity is (it_i). To facilitate notation we feel free here to replace the double subscript ij by the single subscript h, with $1 \leqslant h \leqslant K\ (=\Sigma_i t_i)$. We indicate that an h corresponds to a particular task ij by the symbolism $h \leftrightarrow ij$.

We assume that, in general, there are R scarce resources, and that for each task h there corresponds a vector $\mathbf{c}_h = (c_{h1}, c_{h2}, \ldots, c_{hR})$, where c_{hr} is the amount of resource r consumed by task h (in a unit time), $1 \leqslant r \leqslant R$. To each subset A_s, formed as indicated in the previous discussion, $s = 1, 2, \ldots, S$; there corresponds a vector $\mathbf{G}_s = (G_{s1}, G_{s2}, \ldots, G_{sM})$, where $G_{sm} = \Sigma_{h \in A_s} c_{hr}$, the total consumption of resource r by the tasks specified by the subset of tasks in A_s.

The other side of the balance sheet is, of course, the availability of these very resources. Let $\mathbf{L}_\tau \triangleq (l_{1\tau}, l_{2\tau}, \ldots, l_{R\tau})$ denote the limit availability of resources 1 through R in period (e.g., day)τ, where τ ranges over the planning horizon $1 \leqslant \tau \leqslant H$. If the availability of the resources does not vary with time, as we assume henceforth for the sake of clarity of exposition, we simply drop the subscript τ and have the availability vector be given by $\mathbf{L} = (l_1, l_2, \ldots, l_R)$.

The important fact to remember is that the "tasks" defined above may have a stronger relationship to each other than mere precedence. If the tasks are, in fact, segments of the *same activity*, they must occur in *contiguous time periods* (assuming the activities are indivisible). This gives birth to the notion of one task that *must follow* another. Hence, if the earlier task is contained in subset A_s, the later task must be contained in subset A_{s+1}, which should *immediately follow* A_s on the time scale (this is indicated on the network by a heavy line). With these two important distinctions of ANs to bear in mind, we are ready to state the modification as suggested by Black.

Denote, as before, the set of all tasks by J. Then we desire assignments $\{A_s\}$ formed as subsets of J and represented as nodes of a graph, and directed arcs (s, u) joining subset A_s to subset A_u such that the following conditions are satisfied:

1. $A_0 = \phi$, the empty set,
   $A_S = J$, the set of all tasks,
   and any subset A_s has the property that if $h' \in A_s$ and $h \prec h'$, then $h \in A_s$ also.
2. The contiguity and precedence constraints are satisfied:
   (a) if segment $ij \leftrightarrow h$ is in subset A_s, then segment $i, j+1 \leftrightarrow h+1$ must be in A_{s+1}, and an arc exists from A_s to A_{s+1};
   (b) if segment $i_1 t_{i_1} \leftrightarrow h$ is in subset A_s, then segment $i_2 1 \leftrightarrow h'$ must be in subset A_u, with $s \leqslant u$, iff $i_1 \prec i_2$.

3. An arc (s,u) is directed from node A_s to node A_u, $s<u$, iff
   (a) $A_s \subset A_u$
   (b) $\mathbf{G}_u - \mathbf{G}_s \leqslant \mathbf{L}$.

The alert reader will note that the contiguity constraints (for segments of the same job) gave rise to condition (2a). On the other hand, the resource limitations gave rise to condition (3b). It is easy to see that each arc of the A-network represents an assignment of a set of tasks to one time period. Consequently, such assignment is feasible only if it does not demand in excess of the available resources.

Consequently, the generation of assignments $\{A_s\}$ and the construction of the A-network proceed exactly in the same manner as described in the ALB problem discussed before, with the added provisos of conditions (2a) and (3b) stated above.

As an example of the procedure, consider the simple project network of Fig. 3.12. The subdivision of activities into segments gives rise to the network of Fig. 3.13. The generation of the assignments $\{A_s\}$ is detailed in

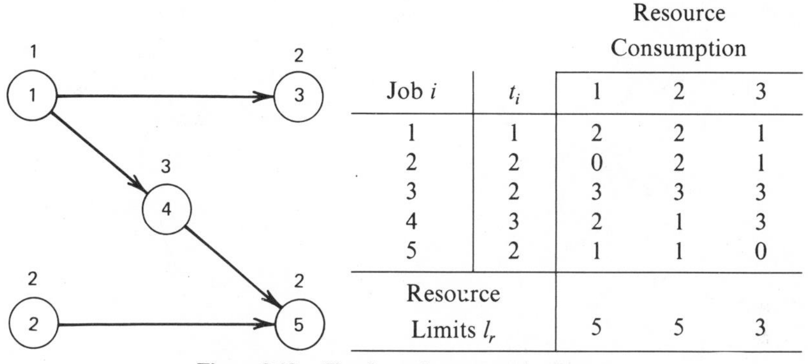

Job i	t_i	Resource Consumption 1	2	3
1	1	2	2	1
2	2	0	2	1
3	2	3	3	3
4	3	2	1	3
5	2	1	1	0
Resource Limits l_r		5	5	3

Figure 3.12. Simple project network (A-on-N)

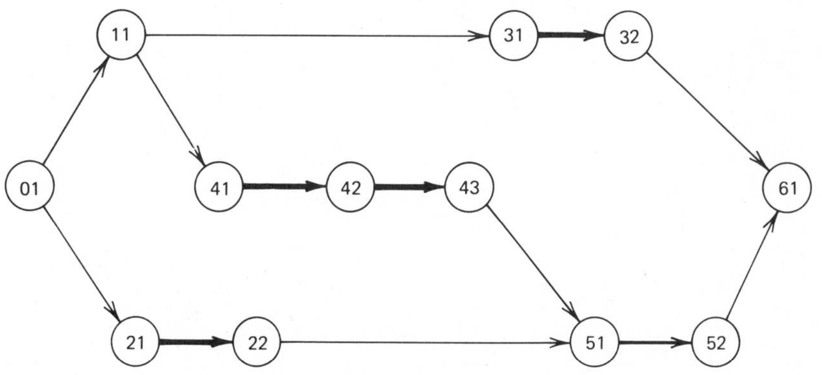

Figure 3.13. Nodes (0, 1) and (6, 1), pseudotasks; a heavy arrow represents "must follow."

Table 3.4. The generation of subsets $\{A_s\}$ for Fig. 3.13

Stage	Assignment Number, s	Tasks in Assignment A_s	Resource requirements, R_s	Deleted[a] Tasks	UIF(A_s)
0	0	ϕ	(0,0,0)		(11,21)
1	1	(11)	(2,2,1)	(11)	(31,41)
	2	(21)	(0,2,1)	(21)	(22)
	3	(11,21)	(2,4,2)		(22,31,41)
2	4	(11,31)	(5,5,4)	(22)	(32)
	5	(11,41)	(4,3,4)	(31)	(42)
	6	(11,31,41)	(7,6,7)	(41)	(32,42)
	7	(21,22)	(0,4,2)		ϕ
	8	(11,21,22)	(2,6,3)		ϕ
	9	(11,21,31)	(5,7,5)		(32)
	10	(11,21,41)	(4,5,5)		(42)
	11	(11,21,22,31)	(5,9,6)		(32)
	12	(11,21,22,41)	(4,7,6)		(42)
	13	(11,21,31,41)	(7,8,8)		(32,42)
	14	(11,21,22,31,41)	(7,10,9)		(32,42)
3	15	(11,31,32)	(8 8,7)	(32)	ϕ
	16	(11,41,42)	(6,4,7)	(42)	(43)
	17	(11,31,32,41)	(10,9,10)		ϕ
	18	(11,31,41,42)	(9,7,10)		(43)
	19	(11,31,32,41,42)	(12,10,13)		(43)
	20	(11,21,31,32)	(8,10,8)		ϕ
	21	(11,21,41,42)	(6,6,8)		(43)

	22	(11, 21, 22, 31, 32)	(8, 12, 9)		ϕ
	23	(11, 21, 22, 41, 42)	(6, 8, 9)		(43)
	24	(11, 21, 31, 32, 41)	(10, 11, 11)		ϕ
	25	(11, 21, 31, 41, 42)	(9, 9, 11)		(43)
	26	(11, 21, 31, 32, 41, 42)	(12, 12, 14)		(43)
	27	(11, 21, 22, 31, 32, 41)	(10, 13, 12)		ϕ
	28	(11, 21, 22, 31, 41, 42)	(9, 11, 12)		(43)
	29	(11, 21, 22, 31, 32, 41, 42)	(12, 14, 15)		(43)
4	30	(11, 41, 42, 43)	(8, 5, 10)	(43)	ϕ
	31	(11, 31, 41, 42, 43)	(11, 8, 13)		ϕ
	32	(11, 31, 32, 41, 42, 43)	(14, 11, 16)		ϕ
	33	(11, 21, 41, 42, 43)	(8, 7, 11)		ϕ
	34	(11, 21, 22, 41, 42, 43)	(8, 9, 12)		(51)
	35	(11, 21, 31, 41, 42, 43)	(11, 10, 14)		ϕ
	36	(11, 21, 31, 32, 41, 42, 43)	(14, 13, 17)		ϕ
	37	(11, 21, 22, 31, 41, 42, 43)	(11, 12, 15)		(51)
	38	(11, 21, 31, 32, 41, 42, 43)	(14, 15, 18)		(51)
5	39	(11, 21, 22, 41, 42, 43, 51)	(9, 10, 12)	(51)	(52)
	40	(11, 21, 31, 41, 42, 43, 51)	(12, 13, 15)		(52)
	41	(11, 21, 22, 31, 32, 41, 42, 43, 51)	(15, 16, 18)		(52)
6	42	(11, 21, 22, 41, 42, 43, 51, 52)	(10, 11, 12)		ϕ
	43	(11, 21, 22, 31, 41, 42, 43, 51, 52)	(13, 14, 15)		ϕ
	44	(11, 21, 22, 31, 32, 41, 42, 43, 51, 52)	(16, 17, 18)		ϕ

[a]When all feasible subsets in stage k have been considered, each task in the list of UIF of stage k is "marked" in stage $k+1$. "Marking" a task prohibits its inclusion again in the set of UIF of stage $k+1$. This eliminates duplication of assignments.

Table 3.4, and the A-network is shown in Fig. 3.14. Since there are seven arcs in the A-network, we conclude that the minimum (i.e., optimum) duration of this project subject to the given resource constraints is 7 days.

The schedule itself is obtained from the sequence of assignments $A_0, A_3, A_8, A_{12}, A_{23}, A_{34}, A_{40}, A_{44}$ by simple subtraction. For economy of notation we let the vector (s_i, i) indicate that task i is *started* in period s_i. Hence a schedule is given by the set $\{(s_i, i)\}$ for all tasks $\{i\}$. We have:

$$A_3 - A_0 = (11, 21) \Rightarrow \{(1, \mathit{1}), (1, \mathit{2})\};\ \text{start } \mathit{1} \text{ and } \mathit{2} \text{ period 1}$$

$$A_8 - A_3 = (22) \quad \Rightarrow \text{continue } \mathit{2} \text{ in period 2}$$

$$A_{12} - A_8 = (41) \quad \Rightarrow \text{add } (3, \mathit{4});\ \text{start } \mathit{4} \text{ in period 3}$$

$$A_{23} - A_{12} = (42) \quad \Rightarrow \text{continue } \mathit{4} \text{ in period 4}$$

$$A_{36} - A_{23} = (43) \quad \Rightarrow \text{continue } \mathit{4} \text{ in period 5}$$

$$A_{40} - A_{34} = (31, 51) \Rightarrow \text{add } \{(6, \mathit{3}), (6, \mathit{5})\};\ \text{start } \mathit{3} \text{ and } \mathit{5} \text{ in period 6}$$

$$A_{44} - A_{40} = (32, 52) \Rightarrow \text{continue } \mathit{3} \text{ and } \mathit{5} \text{ in period 7.}$$

Therefore, the complete schedule is $\{(1, \mathit{1}), (1, \mathit{2}), (3, \mathit{4}), (6, \mathit{3}), (6, \mathit{5})\}$. We leave it as an exercise to the reader to draw the Gantt chart corresponding to this schedule and verify that both precedence and resource availabilities are respected (see also Fig. 3.16*b*).

Although the extension of the original ALB procedure to ANs subject to scarce resources is "straightforward," it suffers from the possibly very large size of the A-network. This possibility was noted by Gutjahr and Nemhauser (1964) in the context of line balancing. Conditions deteriorate in ANs and a serious attempt at a resolution was made by Davis (1968) and Davis and Heidorn (1971). We briefly outline their line of reasoning.

Reduction in the size of the A-network is effected through a conscious effort to maintain a file of only those assignments that cannot be eliminated by considerations of feasibility, dominance, and bounding. Central to the efforts in this direction are the following remarks:

1. A task h appears for the first time in feasible subsets generated at stage k when k is the earliest start time (EST) of task h. The validity of this observation follows by contradiction; if it were not true, then either $k < (\text{EST})_h$, which means that a segment can be done before it can be done, an obvious absurdity, or $k > (\text{EST})_h$, which would violate the manner in which the "undeleted immediate followers" (see Table 3.4) lead to new assignments.

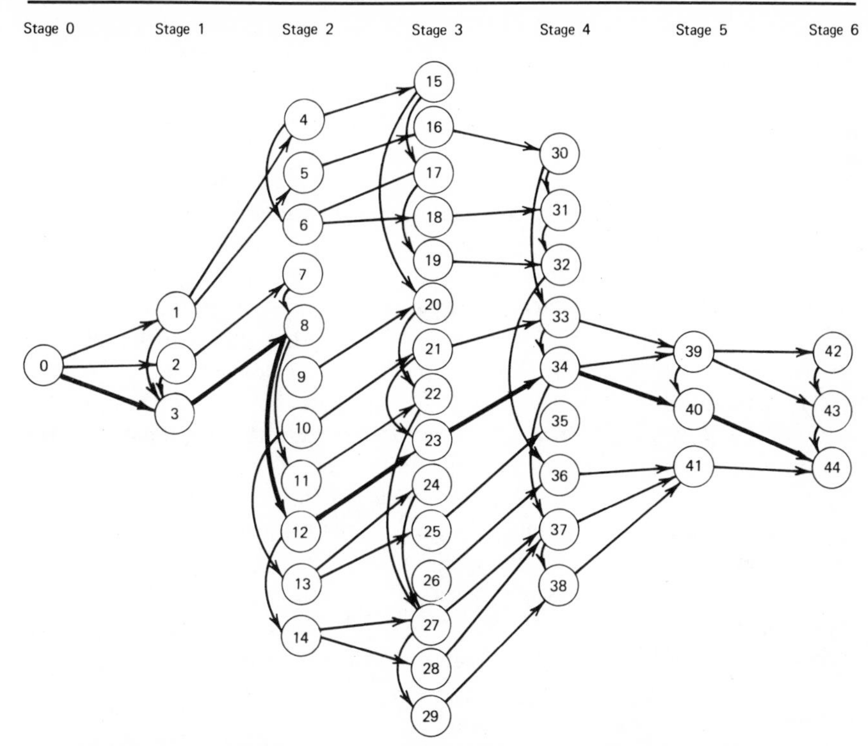

Figure 3.14. The A-network of example (heavy lines indicate shortest path).

2. The number of stages in the generation of the A-network is equal to the number of time periods in the longest path through the network when resource availabilities are taken into consideration. For if not, either there are stages with no generation of new assignments, or there are tasks which appear in no stage, both of which eventualities are impossible.
3. Directed arc (s,u) can exist only if $s<u$; furthermore, when A_s is in stage k and A_u in stage $k+1$, then it must be true that the subset used to generate A_s must bear a number not larger than the number of the subset used to generate A_u. In other words, two assignments generated in two consecutive (or succeeding) stages must be generated by assignments which are themselves consecutively numbered.

The third property follows immediately from the manner in which the A-network is constructed. All three of these properties are used to eliminate some nodes (i.e., assignments) of the A-network and thus save the trouble of investigating all their "descendents." Naturally, the earlier this elimination occurs, the better off we are computationally. In particular, two conditions are suggested.

A Time-Based Elimination Condition:

If for some reason we require that task h be performed on or before schedule period d, then we can eliminate from consideration feasible subsets generated at stages $k \geqslant d$ that do not contain task h.

A natural question in this context is how these "due dates" $\{d\}$ are determined in the first place. Clearly, if they are "tight" they lead to infeasible programs, and if they are "loose" they are worthless! One obvious approach, which is also "foolproof" in the sense that it must lead to the optimum duration, is to assign the $\{d\}$ values based on the duration of the CP *with no regard to the resource constraints*, and keep increasing the duration of the CP by unity until feasibility is achieved.† The drawback of this stepwise approach to the optimum is its possible difficulty in computation, especially if the difference between the duration of the original CP and the optimum duration is large.

Alternatively, one may apply a set of heuristics to establish a feasible schedule, which would yield a feasible project target date. Then one can move "backward" in a stepwise fashion to the optimum.

A Resource-Based Elimination Condition:

Since we defined A_S to be the assignment containing *all* the tasks, then $\mathbf{G}_S$ is the vector of total resources consumption. Obviously, if the project is completed in period T, and the resources available per unit time are given by the vector $\mathbf{L}=(l_1, l_2, \ldots, l_R)$, then the total consumption of resources through period τ must be no less than $\mathbf{W}_\tau \equiv (w_{1\tau}, w_{2\tau}, \ldots, w_{R\tau})$; where

$$w_{r\tau} = \max(0; G_{Sr} - (T-\tau)l_r)$$

For instance, in period $T-1$, *at least*

$$w_{r,T-1} = \max(0; G_{Sr} - l_r), \quad r=1,2,\ldots,R$$

must have been consumed of each resource r. The disadvantage of this condition is that in many instances it may be ineffectual because $w_{r\tau}=0$. Nevertheless, it is valuable whenever applicable.

Apart from these two conditions, Davis (1968) suggests other elimination conditions of dubious effectiveness. Experimentation with some 65 diffe-

†This should be reminiscent of the Patterson–Huber approach described in Section 4.3, in which the optimization problem (of ILP) is changed into a sequence of tests for feasibility.

rent artifically generated projects indicates the inconclusive nature of the approach, especially in carrying the calculations to completion with even small networks of 27 activities, approximately 80 tasks and three scarce resources. This is one more indication that medium- and large-scale projects must be either partitioned into smaller and more manageable parcels, or else one must rely on heuristics to reach "good" or optimal solutions. When one insists on achieving optimality, the class of heuristics used fall in the class of search procedures commonly known as "implicit enumeration" or B&B methods. We devote the next section to a brief description of such an approach and its application to ANs.

§6. A BRANCH-AND-BOUND (B&B) APPROACH

§6.1 Problem Solving Through B&B

We have previously encountered some of the basic notions of B&B theory, particularly in Section 5 of Chapter 2, but always in a disguised form. Now we render these concepts explicit, since B&B may prove to be the only practical means for achieving optimality in any but the most trivial of projects.

The approach of B&B is basically a heuristic tree search in which the space of feasible solutions, which may contain a finite or an infinite number of points, is systematically searched for the optimum.

According to Mitten and Warburton (1973), "the search proceeds iteratively by alternately applying two operations: subset formation and subset elimination. In the former, new subsets of alternatives are formed, while in the latter some subsets of alternatives may be eliminated from further consideration. The procedure terminates when a collection recognized to contain only optimal solutions is reached."

The search has two guiding principles: (a) that every point in the space is enumerated either explicitly or implicitly, and (b) that the minimum number of points be explicitly enumerated. (We view B&B as an approach for implicit enumeration, although we concede that, mainly due to historical coincidences, the label "implicit enumeration" has been applied to approaches that need not employ the "bounding" feature of B&B.)

The implicit enumeration of feasible points is accomplished through *dominance* (which may employ bounding), *redundancy*, and *feasibility* considerations. Each of these concepts is discussed and illustrated below, but first we give a laconic description of them that offers the uninitiated reader a general grasp of the subject.

The basic idea in B&B is to divide the feasible space, denoted by Ω, into subsets $\Omega_1, \Omega_2, \ldots, \Omega_K$ which may or may not be mutually disjoint. Assuming that the optimum falls in subset Ω_k, a bound on its value is determined: an upper bound in the case of maximization, and a lower bound in the case of minimization, or better still, both an upper and a lower bound in either case. (Recall the bounds established on the optimum cost in Section 5 of Chapter 2.) Based on such bounds two actions may take place:

1. A particular subspace, Ω_k, is selected for more intensive search by further partitioning into *its* subsets (this is the branching, or "formation" function).
2. Some feasible points (i.e., subspaces) are declared "noncandidates" for the optimum and are thus eliminated from further consideration (this is the "elimination" function).

The latter concept (2) is a consequence of *dominance* since it is based on the determination that any element of a particular subset Ω_i is better (or worse)—in the sense of the criterion function—than any element in another subset Ω_j. Then, indeed, we may declare the points in Ω_j (or in Ω_i) as noncandidates for the optimum and eliminate them from further analysis.

While dominance may be established on the basis of the bounds evaluated on subsets Ω_k, it is also true that dominance can be established independent of any bounding considerations. In some circles (especially in the scheduling literature) this is referred to as "elimination" procedures. The final result is the same, namely, to establish that certain subsets cannot contain the optimum because they are dominated by other subsets.

A similar idea underlies the *feasibility* considerations. They arise because in the majority of cases one is forced to hypothesize a rather "rich" original space Ω, in the sense of containing more points than the original feasible space as defined by the set of constraints, through the *relaxation* of some of the original problem constraints. For example, one may relax the integer requirements and treat a LP instead of an ILP, or one may ignore the "noninterference" constraints and treat an independent set of activities instead of strongly dependent ones. At some stage of analysis, it may be possible to establish that certain subsets of Ω are actually infeasible, in the sense of violating some (previously relaxed) constraint of the problem, in which case such subsets can also be eliminated from further study.

The concept of *redundancy* is closely related to those of irrelevance and/or inferiority with respect to what is already in hand or achievable.†

†The *Webster Collegiate Dictionary* defines redundancy as "the part of a message that can be eliminated without loss of essential information" and cites the synonyms, *superfluity* and *prolixity*.

For instance, a schedule that delays an activity while resources are available for its undertaking is evidently a redundant schedule, albeit it is feasible, under the objective of minimizing the total project duration.

Heuristics enter the tree search in all three basic phases of the approach: (a) in the definition of the partitioning procedure, (b) in the calculation of the bounds, and (c) in the philosophy of searching the three.

Here we wish to draw the reader's attention to the following important, and rather crucial distinction: the formal structure of B&B admits the use of heuristics,† but these are "reliable heuristics" in the sense that if the procedure is permitted to run to completion, the optimum will be achieved. Furthermore, if the procedure is terminated before it has achieved the optimum, it yields a bound on the error committed. This is in sharp contrast to "heuristic problem-solving procedures" that lay no claim to either optimality or a measure of the error committed at premature abortion (see Section 2 of this chapter).

The following remarks pertain to the use of heuristics in the three phases of B&B:

1. As mentioned above, the definition of the partitioning procedure is synonymous with the definition of the branching procedure in the tree search. In the majority of cases, there are several ways in which the space of feasible solutions Ω may be partitioned and the choice of a particular partitioning rule is at the discretion of the analyst. [In Problem 19 we ask the reader to suggest at least two partitioning schemes for a particular ILP derived from Eqs. (3.4)–(3.6).]
2. The calculation of the lower bound is equally open to a variety of heuristic approaches. Naturally, we are always interested in the procedure that yields the greatest lower bound, because then the bounds obtained are uniformly the most powerful. Unfortunately, such power of discrimination is either not known or not achievable with reasonable computing effort. Then heuristics are called upon to obtain "good" or "tight" lower bounds. Usually, such "tightness" is obtained at the expense of increased computing effort, and it may be better to be content with a less powerful procedure which is very easy to calculate. Alternatively, the procedure need not be the same from the start of the search to its termination: intuition suggests that the higher up the search tree, the more worthwhile may be the additional computation time necessary to obtain tighter lower bounds.
3. Finally, the philosophy of searching the tree is also open to the use of heuristics. Basically, there are two extreme philosophies, with an infinity of intermediate ones. On the one end of the spectrum, there are

†So does the simplex algorithm of LP.

heuristics (e.g., branch from the node with the smallest lower bound) that favor the nodes higher up in the tree. In this case, the construction of the search tree will proceed "horizontally"—the so-called *jump-tracking.* On the other end of the spectrum, there are heuristics that favor the pursuit of search in one subset of the space of search Ω until it is "fathomed," that is, until either all nodes become final or the lower bound of each exceeds the actual cost of a solution previously obtained —the so-called *back-tracking* approach. The advantage of a procedure of this type is that certain information is readily available when branching is from the node (i.e., subset) just created but otherwise would need to be recomputed. Furthermore, the procedure goes directly to a feasible solution so that if calculations are stopped before optimality is achieved there is available a feasible solution as well as an upper bound on the optimal solution. Other procedures may be adopted that fall between these two extremes, such as the so-called "*choosing-up-the-tree*" procedure. In this case branching always continues from one of the nodes just created until eventually either final nodes are obtained, all nodes are infeasible, or their lower bounds exceed the actual cost of a known solution. When this occurs, the next intermediate node is chosen as follows. Trace back up the tree until a node N is found which has the property that one of the $m-1$ other nodes created when branching took place from N is still an intermediate node. Branching is then continued from this intermediate node. Evidently, the choice of the search pattern determines the computing effort as well as the storage requirements.

Ingenious devices are used for the implementation of these ideas. We do not devote any more space to elaborate on them. The interested reader is directed to the review paper by Elmaghraby and Elshafei (1976) and the references cited therein. Rather, we return to the problem of minimum project duration subject to resource constraints, utilizing the concepts discussed above.

§6.2 Application to ANs

As always, let A denote the *set* of activities in the project as well as their number.[†] We assume the structure of the network given, and denote the set of activities *immediately* preceding (i.e., immediately occurring before) activity i by $\mathcal{B}(i)$, the set of activities preceding i by $\mathcal{P}(i)$, the set of activities immediately succeeding (i.e., immediately occurring after) activity i by $\mathcal{A}(i)$; and the set of activities succeeding i by $\mathcal{S}(i)$. These various sets are illustrated in Fig. 3.15.

[†]It would be helpful to think in terms of A-on-N representation.

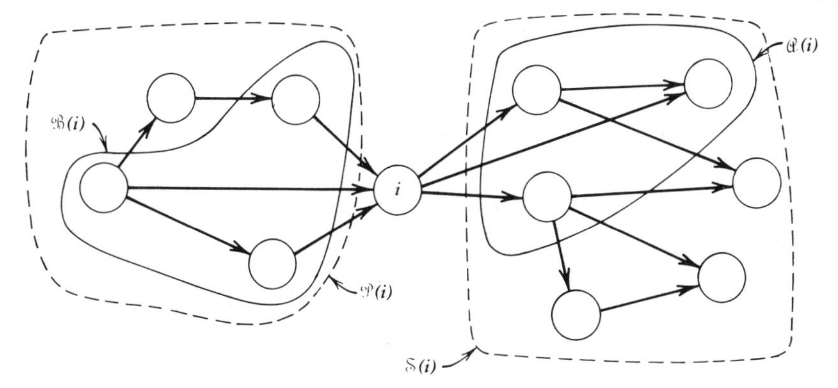

Figure 3.15. The sets $\mathcal{A}$, $\mathcal{B}$, $\mathcal{P}$, and $\mathcal{S}$.

We assume that there are R resources, and we denote the start time of activity i by s_i. In what follows, a *schedule* is understood to be the specification of the start times $\{s_i\}$ of the activities in A satisfying the precedence constraints. A *feasible schedule* satisfies, in addition, the resource-availability constraints. A *partial schedule* and a *partial feasible schedule* then refer to a subset of A. A feasible schedule is *ineligible* if there exists some activity that can be started earlier without changing the start times of any other activities and still maintain feasibility. Schedules that are not ineligible are defined to be *eligible schedules*. A similar definition applies to *eligible partial schedules*.

By the terms *schedule activity* i we mean, for a given partial schedule not yet including i, to determine its earliest permissible start time, s_i, and add activity i to the partial schedule with this start time. In the sequel we refer to the vector (s_i, i). Clearly, i can be scheduled only if $\mathcal{P}(i)$ has already been scheduled. Furthermore, if the original partial schedule were eligible, the new schedule, which includes i, is also eligible.

Perhaps the most convenient way to "remember" a partial schedule is through its corresponding *permutation*. An *eligible partial permutation*, denoted by $\mathfrak{P}$, is a permutation of a subset of the indices $1, 2, \ldots, A$ for which if $i \prec j$ in the network, then i appears before j in the partial permutation. In other words, an eligible partial permutation respects, in a sense to be defined more precisely in the next paragraph, the precedence relationship among the activities.

It is not difficult to see that there is a one-to-one correspondence between an eligible partial schedule and an eligible partial permutation, if we insist on the following "correspondence" procedures:

1. The eligible partial permutation is obtained from the eligible partial

schedule by ordering the indices of the activities in the eligible partial schedule lexicographically by the vector (s_i, i).

2. The activities are scheduled as early possible (respecting precedence) in order of their appearance in the eligible partial permutation. Thus we obtain a compact means of representing the eligible partial schedule. For example, consider the (miniscule) project of Figure 3.12 (repeated in Fig. 3.16*a*). A schedule S given by

$$S = \{(1,\mathit{1}),(1,\mathit{2}),(3,\mathit{3}),(5,\mathit{4}),(8,\mathit{5})\}$$

yields a completion time $=9$. This is a feasible albeit ineligible (hence nonoptimal) schedule (why?) and, therefore, has no corresponding permutation.[†] As we know from Figure 3.14, the optimal schedule is given by

$$S^* = \{(1,\mathit{1}),(1,\mathit{2}),(3,\mathit{4}),(6,\mathit{3}),(6,\mathit{5})\}$$

for a total duration $=7$, and the corresponding (optimal) permutation is $\mathfrak{P}^* = (\mathit{1,2,4,3,5})$. The Gantt chart for both schedules, and the resource requirements are shown in Figure 3.16.

In order to avoid repeating the enumeration of what is essentially the same eligible partial permutation (and its completions) over and over again, we wish to eliminate from consideration the so-called *redundant* permutations, defined as follows: an eligible partial permutation is redundant if there exist two activities i and j so that i appears before j in the partial permutation, but when the activities are scheduled in the order of their appearance in the partial permutation, then either $s_i > s_j$, or $s_i = s_j$ and $i > j$. The term *redundant* is used advisedly since no loss is incurred if such a permutation is ignored. Consequently, to generate all eligible schedules we need only concern ourselves with generating nonredundant eligible partial permutations. For instance, relative to the network of Figure 3.16, we may conclude that the permutation $(\mathit{2,1,3,4,5})$ is a redundant permutation (why?).

To implicitly enumerate only nonredundant eligible partial permutations is a laudable objective, albeit quite difficult to achieve. (This is illustrated below.) It is approached in the B&B scheme through having as our starting point of search the eligible schedules and their representation as eligible partial permutations. (Thus, the relaxation is seen to be relative to satisfying the resource constraints.)

We now discuss the determination of the lower bound. The approach is

[†]Note that we have defined permutations only relative to *eligible* schedules and partial schedules.

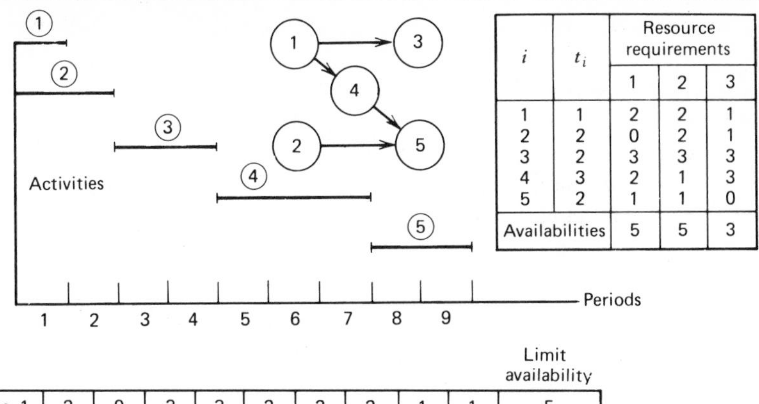

i	t_i	Resource requirements		
		1	2	3
1	1	2	2	1
2	2	0	2	1
3	2	3	3	3
4	3	2	1	3
5	2	1	1	0
Availabilities		5	5	3

										Limit availability
Resource 1	2	0	3	3	2	2	2	1	1	5
Resource 2	4	2	3	3	1	1	1	1	1	5
Resource 3	2	1	3	3	3	3	3	0	0	3

(a) Feasible schedule

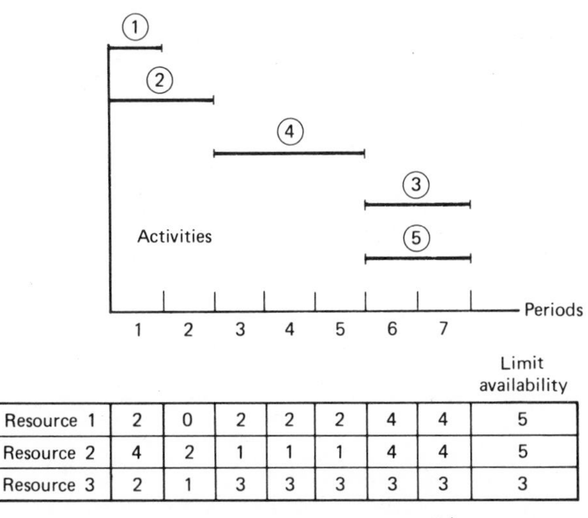

								Limit availability
Resource 1	2	0	2	2	2	4	4	5
Resource 2	4	2	1	1	1	4	4	5
Resource 3	2	1	3	3	3	3	3	3

(b) Optimal permutation $\mathfrak{P}^* = (1,2,4,3,5)$

Figure 3.16. Two eligible permutations of the project of Fig. 3.12.

a natural follow-up of the "resource-based elimination condition" described above. As usual, we need to clarify our terminology.

As before, let l_{rt} denote the limit availability of resource r in period t, $r = 1, 2, \ldots, R$. The vector $\mathbf{L}_t = (l_{1t}, l_{2t}, \ldots, l_{Rt})$ then represents the vector of limit availabilities in period t. Also let $c_i(r,t)$ denote the consumption of resource r in period t by activity i.

New terminology is introduced as follows. For purposes of lower-bound calculations, we need to identify two schedules: (a) the earliest start schedule (ESS) and (b) latest start schedule (LSS). The consumption of resource r in period t is denoted by $C_{rt}(E)$ for the ESS and $C_{rt}(L)$ for the LSS. These C values correspond to the "skyline" mentioned in Section 1.2 (this chapter). The resource consumption under any schedule S is denoted by $C_{rt}(S)$. The *cumulative* consumption of resources is denoted by $\mathcal{C}_{rt}(\cdot)$, where

$$\mathcal{C}_{rt}(\cdot) = \sum_{\tau=1}^{t} C_{r\tau}(\cdot)$$

The cumulative consumption "from the end" is denoted by

$$\bar{\mathcal{C}}_{rt}(\cdot) = \sum_{\tau=t+1}^{T} C_{r\tau}(\cdot)$$

The cumulative availability through period t is denoted by $\mathcal{L}_{rt}$ and is given by

$$\mathcal{L}_{rt} = \sum_{\tau=1}^{t} l_{r\tau}$$

We define $\bar{\mathcal{L}}$ similar to $\bar{\mathcal{C}}_{rt}(\cdot)$; in other words,

$$\bar{\mathcal{L}}_{rt} = \sum_{\tau=t+1}^{T} l_{r\tau}$$

These definitions are illustrated in Figs. 3.17 and 3.18, which are drawn for the network of Fig. 3.16*a*. Here, Fig. 3.17 contains the ESS and LSS that correspond to the particular schedule of Fig. 3.16*a*. In Fig. 3.17 we have circled any entry $C_{rt}(\cdot) \geqslant l_{rt}$ for later use. Figure 3.18 illustrates the quantities $\mathcal{C}_{rt}(E)$, $\mathcal{C}_{rt}(L)$, $\mathcal{L}_{rt}$, and $\bar{\mathcal{L}}_{rt}$ for resource $r=3$.

The following two observations are immediate, and they assist in discarding a project duration T as infeasible:

OBSERVATION 1. If $\mathcal{C}_{rt}(L) > \mathcal{L}_{rt}$ for any $t = 1, 2, \ldots, T$, then the duration T is infeasible.

This follows from the fact that, for any arbitrary schedule S, we must have

$$\mathcal{C}_{rt}(S) \geqslant \mathcal{C}_{rt}(L) \text{ by virtue of the structure of LSS,}$$

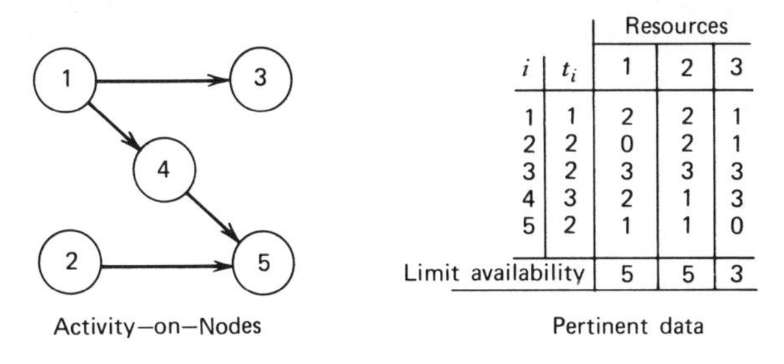

i	t_i	Resources 1	2	3
1	1	2	2	1
2	2	0	2	1
3	2	3	3	3
4	3	2	1	3
5	2	1	1	0
Limit availability		5	5	3

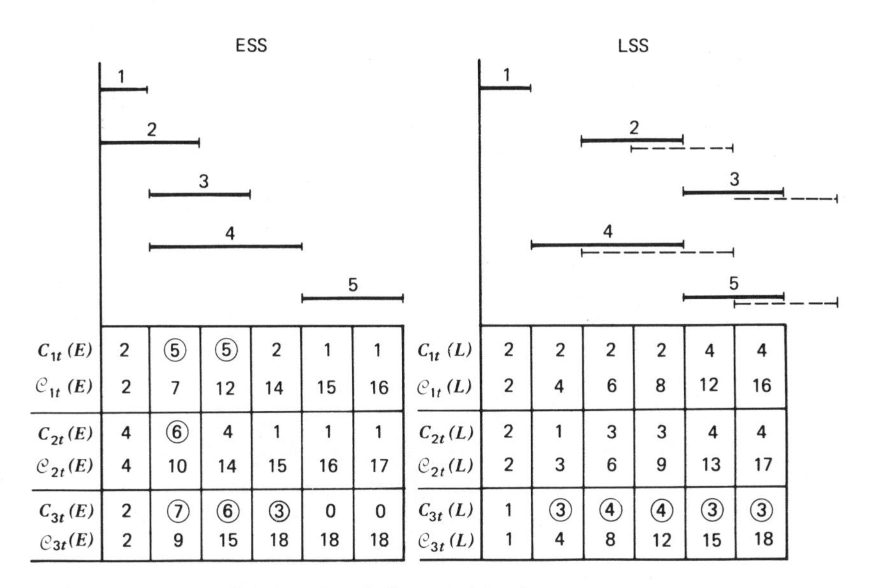

$C_{1t}(E)$	2	⑤	⑤	2	1	1
$\mathcal{C}_{1t}(E)$	2	7	12	14	15	16
$C_{2t}(E)$	4	⑥	4	1	1	1
$\mathcal{C}_{2t}(E)$	4	10	14	15	16	17
$C_{3t}(E)$	2	⑦	⑥	③	0	0
$\mathcal{C}_{3t}(E)$	2	9	15	18	18	18

$C_{1t}(L)$	2	2	2	2	4	4
$\mathcal{C}_{1t}(L)$	2	4	6	8	12	16
$C_{2t}(L)$	2	1	3	3	4	4
$\mathcal{C}_{2t}(L)$	2	3	6	9	13	17
$C_{3t}(L)$	1	③	④	④	③	③
$\mathcal{C}_{3t}(L)$	1	4	8	12	15	18

Circled numbers indicate $C_{rt}(\cdot) \geqslant l_{rt}$

Figure 3.17. The ESS and LSS for the network of Fig. 3.12 and corresponding resource consumptions.

hence $> \mathcal{L}_{rt}$ for some t; $1 \leqslant t \leqslant T$; under the hypothesis, and S is infeasible. Incidentally, we immediately conclude that $T^* \geqslant T+1$; that is, that $T+1$ is a lower bound on the optimal duration T^*; and that $\mathcal{C}_{rt}(L)$ for $t = 1, 2, \ldots, T$, is a lower bound on the resource requirements.

OBSERVATION 2. If $\bar{\mathcal{C}}_{rt}(E) > \bar{\mathcal{L}}_{ri}$, for any t; $1 \leqslant t \leqslant T$; then the duration T is infeasible.

This statement is equivalent to Observation 1 when the resource availability and consumption are viewed "backward" from the end T. For, indeed,

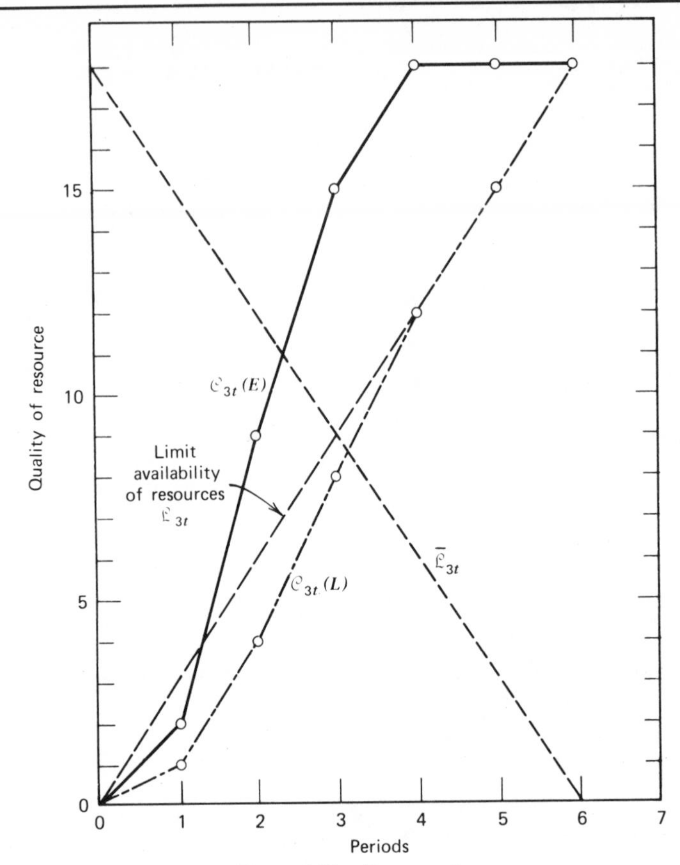

Figure 3.18. Resource 3.

any arbitrary schedule S must require resources no less than the requirement of ESS in the interval $[T-t, T]$. The value of $\bar{\mathcal{C}}_{r,t}(E)$; $t = T-1, \ldots, 1$; is easily seen to be a lower bound on the resource requirements over that interval.

These two observations together lead to the improved lower bound on the duration T^*:

$$T^* \geqslant \max\left\{\left(T|\mathcal{C}_{rt}(L) > \mathcal{L}_{rt} \text{ for some } t\right); \left(T|\bar{\mathcal{C}}_{rt}(E) > \bar{\mathcal{L}}_{r,t} \text{ for some } t\right)\right\} + 1 \tag{3.14}$$

Further sharpening of the lower bound of (3.14) may be achieved through considerations of the pattern of resource consumption. To simplify the discussion, assume *constant resource availability*, l_r, for all t. Consider the two (hypothetical) skylines of a resource r shown in Fig. 3.19 drawn for ESS and LSS.

Let k_r, if it exists, be the first period under ESS in which $C_{rt}(E) \geqslant l_r$, otherwise, put $k_r = 1$. Let K_r, if it exists, denote the last period under LSS in which $C_{rt}(L) \geqslant l_r$; otherwise, put $K_r = T$. It is easy to see that, for any arbitrary schedule S

$$\mathcal{C}_{r,k_r-1}(S) \leqslant \mathcal{C}_{r,k_r-1}(E)$$

Hence, the resource-period availability represented by the hatched area in Fig. 3.19*a* is unutilizable. In other words, any schedule "wastes" *at least* $[l_r(k_r - 1) - \mathcal{C}_{r,k_r-1}(E)]$ of the resource availability (since such availability is not "transferable" from one period to another).

A similar argument applies to the hatched area of Fig. 3.19*b*, from which we deduce that any schedule must "waste" *at least* $[l_r(T - K_r) - \bar{\mathcal{C}}_{rK_r}(L)]$ of the resource availability.

Consequently, the *net* resource-period availability is *at most*

$$\eta_r = l_r T - \left[l_r(k_r - 1) - \mathcal{C}_{r,k_r-1}(E) \right] - \left[l_r(T - K) - \bar{\mathcal{C}}_{r,K_r}(L) \right] \quad (3.15)$$

The limit availability of (3.9) may be used to exclude a project duration T because of infeasibility. Clearly, if the total resource requirements, $\mathcal{C}_{rT}(E) = \bar{\mathcal{C}}_{r0}(L)$ are $> \eta_r$, then the duration T is infeasible. However, if $\mathcal{C}_{rT}(E) \leqslant \eta_r$, then T cannot be eliminated based on feasibility. We then declare it

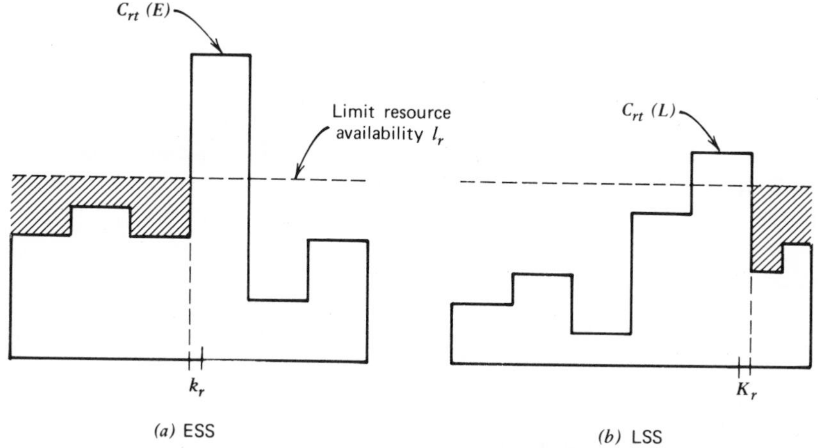

Figure 3.19. Two hypothetical skylines.

"feasible"; but it is understood that this is a shorthand notation that does not imply the realization of a feasible schedule in the given horizon T. Evidently, in that case any $T' \geqslant T$ is also "feasible" relative to resource r.

Example 3.2. The above infeasibility and bounding techniques are now applied to the simple network of Fig. 3.17, with the resource requirements and availabilities as shown in the table of that figure.

At the outset we seek an upper bound on T^* by applying any reasonable heuristic to achieve a feasible schedule. Little reflection (or reference to Fig. 3.16*a*) yields the following schedule

$$S = \{(1,1),(1,2),(3,3),(5,4),(8,5)\}$$

which is read as follows: in period 1 start activity *1*, in period 1 start activity *2*, in period 3 start activity *3*, and so on. This schedule is clearly feasible with duration $T=9$. Hence, we know that T^* is bounded as follows: $6 \leqslant T^* \leqslant 9$. (The lower bound of 6 is obtained from the CP and resource requirements.) This, in turn, immediately restricts the possible start times of the activities, in the spirit of the a-bounds of Section 4.2, as follows:

Activity i	a_i	$\bar{a}_i$
1	1	4
2	1	6
3	2	8
4	2	6
5	5	8

The set of nodes (i.e., activities) with no predecessors is $\{1,2\}$. Let the starting partial permutation be $\mathfrak{P}(1)=\{1\}$. The first branching is on activity *1* being either started in period 1 or not, indicated by $s_1=1$ or $s_1 \neq 1$. In the latter case, the earliest start time of *1* is period 2. The ESS and LSS under $s_1=1$ have already been evaluated in Fig. 3.17.

$s_1 = 1$

Relative to Fig. 3.17 we see that the CP$=6$. We obtain

$$k_1=2; K_1=6 \Rightarrow \eta_1 = 27 > \mathcal{C}_{1,6}(E) \Rightarrow T=6 \text{ feasible}$$

$$k_2=2; K_2=6 \Rightarrow \eta_2 = 29 > \mathcal{C}_{2,6}(E) \Rightarrow T=6 \text{ feasible}$$

$$k_3=2; K_3=6 \Rightarrow \eta_3 = 17 < \mathcal{C}_{3,6}(E) \Rightarrow T=6 \text{ infeasible} \Rightarrow T^* \geqslant 7$$

Pursuing the matter one step further, put $T=7$ *with s_l still fixed at* 1. It is easy to see that the ESS of Fig. 3.17 remains the same, but that the LSS is shifted in toto one period to the right (the dotted lines). Consequently, we have: $k_3'=2; K_3'=7 \Rightarrow \eta_3'=20 > \mathcal{C}_{3,7}(E) \Rightarrow T=7$ feasible, from which we conclude that the lower bound for $s_l=1$ is 7.

$s_l = 2$

The ESS and LSS under $s_l=2$ are shown in Fig. 3.20. We conclude that the CP$=7$, and

$$k_1=3; K_1=7 \Rightarrow \eta_1=27 > \mathcal{C}_{1,7}(E) \Rightarrow T=7 \text{ feasible}$$

$$k_2=1; K_2=7 \Rightarrow \eta_2=35 > \mathcal{C}_{2,7}(E) \Rightarrow T=7 \text{ feasible}$$

$$k_3=3; K_3=7 \Rightarrow \eta_3=14 < \mathcal{C}_{3,7}(E) \Rightarrow T=7 \text{ infeasible} \Rightarrow T^* \geqslant 8$$

Pursuing the analysis as above, put $T=8$ with s_l still fixed at 2. Then,

$$k_3'=3; K_3'=8 \Rightarrow \eta_3'=20 > \mathcal{C}_{3,8}(E) \Rightarrow T=8, \text{ feasible}$$

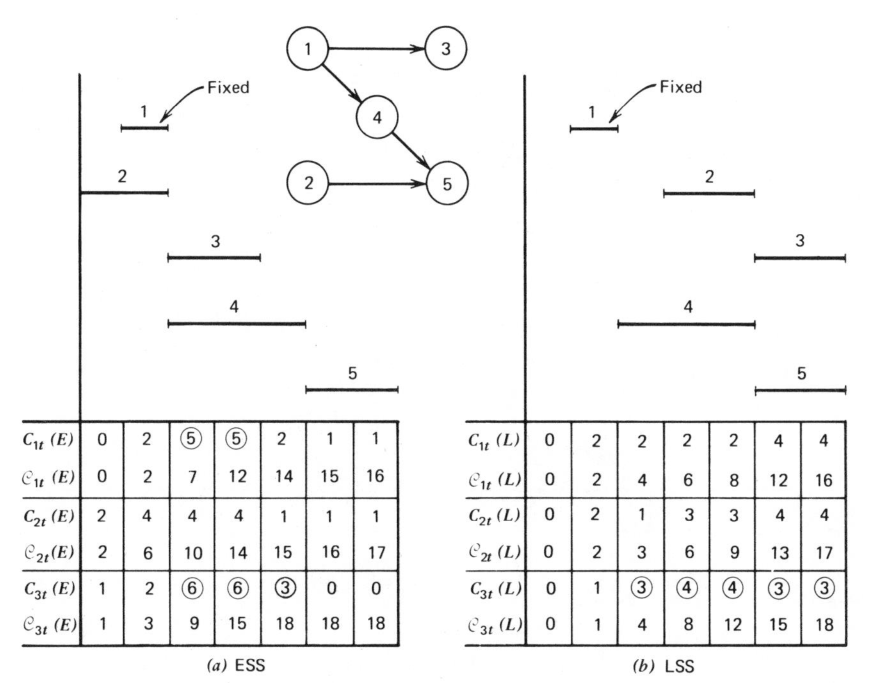

C_{1t} (E)	0	2	⑤	⑤	2	1	1
$\mathcal{C}_{1t}$ (E)	0	2	7	12	14	15	16
C_{2t} (E)	2	4	4	4	1	1	1
$\mathcal{C}_{2t}$ (E)	2	6	10	14	15	16	17
C_{3t} (E)	1	2	⑥	⑥	③	0	0
$\mathcal{C}_{3t}$ (E)	1	3	9	15	18	18	18

(*a*) ESS

C_{1t} (L)	0	2	2	2	2	4	4
$\mathcal{C}_{1t}$ (L)	0	2	4	6	8	12	16
C_{2t} (L)	0	2	1	3	3	4	4
$\mathcal{C}_{2t}$ (L)	0	2	3	6	9	13	17
C_{3t} (L)	0	1	③	④	④	③	③
$\mathcal{C}_{3t}$ (L)	0	1	4	8	12	15	18

(*b*) LSS

Figure 3.20. The node $s_l=2$.

The branching of Fig. 3.21 shows the two lower bounds just obtained.

Note that, as of this point in analysis, we have only an indication of the availability of sufficient resources over the horizon $T=7$ when $s_1=1$. However, we do not possess a *feasible* schedule; if one were in hand it would be *optimal* (why?).

Consequently, we must proceed with the search. Since the two lower bounds of Fig. 3.21 are unequal, and since activity *1* is on the CP, we branch from node $s_1=1$ to nodes $1 \leqslant s_2 \leqslant 6$ (as shown in Fig. 3.22). Now $\mathfrak{P}(2)=\{1,2\}$.

$s_2=1$.

The ESS and LSS are given in Fig. 3.24. Since activities *1* and *2* are fixed, we need only be concerned with the remaining activities. It is easily seen that scheduling activities *3*, *4*, and *5* at their earliest start times will duplicate the ESS of Fig. 3.17, a redundant eligible partial schedule, resulting in a CP of duration 6, which we know is infeasible. This unwarranted duplication of effort could be eliminated by maintaining a file of previously generated eligible partial schedules. On the other hand, searching a long file and effectuating the comparisons may be too time

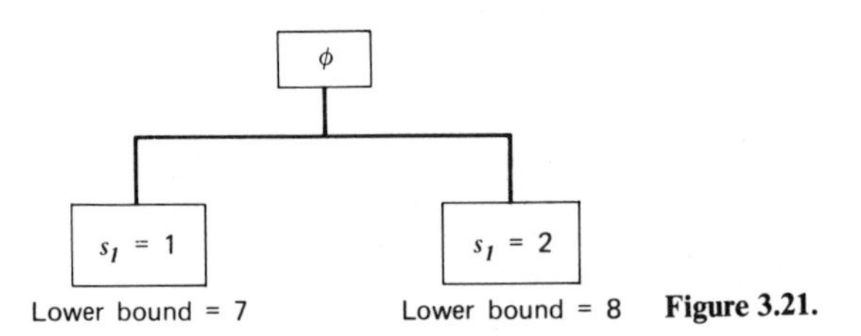

Figure 3.21.

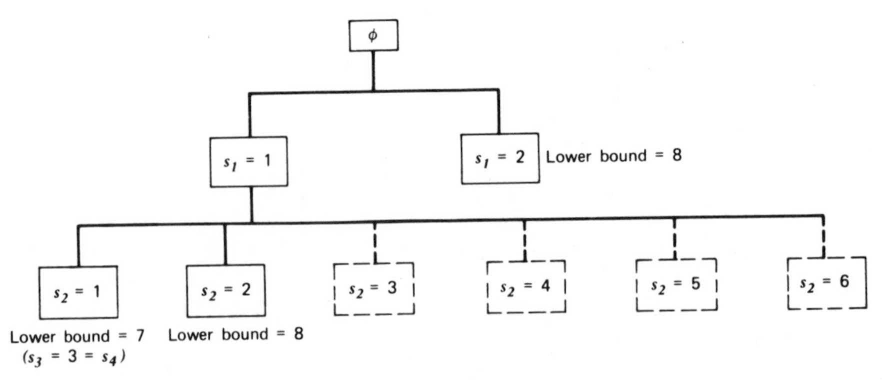

Figure 3.22.

consuming, in which case it may be faster to go ahead with enumerating eligible partial schedules redundantly. This illustrates our contention above of the difficulty in adhering to the dictum of enumerating only nonredundant eligible partial schedules (see p. 206).

In any event, for our expository purposes we can certainly avoid such duplication and proceed with $s_3=3=s_4$, which leads to a CP of length 7. We have (see Fig. 3.24)

$$k_1=3, K_1=7 \Rightarrow \eta_1=29>\mathcal{C}_{1,7}(E) \Rightarrow T=7 \text{ feasible}$$

$$k_2=1, K_2=7 \Rightarrow \eta_2=35>\mathcal{C}_{2,7}(E) \Rightarrow T=7 \text{ feasible}$$

$$k_3=3, K_3=7 \Rightarrow \eta_3=18=\mathcal{C}_{3,7}(E) \Rightarrow T=7 \text{ feasible}$$

From which we conclude that for $s_2=1$ (and $s_3=3=s_4$), the lower bound is $=7$. From this point on the same reasoning is applied to the various nodes of the search tree, which is developed sequentially as follows. We branch to $s_2=2$ to obtain lower bound $=8$ (see Fig. 3.25), and we forfeit for the time being the branching to other feasible values of s_2 (viz., $s_2=3,4,5,6$) and obtain the tree shown in Fig. 3.22. From the node $s_2=1$ branch to the feasible values of s_3, namely $s_3=4$, $s_3=5$, and $s_3=6$ as shown in Figure 3.23. The analysis for $s_3=6$ is shown in Fig. 3.26, from which we see that the LSS is *feasible.* Hence it is optimal, with the following schedule.

$$S^*=\{(1,1),(1,2),(3,4),(6,3),(6,5), \text{ with } T^*=7$$

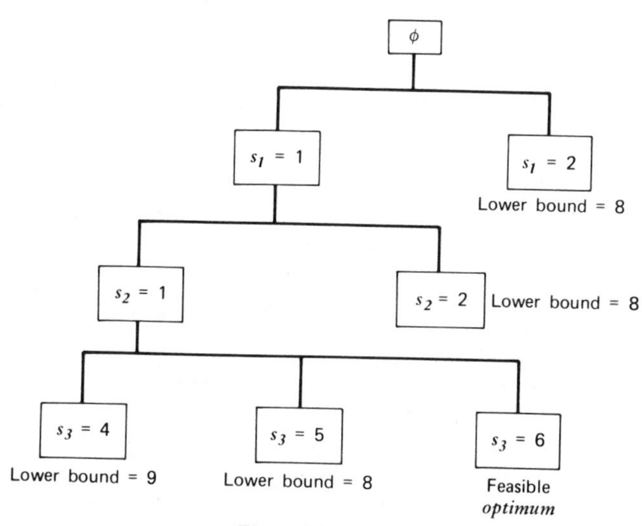

Figure 3.23.

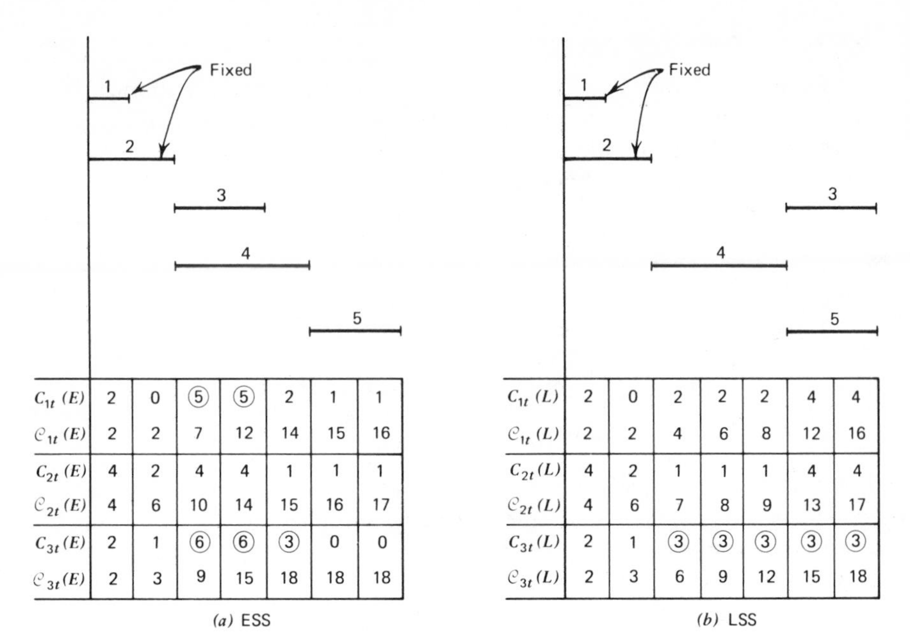

$C_{1t}(E)$	2	0	⑤	⑤	2	1	1
$\mathcal{C}_{1t}(E)$	2	2	7	12	14	15	16
$C_{2t}(E)$	4	2	4	4	1	1	1
$\mathcal{C}_{2t}(E)$	4	6	10	14	15	16	17
$C_{3t}(E)$	2	1	⑥	⑥	③	0	0
$\mathcal{C}_{3t}(E)$	2	3	9	15	18	18	18

(*a*) ESS

$C_{1t}(L)$	2	0	2	2	2	4	4
$\mathcal{C}_{1t}(L)$	2	2	4	6	8	12	16
$C_{2t}(L)$	4	2	1	1	1	4	4
$\mathcal{C}_{2t}(L)$	4	6	7	8	9	13	17
$C_{3t}(L)$	2	1	③	③	③	③	③
$\mathcal{C}_{3t}(L)$	2	3	6	9	12	15	18

(*b*) LSS

Figure 3.24. The node $s_2 = 1$.

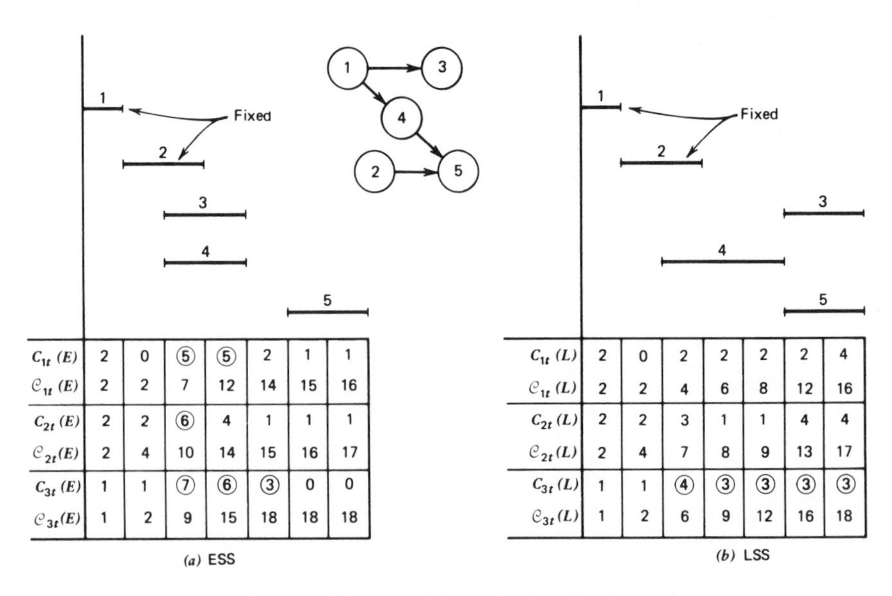

$C_{1t}(E)$	2	0	⑤	⑤	2	1	1
$\mathcal{C}_{1t}(E)$	2	2	7	12	14	15	16
$C_{2t}(E)$	2	2	⑥	4	1	1	1
$\mathcal{C}_{2t}(E)$	2	4	10	14	15	16	17
$C_{3t}(E)$	1	1	⑦	⑥	③	0	0
$\mathcal{C}_{3t}(E)$	1	2	9	15	18	18	18

(*a*) ESS

$C_{1t}(L)$	2	0	2	2	2	2	4
$\mathcal{C}_{1t}(L)$	2	2	4	6	8	12	16
$C_{2t}(L)$	2	2	3	1	1	4	4
$\mathcal{C}_{2t}(L)$	2	4	7	8	9	13	17
$C_{3t}(L)$	1	1	④	③	③	③	③
$\mathcal{C}_{3t}(L)$	1	2	6	9	12	16	18

(*b*) LSS

$k_1=3, K_1=7 \rightarrow \eta_1=27 > C_{1,7}(E) \rightarrow T=7$ feasible

$k_2=3, K_2=7 \rightarrow \eta_2=29 > C_{2,7}(E) \rightarrow T=7$ feasible

$k_3=3, K_3=7 \rightarrow \eta_3=17 < C_{3,7}(E) \rightarrow T=7$ infeasible $\rightarrow T^* \geqslant 8$

Figure 3.25. The node $s_2 = 2$.

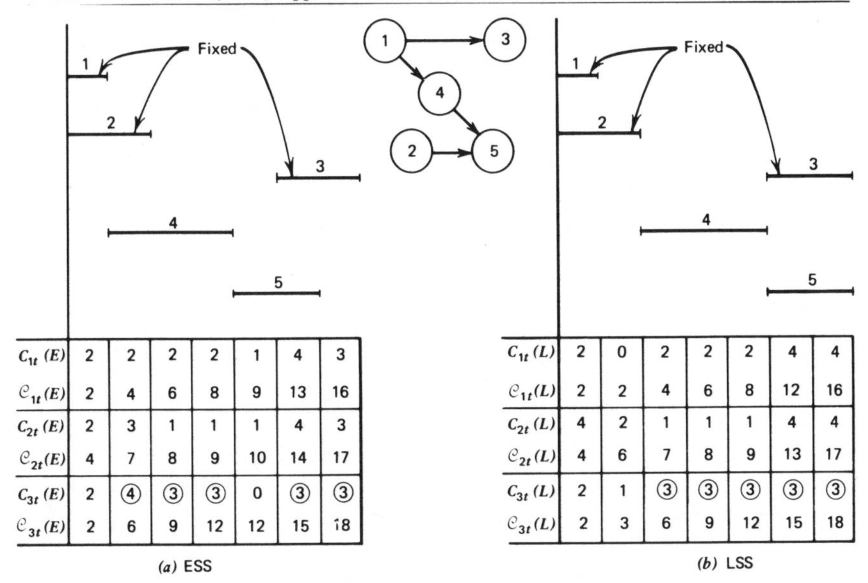

$C_{1t}(E)$	2	2	2	2	1	4	3
$\mathcal{C}_{1t}(E)$	2	4	6	8	9	13	16
$C_{2t}(E)$	2	3	1	1	1	4	3
$\mathcal{C}_{2t}(E)$	4	7	8	9	10	14	17
$C_{3t}(E)$	2	④	③	③	0	③	③
$\mathcal{C}_{3t}(E)$	2	6	9	12	12	15	18

(a) ESS

$C_{1t}(L)$	2	0	2	2	2	4	4
$\mathcal{C}_{1t}(L)$	2	2	4	6	8	12	16
$C_{2t}(L)$	4	2	1	1	1	4	4
$\mathcal{C}_{2t}(L)$	4	6	7	8	9	13	17
$C_{3t}(L)$	2	1	③	③	③	③	③
$\mathcal{C}_{3t}(L)$	2	3	6	9	12	15	18

(b) LSS

Figure 3.26. Activities *1*, *2*, and *3* fixed at 1, 1, and 6, respectively; schedule LSS feasible⇒ optimal.

Indeed, it is the schedule obtained by the approach of Section 5 (see p. 198 or Fig. 3.16*b*).

The B&B procedure is now complete. The reader may be awed by the large amount of calculations that were required to solve such a simple project of only six activities. Unfortunately, that is the price to be paid for insisting on achieving optimality. The problem of scheduling activities on multiple resources when the activities are subject to precedence constraints and the availability of the resources is limited is known to be "NP hard," see Lenstra (1976). This implies that, in all probability, there shall be no "efficient" algorithm for solving this problem (in the sense of achieving optimality). One is then obliged to resort to B&B procedures, which are inherently demanding in their computational requirements.

PROBLEMS

1. The following network gives the precedence relationship among activities, and the table gives the expected duration of each activity together with its demands on two scarce resources.
   a. Determine the critical path.
   b. Determine resource requirements assuming primary ranking on i and secondary ranking on j [for activities $(i,j_1),(i,j_2)\cdots$ starting from node i].

		Resources				Resources	
Activity	Duration	1	2	Activity	Duration	1	2
(1,2)	12	3	—	(5,10)	19	17	—
(1,3)	10	2	—	(6, 9)	30	6	13
(1,4)	11	—	4	(6,11)	22	10	—
(2,5)	6	—	—	(7,10)	7	—	—
(2,7)	5	2	—	(8, 9)	2	—	1
(3,5)	4	—	1	(8,12)	14	6	6
(3,6)	15	2	2	(9,10)	14	—	7
(3,8)	31	3	14	(9,11)	10	2	2
(4,6)	6	1	—	(10,12)	16	8	1
(4,7)	18	5	5	(11,12)	15	5	5

c. Attempt leveling of both resources by any technique you develop. State your procedure explicitly (i.e., in algorithm format) and the final results of leveling using your procedure.

2. Consider the project network shown (A-on-A representation) that demands resources (numbered 1, 2, 3, 4). The attached table summarizes the activity durations and the resource requirements for each activity:

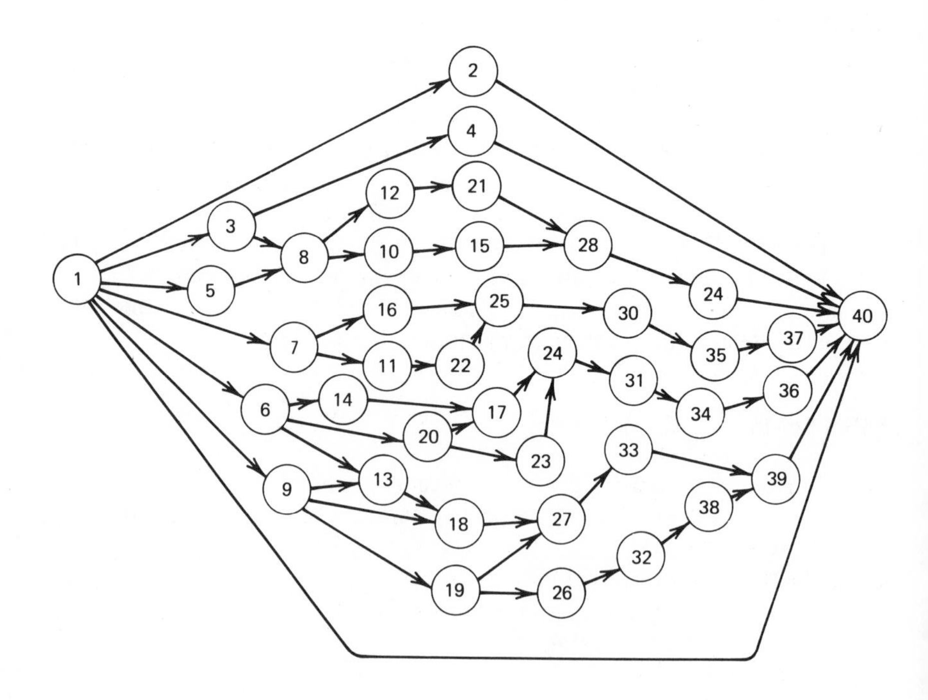

Data for Problem 2

Activity No.	Initial Node i	Final Node j	Duration	Requirement	Resource	Activity No.	Initial Node i	Final Node j	Duration	Requirement	Resource
1	1	2	4	2	3	28	14	17	2	7	3
2	2	40	3	8	1	29	17	24	3	3	3
3	1	3	2	5	1	30	6	20	4	1	2
4	3	4	2	8	3	31	17	20	2	4	4
5	4	40	2	3	4	32	20	23	1	7	2
6	1	5	1	2	2	33	23	24	2	3	1
7	5	8	1	7	3	34	24	31	3	8	4
8	8	10	2	5	4	35	31	34	3	8	4
9	8	12	3	8	3	36	34	36	5	4	4
10	10	15	5	8	2	37	36	40	4	2	4
11	12	21	3	4	1	38	1	9	3	7	1
12	15	28	1	4	1	39	9	13	4	3	3
13	21	28	4	9	3	40	9	19	2	4	1
14	28	29	1	5	2	41	13	18	3	2	4
15	29	40	3	6	2	42	18	27	5	6	2
16	1	7	4	3	4	43	19	27	4	4	2
17	7	16	3	4	1	44	19	26	1	3	1
18	16	25	2	4	2	45	26	32	1	1	3
19	7	11	1	6	3	46	27	33	3	2	4
20	11	22	5	2	2	47	32	33	3	2	3
21	22	25	1	9	3	48	32	38	2	9	1
22	25	30	5	8	4	49	38	39	5	8	4
23	30	35	4	4	2	50	33	39	4	9	3
24	35	37	1	2	3	51	39	40	1	3	1
25	37	40	2	2	3	52	1	40	10	8	1
26	1	6	2	1	3	53	6	13	1	4	4
27	6	14	1	1	1	54	9	18	2	5	4
						55	3	8	4	1	4

You are asked to determine the schedule of the activities that minimizes the project duration subject to resource availabilities of 9, 8, 9, 9 in the four shops. (*Hint*: Devise a heuristic procedure.)

3. Consider the problem of project scheduling under scarce resources in which it is desired to "smooth" the resource utilization (see Section 4.1). Suppose you have devised a heuristic procedure that you believe to be "good" and are interested in demonstrating its "superiority" to other heuristic procedures. You are willing to compare its performance to the performance of others when all are applied to a "wide variety" of projects.
   a. Enumerate the different parameters that you believe control the performance of any procedure. Determine the range of variation of each parameter. Subsequently, suggest an experimental design that you believe would convince any skeptic. (Such design must include the number of experiments you plan to conduct and the level of confidence to which you aspire.)
   b. State the various measures of performance you would suggest for your procedure as well as others.
4. Complete the construction of the smoothing model of Section 4.1 with the structural variables $\{y_{ij}\}$ of Eqs. (3.1)–(3.3). In particular:
   a. Suppose that resource r is available in quantities m_j^r in period j. State the resource-availability constraints.
   b. Suppose that the cost parameters are as follows: (i) c_{rj} is the cost per unit increase in the use of resource r, (ii) d_{rj} is the cost per unit decrease, and (iii) p is the penalty per period of delay beyond D. State the criterion function that minimizes the total cost. State the number of variables and constraints, assuming N activities, R resources, and a maximum duration of H.
5. Expand the ILP of Problem 4 to recognize the possibility of overtime work at increased cost. The capacity of such work is a fraction of the regular time capacity (varying between 0.20 and 0.50).
6. In the ILP of Problems 4 and 5 redefine the structural variables to represent the increase and decrease in the level of the activity in each period (rather than its absolute level). State the resulting ILP, and compare it to that of the previous two problems (especially compare the number of variables and constraints). Comment briefly.
7. Complete the construction of the scarce resources planning model with the structural variables given in Eqs. (3.4)–(3.6). In particular;
   a. Indicate explicitly how the parameters a_i and $\bar{a}_i$ can be determined. Illustrate your proposed procedure on the following network (A-on-N representation, with the activity duration as the number next to each node).

b. Suppose that a unit of resource r can be acquired at a cost c_r, independent of period, and that a penalty p is incurred for each period of delay beyond the due date D. State the criterion function that minimizes the combined costs of resources acquisition and project delay. Comment briefly on the size of the resulting ILP.

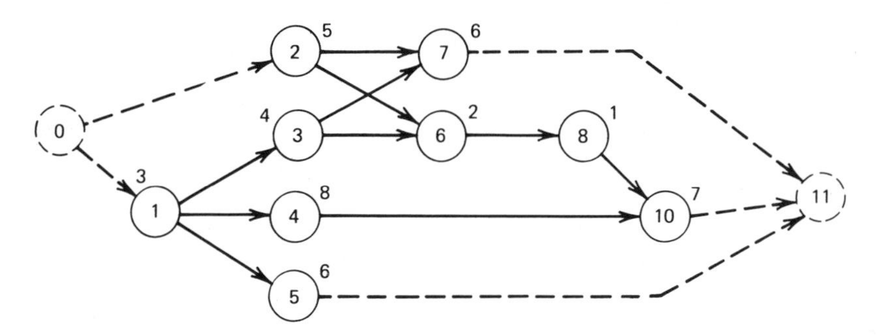

8. A frequently occurring restriction in scheduling tasks (or activities) over a single processor is to require that in any period of time t the processor is devoted to at most one task. (This is the *noninterference constraint*). Let i and j be two tasks unrelated to each other by any precedence relations. Let p_i denote the processing time of i and p_j denote the processing time of j. Write the constraint (or constraints) that model the noninterference condition, namely, that tasks i and j should not be scheduled simultaneously on the processor in any period t in the planning horizon H.

9. In the model of Section 4 it was necessary to "price" delays in the completion of the project. To this end it was necessary to determine the completion time of the project, which is given by the variable z of Eq. (3.11), and the shortest possible project duration D in order to write the last term in Eq. (3.12).

   Now suppose D is not known or that it is known but that the penalty does not apply except if the project duration is longer than a contractual date D. (The project *may* be finished prior to that date.) What modifications in the model of Eqs. (3.7)–(3.12) would you introduce to reflect this new characterization?

10. Develop a model for the following problem of multiproject resource planning.

    Denote a project by q; $q=1,2,\ldots,Q$. A project will contain N_q activities related to each other by some precedence network and will have a specified "starting date"—its completion date is dictated by its CP. A project "fits" among others by virtue of precedence relations between some of its activities and activities in other projects.

Each activity a (in each project) has duration $y_a \geqslant 0$ and consumes some resources throughout its duration. Let a resource be denoted by $r, r=1,\ldots,R$, and denote the consumption of resource r by activity a in project q by b_{aqr}. Assume the availabilities of the resources to be limited at A_r. The resources are available to all projects simultaneously.

It is desired to respect the resources' availabilities and not to exceed them. It is also desired to complete all projects "on time" where "time" is specified by their respective completion dates. If a conflict exists, then a whole project will be subcontracted, at a cost c_q. This would free all the resources consumed by that project simultaneously. It is desired to minimize the total cost of subcontracting.

Discuss any computing suggestions you may have to achieve numerical answers in real-life situations, where each project may have from 60 to 80 activities and management may be handling some 400 projects simultaneously.

11. Based on the ILP models developed in Section 4, construct a model to minimize the total costs in the following jobshop scheduling problem:
    a. Six jobs numbered $1,2,\ldots,6$
    b. Four machine groups $A(2), B(1), C(4), D(4)$. The number in parentheses is the number of machines, n_m, in machine group m.
    c. Planning horizon $s=$ four 8-hr shifts $=32$ hr $=T$. Every job i has its own:
        i. route, as given in the sequence of machine groups it is supposed to visit. Denote the total set of groups visited by $M_i; m=A, B, C,$ or D and $i=1,\ldots,6$; and denote the route by $[i_1],[i_2],\ldots,[i_{M_i}]$;
        ii. processing time, p_{im} on machine $m \in M_i$;

Data for Problem 11

Machine Groups	A	B	C	D	Route	M_i	Delay Cost	Holding Cost
Number of Machines	2	1	4	4			c_{ij}	$h_{[i_k,l]}$
Capacity	64	32	128	128				
	p_{iA}	p_{iB}	p_{iC}	p_{iD}				
$i=1$	6	2	8	6	A,B,C,D	4	$c_{1j}=3j-66; j>22$	0.5
2	7		10	7	A,C,D	3	$c_{2j}=5j-120; j>24$	1.0
3	4	1	5	8	B,A,C,D	4	$c_{3j}=j-18; j>18$	2.0
4		2	4	9	B,D,C	3	$c_{4j}=j-20; j>20$	1.0
5	5	1	8	5	C,A,D,B	4	$c_{5j}=6j-114; j>19$	0.5
6	6	2	6	6	D,C,B,A	4	$c_{6j}=6j-120; j>20$	3.0
Total Demand	28	8	41	41				

iii. delay cost c_{ij}, a function of the time of completion of the job, and is dependent on the due date, d_i, and the period j in which the job is completed;

d. in-waiting storage cost, $h_{[i_{k,l}]}$, incurred per unit time of waiting between machine group $[i_k]$ and machine group $[i_l]$, $1 \leqslant k < l \leqslant M_i$.

12. Construct an ILP model for the problem of simultaneous determination of optimal activity durations *and* the explicit recognition of the individual resources. The particulars of the problem are as follows.

An activity, say i, may require a set of resources; its total content of resource k is given as b_{ik}. In any period, one or more resources may be allocated to the activity, within specified lower and upper limits (e.g., limits on crew size). The availability of the resources are known for each period and are denoted by a_k^t. It is desired to minimize the total project durations, T.

[*Hint*: Translate the objective into a test for feasibility for various (decreasing) values of T, starting from a known feasible solution. Define x_{ik}^t to be the quantity of resource k allocated to activity i in period t; and let $l_{ik} \leqslant x_{ik}^t \leqslant u_{ik}$. There are three types of constraints:

a. The total allocation of each resource to each activity must be equal to the total content, b_{ik}.
b. The availabilities of resources should not be exceeded in any period.
c. The precedence constraints must be respected.]

The interaction between resource availabilities and project "compression" is more intimate than has been hitherto realized. The following two problems state some practical considerations that, to the best of our knowledge, have not been treated previously.

13. In many instances the availability of resources is almost unlimited, but at a price. Furthermore, the cost of an activity is composed of not only the cost of the utilization of these resources, but also of the nonutilization (i.e., idle time) of the resources. Implicit in this situation is the assumption that resources are available throughout the duration of the project; hence, their cost must involve some opportunity cost when they are not fully utilized (e.g., resources made available for the Alaskan pipeline). Since the duration of the project is itself a function of the resource availabilities, we essentially have a compound function of resource availabilities (resource availabilities determine project duration which, in turn, determine resource availabilities through costs incurred).

It is desired to formulate a model whose objective is the minimization of the total project cost through completion, taking such opportunity cost into account. Apply the model to the following network:

Activity	Duration	Cost of Utilization	Opportunity cost per period (when Resource Idle)
(1,2)	3	2100	500
	4	2000	400
	5	2100	300
(1,3)	1	9000	6000
	2	8000	3000
	3	8400	3000
(2,3)	6	6600	800
	8	6400	600
	9	7200	300
(2,4)	8	1200	100
	9	1080	80
	11	1320	90
(3,4)	4	5200	1000
	5	4500	800
	6	4800	700

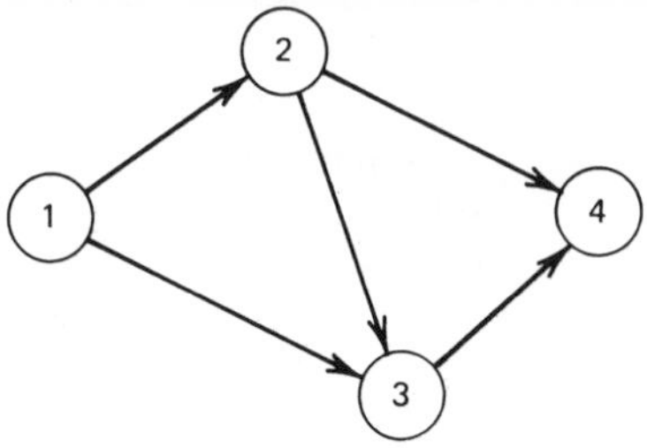

14. Suppose that resources are available in fixed quantities and are not to be exceeded. However, an activity may be compressed, at a cost, utilizing the *same* total resources but in shorter time (i.e., $y_a z_{ar} = \text{const}$, where y_a is the duration of activity a and z_{ar} is the rate of consumption of resource r by activity a). Thus, an early activity may be compressed in order to free resources later on and thereby achieve earlier project completion. The parameters of the problem are: (a) the resource availabilities, (b) the upper and lower bounds on the activity durations, (c) the cost-duration functions, (d) the total resource requirements for each activity, and (e) the penalty for project delay beyond a target due date. You are asked to formulate a model of this scenario.
15. Consider the project of Problem 1 and suppose that the objective of analysis is to determine the minimum project duration (rather than to smooth the resources utilization) subject to fixed availabilities:
    a. Resource 1: 30 units
    b. Resource 2: 25 units

    Determine the minimum duration utilizing a B&B approach.
16. Explain why the B&B approach is differentiated from other procedures that use heuristic rules by referring to it as a procedure of "reliable heuristic." Illustrate how one can obtain a bound on the error if one halts the B&B calculations before the optimum has been secured (and verified).

17. In Section 1 of Chapter 1 we introduced the concepts of "overlapping" activities, that is, activities that can be partially performed simultaneously. Consider a project in which a lag restriction, if it exists, is of the form of a SS-lag type.

    Under such conditions the precedence relations bear a very special interpretation. In particular, an arrow from activity i to activity j implies that i starts before j and that the start of j is governed by the "lag" restrictions. The absence of such restrictions signifies the ordinary interpretation of the arrow, namely, that i must be completed before j is started.

    Formulate an ILP model that minimizes the completion time of the project in the presence of resource constraints. Then apply your model to the following project (also see Fig. 3.12):

A-on-N

Activities	SS lag
$1 \to 3$	2
$4 \to 5$	3

Activity i	t_i	Resource consumption 1	2	3
1	2	2	2	1
2	4	0	2	1
3	4	3	3	3
4	6	2	1	3
5	4	1	1	0
Resource Availability l_r		5	5	3

18. Consider the following problem of scheduling the activities of a project under the constraint of one scarce resource. You are given the project network; that is, the structure of the AN is given. Furthermore, the duration of each activity is known, say y_i for activity i, as well as the availability of the resource, say m_t in period t. Assume for simplicity that each activity requires 1 unit of the resource throughout its duration. (The resource may be a special type of machine or facility, of which there are several units, and an activity occupies the machine or facility for its duration.) It is desired to schedule the activities so as to minimize the completion time of the project. Construct either an ILP or a DP model of the problem and comment on its computational requirements. Would your model or conclusions change at all if there were only one unit of the resource available at all times (i.e., if $m_t = 1$ for all periods t)? State your answer as explicitly as possible.
19. Consider the ILP of Eqs. (3.4)–(3.6) in which the desired objective is to minimize the project duration z subject to the stated resource availabil-

ities. Suppose that it is desired to solve this ILP by B&B. You are asked to devise two possible heuristic rules for partitioning the space of feasible solutions. Comment briefly on their respective characteristics. [*Hint*: Since each x_{ij} has value either 0 or 1, this property may constitute a base for the first rule. On the other hand, in any period j a subset of activities may be eligible for being started. This may constitute a base for the second rule.]

REFERENCES

Arthanari and Ramamurthy, 1970.
Balas, 1970.
Bennett, 1966.
———, 1968.
Bennington and McGinnis, 1973.
Berman, 1964.
Black, 1965.
Blair, 1963.
Burgess and Killebrew, 1962.
Butcher, 1967.
Calica, 1965.
Carruthers, 1968.
Chapman, 1970.
Clark, 1961.
Cooper, 1976.
Davis, 1966.
———, 1968.
———, 1973.
———, 1974.
——— and Heidorn, 1971.
——— and Patterson, 1975.
DeWitte, 1964.
Dike, 1964.
Dogrusoz, 1961.
Elmaghraby, 1968.
———, 1969.
——— and Elshafei, 1976.
Emmons, 1968.
Fendley, 1968.
Fisher, 1973.
———, 1973.
Freeman, 1960a.
———, 1960b.
Freeman and Tucker, 1967.
Galbreath, 1956.
Ghare, 1965.
———, 1965.
Gonguet, 1969.
Gorenstein, 1972.
Gutjahr, 1963.
Gutjahr and Nemhauser, 1964.
Hastings, 1972.
Hayes, Koman, and Byrd, Jr., 1973.
Hegelson and Birnie, 1961.
Hooper, 1965.
Hsu, 1969.
Johnson, 1967.
Kapper, 1966.
Kapur, 1973.
Karush, 1964.
———, 1967.
Kelley, 1963.
Kilbridge and Wester, 1962.
Klein, 1967.
Lambourn, 1963.
Lave, 1968.
Lenstra, 1976.
Levy, Thompson, and Wiest, 1962.
Litsois, 1965.
Lofts, 1974.
McCoy, 1972.
McGee and Markarian, 1962.
McGinnis, 1973.
Mansoor, 1964.
Martino, 1965.
Mason and Moodie, 1971.
Meyer and Shaffer, 1963.
Mitten and Warburton, 1973.
Mize, 1964.
Moder, 1975.
Moodie and Mandeville, 1966.
Moshman, Johnson, and Larsen, 1963.
Moore, 1968.
Nepomiastchy, 1974.

Norden, 1963.
Parikh and Jewell, 1965.
Pascoe, 1965.
———, 1966.
Patterson, 1973.
———and Huber, 1974.
———and Werne, 1972.
Patton, 1968.
Petrovic, 1968.
Polya, 1957.
Pritsker, Watters, and Wolfe, 1969.
Richards, 1973.
Rosenbloom, 1964.
Rothkopf, 1966.
Roy, 1964a.
Roy, 1964b.
Schrage, 1970.
———, 1972.
Shwimer, 1972.
Simon and Newell, 1958.
——— and ———, 1964.
Sunaga, 1970.
Trilling, 1966.
Verhines, 1963.
Voronov and Petrushinin, 1966.
Wiest, 1964.
———, 1966.
———, 1967.
Wilson, 1964.
Woodgate, 1968.
Woodworth and Dane, 1974.
Zaloom, 1971.

4

PROBABILISTIC ACTIVITY NETWORKS (PANs): A CRITICAL EVALUATION AND EXTENSION OF THE PERT MODEL

The distinctive feature of the original PERT model was claimed to be the capability to take uncertainty (in the estimated durations of the activities) into account and still come up with meaningful results. In particular, through the assumption of Beta PDFs for the duration of an activity, and by making certain approximations (see Section 2.2 of Chapter 1) the model claimed some important features such as:

1. *Flexibility*: The beta DF is really a four-parameter DF that can "fit" any desired shape and location;
2. *Simplicity*: Although the beta DF itself is a complicated expression, its application is simplified through the use of simple approximations. From that point on, calculations involve simple additions.
3. *Potency*: Not only do we possess all the information that the deterministic model of CPM gave, but also we can deduce confidence limits on the duration of the project (or any part thereof) by simple recourse to standard Normal tables.

As was indicated in Section 2.2 of Chapter 1, the assumptions and most of the conclusions based on them are open to serious questions that cast a grave shadow on their validity.

In this chapter we consider in great detail these various assumptions and their subtleties as they reflect in one aspect after another on the PERT model. As the reader will discover, extensive and rather fundamental changes will have to be introduced in order to place the model on a more solid foundation. The final result is a model that is radically different from that originally proposed, hence our preference to refer to such a model, and extensions thereof, as *probabilistic activity network* (PAN).

The discussion in the present chapter proceeds as follows. In Section 1 we give some detailed probabilistic considerations, in Section 2 the determination of the expected value of project duration, in Section 3 its variance, and in Section 4, the problem of determining the complete DF of project duration. Our discussion is mostly analytical in nature; for a nonmathematical overall view of PERT as a management tool, see Pocock (1962).

§1. SOME PROBABILISTIC CONSIDERATIONS

The reader will recall that the PERT model, as presented in Section 2.2 of Chapter 1, made the following assumptions:

1. The activities are independent.
2. The CP contains a "sufficiently large" number of activities so that we can invoke the central limit theorem.
3. The PDF of the duration Y_u of an activity can be approximated by the beta distribution.
4. The mean of the beta distribution can be approximated by

$$\bar{y}_u = \frac{a_u + 4m_u + b_u}{6}$$

and the variance by

$$\sigma_u^2 = \frac{(b_u - a_u)^2}{36}$$

(for the definition of the terms a, m, and b, see Section 2.2 of Chapter 1).

It must be evident from that treatment of the PERT model that assumption (3) concerning the *form of the distribution* of Y_u is not fundamental to the subsequent analysis, since only the first two moments of the DF were ever utilized. In the absence of any empirical evidence on the most appropriate form of the DF of Y_u, there seems to be no compelling reason to adopt the one proposed by the originators of PERT!

The assumption of a beta DF may be "logical" and highly convenient, especially under the stipulation of the three time estimates a, m, and b and the approximations based on them. (We discuss this further shortly.) But it seems that in some instances it would be equally "logical" to assume other forms of DF. For example, if a and b represent the range of the possible durations of an activity, with all intermediate values y, $a \leqslant y \leqslant b$, equally probable, a uniform distribution over the closed interval $[a,b]$ would be adequate, as in Fig. 4.1a. On the other hand, if one is wedded to the idea of three time estimates: a, m, and b, with m being the mode, maybe the simple triangular distribution of Fig. 4.1b is adequate, and its mean and variance are certainly easily obtained from the three given parameters.

Consider next the question of probability statements attached to the realization time of an event—say the last node n. In Section 2.2 of Chapter 1 we also learned that given assumptions (1) and (2) above one can state

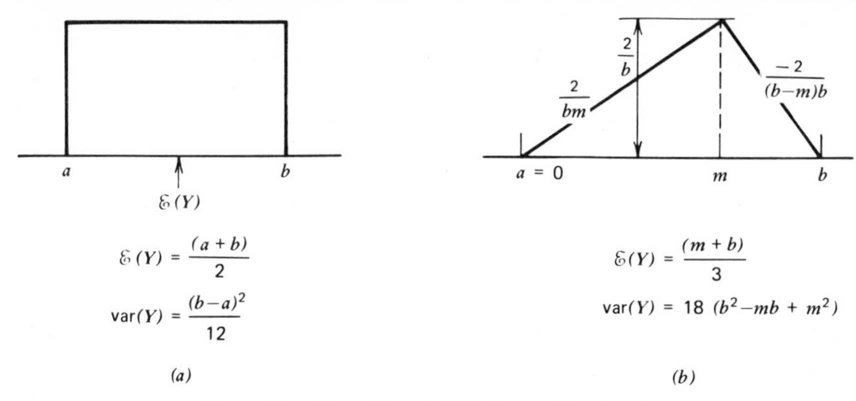

Figure 4.1. Two alternate DFs for activity duration.

that

$$Pr\{\mathrm{T}_n \leqslant t_n(s)\} = \Phi\left\{\frac{t_n(s) - \overline{\mathrm{T}}_n}{\sigma_n}\right\} \tag{4.1}$$

where $\overline{\mathrm{T}}_n = \Sigma_{a \in \pi_C} \bar{y}_a$, π_C is the CP; and $\sigma_n^2 = \Sigma_{a \in \pi_C} \sigma_a^2$, and Φ is the Normal PDF. There are two comments pertinent to such probability statements.

First, consider the case where several paths $\pi_1, \pi_2, \ldots, \pi_r$ lead from the origin, node *1*, to node *i* as shown in Fig. 4.2. Let $T(\pi_j)$ denote the duration of path π_j; then the earliest realization time of node *i* is clearly given by

$$\mathrm{T}_n = \max_{\pi_j} \{T(\pi_j)\} \tag{4.2}$$

If the different paths are assumed independent—which is *not* true if two paths share one or more arcs—then T_n is the maximum of a finite set of

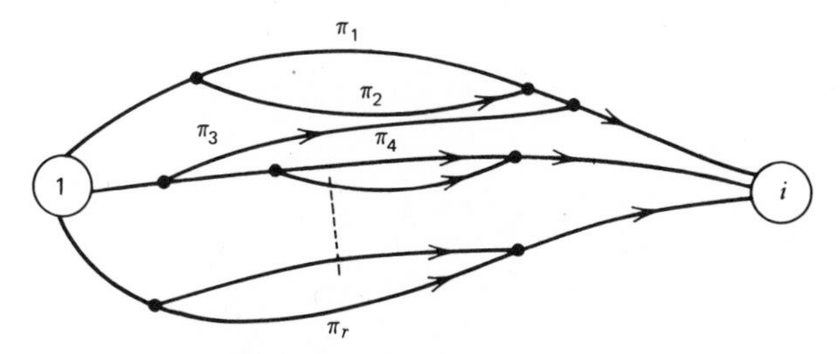

Figure 4.2. Paths in parallel.

random variables. The following remarks refer to the case where $T(\pi_j)$ is assumed Normally distributed with mean $\bar{T}(\pi_j)$ and variance $\sigma^2(\pi_j)$. If $T(\pi_j)$ is *not* Normally distributed, the following remarks hold a fortiori.

It is well known that the maximum of two or more Normally distributed variables is *not* Normally distributed. However, as pointed out by Clark (1961), it can be *approximated* by a Normally distributed variable; but the expected value and variance of this approximation are *not* those given by Eqs. (1.9) of Chapter 1.

In order to appreciate the extent of the error committed in those equations, Table 4.1 is taken from Tippett (1925), whose more comprehensive table gives the mean and standard deviation of the maximum of N standardized Normal deviates. That is, if

$$Z = \max\{X_1, X_2, \ldots, X_N\}$$

and the X values are Normally distributed with mean 0 and variance 1, then the mean and standard deviation of Z for various values of N are as given in Table 4.1.

Interpreting Table 4.1 in the context of ANs, we see that if four paths π_1, π_2, π_3, and π_4 lead from *1* to i in Fig. 4.2, and if the duration of each path is assumed to be Normally distributed with the same mean $\bar{T}$ and the same variance σ^2, then

$$\mathcal{E}(\mathrm{T}_i) = 1.029\sigma + \bar{T} \tag{4.3}$$

and

$$\sigma_i = .7012\sigma \tag{4.4}$$

Had we applied Eqs. (1.9), we would have obtained

$$\mathcal{E}(\mathrm{T}_i) = \bar{T} \text{ and } \sigma_i = \sigma \tag{4.5}$$

It is easy to see that the error committed is not insignificant. For instance, let $\bar{T} = 10$ and $\sigma = 1$; then Eqs. (4.3) and (4.4) yield

$$\mathcal{E}(\mathrm{T}_i) = 11.029;\ \sigma_i = 0.7012$$

Table 4.1. Mean and Standard Deviation of Z

N	1	2	3	4	5	6	7	8	9	10
$\mathcal{E}(Z)$	0	0.5642	0.8463	1.0296	1.1630	1.2672	1.3522	1.4236	1.4850	1.5388
σ_Z	1	0.8256	0.7480	0.7012	0.6690	0.6650	0.6260	0.6107	0.5978	0.5868

Hence, the probability of reaching node i in, say, 11.5 units of time or less is given by

$$Pr[\mathrm{T}_i \leqslant 11.5] \cong \Phi\left\{\frac{11.5 - 11.029}{0.7012}\right\} = \Phi\{0.6717\} = 0.748$$

This is to be compared with

$$\Phi\left\{\frac{11.5 - 10}{1}\right\} = \Phi\{1.5\} = 0.933$$

which would have been obtained using the standard PERT formulas!

If the random variables are correlated, as is the case in almost all ANs,[†] the treatment is more involved. Clark (1961) has also given a stepwise procedure that yields the desired estimates of $\mathcal{E}(\mathrm{T}_i)$ and σ_i taking correlations into account.

To recapitulate the discussion thus far, the duration of the project is the time of realization of the last node, n, given by

$$\mathrm{T}_n = \max(T(\pi_1), T(\pi_2), \ldots, T(\pi_r)); r = \text{number of paths to } n$$

The paths (π_k) are *not* independent because they usually share activities. Even if they are considered "approximately" independent, their duration need not be Normally distributed. Even if the duration of each path is Normally distributed, T_n is *not* normally distributed. In fact, under the assumption of independence, the PDF of T_n is the product of the individual PDFs;

$$\begin{aligned} Pr[\mathrm{T}_n \leqslant \tau] &= Pr[\max\{T(\pi_1), T(\pi_2), \ldots, T(\pi_r)\} \leqslant \tau] \\ &= Pr[T(\pi_1) \leqslant \tau; T(\pi_2) \leqslant \tau; \ldots, T(\pi_r) \leqslant \tau] \\ &= \prod_{k=1}^{r} Pr[T(\pi_k) \leqslant \tau], \text{ by independence} \end{aligned}$$

Finally, even if we are willing to approximate the PDF of T_n by a Normal distribution, it would be a Normal DF with a different mean and different variance from that suggested by PERT!

Consequently, we feel justified in losing faith in the probability statements made according to the PERT procedure, and we trust that the above criticism presents a fair enough warning to the reader.

[†]Activities interact because of competition for scarce resources, especially when the project is running late.

To move on, it was pointed out by Healy (1961) that the probability statement made concerning $t_n(E)$ is a *function of the subdivision of the activities.*

Engineers and industrial managers who deal with real-life problems realize that activities and events can, sometimes, be defined almost at will. That is, the degree of detail in defining the activities and events is not a consequence of unalterable inherent characteristics of the project but rather a result of the manager's tastes and preferences.

It is easy to show that if an activity that was originally defined in a gross manner is subdivided into several activities in series so that the following two conditions are satisfied:

1. The expected duration of the series is equal to the expected duration of the "parent" activity.
2. The range of the small activities add up to the range of the parent activity.

Then the variance of the terminal event is necessarily decreased. Consequently, for the same specified completion time, $t_n(s)$, as before the subdivision, the probability statements would be different. In Problem 10 we ask the reader to verify the statements with respect to a single parent activity that was divided into three smaller activities in series.

The criticism seems to some analysts to be "unfair," and for two good reasons. First, the conclusion is predetermined by the assumptions made; indeed, if one maintains the mean and range, the variance will necessarily be different, and hence the probability statements made thereafter. But why should one fall into that trap in the first place? Second, the subdivision of activities is not done in vacuum. Presumably the manager who defined the gross activity in the first place is also the one who is responsible for its subsequent subdivision (which may turn out to be a small subnetwork), and his time estimates will vary in such a way as to reflect his best judgment on the *total* duration of the activity.

Nevertheless, the criticism serves two useful objectives: it voices a welcome warning on the haphazard definition of activities, and raises two more fundamental questions, namely, *how one can best use the subjective estimates made by managers, and what the best way is to draw out the information from their minds* (e.g., whether the three time estimates of PERT provide the best method of obtaining such information). We return to both these questions in Section 4.2.

In Appendix D we present a review of the salient properties of the beta distribution function. The reader is urged to familiarize himself with that material before proceeding any further, since we wish to study the impact of the PERT approximations [relative to the mean and variance; see assumption (4) above] on the idea of "flexibility" of the beta DF.

The general beta DF is given by

$$f(y)=\frac{1}{(b-a)^{k_1+k_2-1}\beta(k_1,k_2)}(y-a)^{k_1-1}(b-y)^{k_2-1}; \quad a\leqslant y\leqslant b$$

where $\beta(k_1,k_2)=\Gamma(k_1+k_2)/\Gamma(k_1)\Gamma(k_2)$. The standardization of this DF involves changing its domain from $[a,b]$ to the unit interval $[0,1]$. To this end, let $y=a+(b-a)x$; whence $dy=(b-a)dx$. This change of variable yields the standardized beta DF:

$$f(x)=\frac{1}{\beta(k_1,k_2)}x^{k_1-1}(1-x)^{k_2-1}$$

In Appendix D we derive the mean of the standardized beta DF:

$$\mathcal{E}(X)=\frac{k_1}{k_1+k_2};$$

its variance

$$\operatorname{var}(X)=\frac{k_1k_2}{(k_1+k_2)^2(k_1+k_2+1)}$$

and its coefficient of skewness

$$\alpha_3=\frac{\mu_3}{\sigma^3}=\frac{2(k_2-k_1)}{k_1+k_2+2}\left(\frac{k_1+k_2+1}{k_1k_2}\right)^{1/2}$$

Now, using the fact that $y=a+(b-a)x$, we have

$$\mathcal{E}(Y)=a+(b-a)\mathcal{E}(X)$$

$$=a+(b-a)\frac{k_1}{k_1+k_2}$$

$$\operatorname{var}(Y)=(b-a)^2\operatorname{var}(X)$$

$$=(b-a)^2\frac{k_1k_2}{(k_1+k_2)^2(k_1+k_2+1)}$$

Also, the mode m is easily obtained from

$$\frac{df(y)}{dy}=0=\frac{1}{(b-a)^{k_1+k_2-1}\beta(k_1,k_2)}$$

$$\times\left[-(k_2-1)(y-a)+(k_1-1)(b-y)\right]\times(y-a)^{k_1-2}(b-y)^{k_2-2}$$

which yields

$$m=\frac{a(k_2-1)+b(k_1-1)}{k_1+k_2-2}$$

Notice that all three statistics, $\mathcal{E}(Y)$ and var(Y), as well as the mode m, are functions of the four parameters a, b, k_1, and k_2. The parameters a and b fix the location and range, while k_1 and k_2 determine the shape of the curve.

The question of immediate concern is what values of k_1 and k_2 yield the relationships used in the standard PERT approximations:

$\bar{y}\cong(a+4m+b)/6$; connects $\mathcal{E}(Y)$ with m, a, and b

and $\quad\sigma_y^2\cong(b-a)^2/36$; connects var($Y$) with a and b

Substituting in their respective expressions we obtain

$$\frac{a+4m+b}{6}=a+(b-a)\frac{k_1}{k_1+k_2}$$

$$\frac{(b-a)^2}{36}=(b-a)^2\frac{k_1k_2}{(k_1+k_2)^2(k_1+k_2+1)}$$

Substituting for m its value from above, we obtain two nonlinear equations in two unknowns: k_1 and k_2. It is easy to verify that the following three sets of values satisfy these two nonlinear equations:

$$k_1=3+\sqrt{2}\cong4.4142,\quad k_2=3-\sqrt{2}\cong1.5858\Rightarrow\alpha_3=-1/\sqrt{2}$$

or $\quad k_1=3-\sqrt{2}\cong1.5858,\quad k_2=3+\sqrt{2}\cong4.4142\Rightarrow\alpha_3=+1/\sqrt{2}$

or $\quad k_1=4=k_2\Rightarrow\alpha_3=0$

where the implication on α_3 is obtained by direct substitution in its expression.

We are then led to the conclusion that *the simplifying assumptions of PERT in fact restrict the shape of the probability distribution of activity duration to only one of three*, namely those of skewness $\pm 1/\sqrt{2}$ or 0. The contention of total flexibility is illusory.

We pursue the previous comment one step further. In practice we do not have a, m, and b, but rather *subjective estimates* of these three parameters; say $\hat{a}$, $\hat{m}$, and $\hat{b}$. These are *sample values* that are subject to error of *unknown magnitude*, because the sampling is done only once.

If we assume that these estimates are unbiased, which they may not be, they still may have any variability about their true values a, m, and b. Clearly, there is no reason to assume any connection between these variances and the variance of the duration $\sigma_Y^2 = \text{var}(Y)$; the estimated parameters are points on the domain of the RV Y. Still, if we are willing to assume (while recognizing that there is no rationale for such assumption) that the three variances are equal, and all three are the same multiple value of var(Y); in other words, if we assume

$$\sigma_a^2 = \sigma_m^2 = \sigma_b^2 = h^2 \sigma_Y^2 \tag{4.6}$$

then from the approximation $\hat{\bar{y}} = (\hat{a} + 4\hat{m} - \hat{b})/6$, we have that

$$\mathcal{E}(\bar{Y}) = \frac{a + 4m + b}{6},$$

which is fine under the *restricted* beta DF; but

$$\text{var}(\bar{Y}) = \frac{\sigma_a^2 + 16\sigma_m^2 + \sigma_b^2}{36} = \frac{h^2 \sigma_Y^2}{2}; \text{ by Eq. (4.6)}$$

which is *not* fine because the probability statements of PERT are based on $\hat{\bar{y}}$, which has $h^2/2$ the variance of Y. In other words, the probability statements based on $\hat{\bar{y}}$ will give different results from those based on the true, but unknown, DF of Y. It is easy to see that h has to be $= \sqrt{2}$ for the results of PERT analysis to be valid. Alternatively, we may assume that:

$$\hat{a} \text{ estimates } a \text{ with variance} = 10\sigma_y^2$$

$$\hat{b} \text{ estimates } b \text{ with variance} = 10\sigma_y^2$$

$$\text{but} \quad \hat{m} \text{ estimates } m \text{ with variance} = \sigma_y^2$$

in which case the estimate $\bar{y}$ has variance

$$\text{var}(\bar{Y}) = \tfrac{1}{36}\left(10\sigma_y^2 + 16\sigma_y^2 + 10\sigma_y^2\right) = \sigma_y^2$$

as desired. The assumptions on the variances of $\hat{a}$, $\hat{m}$, and $\hat{b}$ are hard to believe!

To sum up, this section has raised serious questions concerning the validity of the three contentions of PERT. We questioned the choice of the beta DF in the first place, as well as the "simplifications" offered to reduce the complexity of the calculations based on it. The PERT probability statements are open to criticism on more than one count, and we gave examples on the seriousness of the errors committed. We now move to a more detailed discussion of three aspects of the PERT model: the mean, the variance, and the PDF of project duration.

§2. EVALUATION OF THE EXPECTED DURATION OF A PROJECT

We assume the AN to be defined by the pair (N, A), where N is the set of nodes and A the set of arcs. In this section we adopt the A-on-A mode of representation. Let $\mathcal{B}_j$ denote the set of arcs (i.e., activities) that join node j to the nodes immediately preceding it (sometimes also referred to as the "bunch" of activities immediately preceding node j), and let $\mathcal{P}_j$ denote the set of arcs preceding j. Notice that $\mathcal{P}_j$ really defines the subnetwork terminating at node j. If we denote the number of elements in a set by the two vertical lines $|\quad|$, then, clearly, $\mathcal{B}_j \subseteq \mathcal{P}_j$ and $|\mathcal{B}_j| \leqslant |\mathcal{P}_j|$. Furthermore, $\mathcal{P}_n$ is the set of all activities of the project; that is, $\mathcal{P}_n \equiv A$.

We assume that for any activity $u \in \mathcal{B}_j$, Y_u may be jointly distributed with other activity durations in $\mathcal{B}_j$, but the set $\mathcal{B}_j$ is independent of all other sets $\mathcal{B}_i$, $i \neq j$. For example, consider node *5* of Fig. 4.3; it is possible that the duration of arcs $(2,5)$ and $(3,5)$ are jointly distributed; but we assume that each arc in $\mathcal{B}_5$ is distributed independently from any arc in any other bunch $\mathcal{B}_i$, $i \neq 5$.

Finally, we have:

Definition: By a realization of a network we mean a result of a random experiment that assigns a duration y_u to each arc $u \in A$, where y_u is chosen according to the PDF of the RV Y_u.

Let $l_j(\mathcal{P}_j)$ denote the CP to node j given any realization of the subnetwork $\mathcal{P}_j$; $j = 2, 3, \ldots, n$. We assume, for the sake of simplicity of this discussion, that Y_u is discretely distributed for all $u \in A$. If Y_u is a

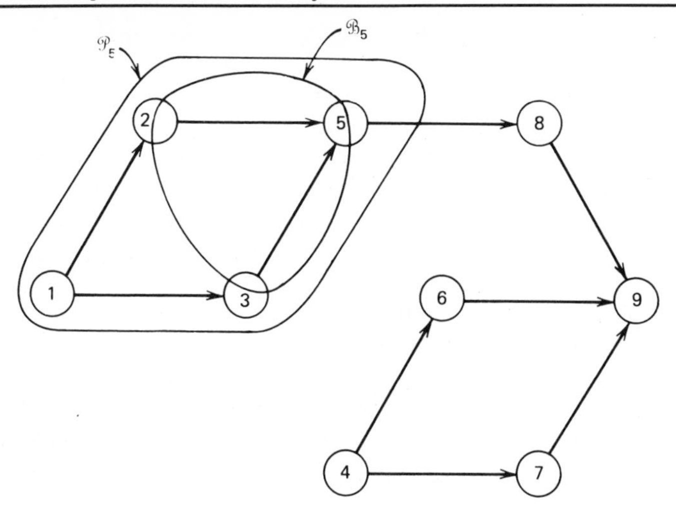

Figure 4.3. The sets $\mathcal{B}$ and $\mathcal{P}$.

continuous RV, obvious changes are made in the following expressions, such as replacing Σ by $\int$, $p(\cdot)$ by $f(\cdot)$, and so on. Let $p[y(\mathcal{B}_j)]$ denote the probability of the realization of the arcs in the bunch $\mathcal{B}_j$, and let $p[y(\mathcal{P}_j)]$ be similarly defined. Then, clearly, the duration of the project is a random variable, previously designated by T_n, whose expected value (which is equal to the average length of the CP) is given by

$$\mathcal{E}(T_n) \triangleq e_n = \sum_{y(\mathcal{P}_n)} l_n(\mathcal{P}_n) p[y(\mathcal{P}_n)], \quad \mathcal{P}_n \equiv A \tag{4.7}$$

The PERT model estimates the expected duration of the project by defining the function g_j recursively as follows:

$$g_1 \equiv 0$$

$$g_j = \max_{i_k \in \mathcal{B}(j)} \{ g_{i_k} + \mathcal{E}(Y_{i_k}) \}; \quad j = 2, 3, \ldots, n \tag{4.8}$$

where

$$\mathcal{E}(Y_{i_k}) \triangleq \text{expected duration of arc } (i_k, j) \text{ for } i_k \in \mathcal{B}_j = \sum_{y_{i_k}} y_{i_k} p(y_{i_k})$$

and $p(y_{i_k})$ is the marginal PF of arc (i_k,j). We assert that

$$g_j \leqslant e_j \text{ for } j = 2,3,\ldots,n \tag{4.9}$$

or, in other words, the maximum based on substituting the expected values for the random variables is no larger than the expected value of the maximum durations. This asserts that the PERT estimate of the expected duration of the project is *biased* on the low side; that is, it is optimistic. (When does *equality* hold between g_j and e_j?)

We do not prove this assertion right now, since it follows as a corollary of a more general inequality to be proven below.†

The immediate question is: realizing that the PERT estimate is biased, can it be improved? The answer is, fortunately, yes. The following improvements are due to Fulkerson (1962) and Elmaghraby (1967), respectively. First, we present Fulkerson's estimate.

§2.1 The First Approach

Define a function f_j recursively as follows:

$$f_1 \equiv 0$$

$$f_j = \sum_{y(\mathcal{B}_j)} \max\{f_{i_1}+y_1;\ f_{i_2}+y_2;\ldots;f_{i_r}+y_{r_j}\} \times p[y(\mathcal{B}_j)];\quad j=2,3,\ldots,n; \tag{4.10}$$

where $y(\mathcal{B}_j)$ is the vector of realization of all arcs in the set $\mathcal{B}_j$ (assumed to contain r_j arcs) and we wrote y_k as a shorthand for y_{i_k}.

In words, f_j is the expected value (over all realizations of $\mathcal{B}_j$) of the maximum length to node j. Note that g_j of PERT, given by Eq. (4.8), is the maximum of the expected values. It is intuitively clear that averaging over the maxima is never less than the maximum of the averages.

Example 4.1. Consider the network of Fig. 4.4, in which each arc bears numbers indicating its possible durations, with all possibilities being equally probable. For instance, the duration of arc $(1,2)$ is a RV, $Y_{(1,2)}$, which can assume three values: 1, 2, and 5 time units. The probability of

†One informative way of proving this inequality (and providing one more link with LP) is to write the problem as a LP with a random right-hand vector, the components of which are simply the activity durations. Then the general results of Madansky (1960) for stochastic LPs can be applied. In particular, for any LP problem with a random r.s. the answer obtained by using expected values is always too optimistic.

$Y_{(1,2)}$ assuming any of these values is the same and is equal to 1/3, and so on for other arcs. We first remark that this (simple) network has 324 different realizations, all equally probable. Just for the sake of comparison, all 324 realizations were enumerated, the CP in each realization evaluated, and then the average evaluated. This turned out to be $e_4 = 12.3148$.† The PERT estimate proceeds as follows: $\bar{y}_{1,2} = 8/3$; $\bar{y}_{1,3} = 5$; $\bar{y}_{1,4} = 11/3$; $\bar{y}_{2,3} = 4$; $\bar{y}_{2,4} = 19/3$, and $\bar{y}_{3,4} = 10/3$; and

$$g_1 \equiv 0$$

$$g_2 = \bar{y}_{1,2} = \frac{8}{3}$$

$$g_3 = \max\left\{0+5;\ \frac{8}{3}+4\right\} = \frac{20}{3}$$

$$g_4 = \max\left\{0+\frac{11}{3};\ \frac{8}{3}+\frac{19}{3};\ \frac{20}{3}+\frac{10}{3}\right\} = 10$$

To be sure, $g_4 = 10 < 12.315 = e_4$. We now evaluate the functions (f_j).

$$f_1 \equiv 0$$

$$f_2 = \bar{y}_{1,2} = \frac{8}{3}$$

$$f_3 = \frac{1}{4}\left(5\frac{2}{3} + 7\frac{2}{3} + 2\times 8\right) = \frac{22}{3}$$

$$f_4 = \frac{1}{27}\left(2\times 8\frac{1}{3} + 2\times 8\frac{2}{3} + 2\times 10 + 6\times 11\frac{1}{3} + 6\times 12\frac{1}{3} + 9\times 14\frac{2}{3}\right)$$

$$= \frac{328}{27} = 12.148$$

Indeed, $g_4 = 10 < f_4 = 12.148 < e_4 = 12.315$. Formally, we have

Theorem 4.1: $g_j \leqslant f_j \leqslant e_j \forall j = 1,2,3,\ldots,n$. That is to say, while f_n is still a biased estimate of e_n, it is nevertheless an improvement on g_n.

Proof. (by induction) 1. To show that $g_j \leqslant f_j$. The statement is true for $j = 1,2$ (recall that, as always, the nodes are numbered such that an arc leads from a small numbered node to a higher numbered node.) Now to

†See Tables 4.7 and 4.8 for more detailed tabulations of the relative frequencies of path and arc occurrences on the CP.

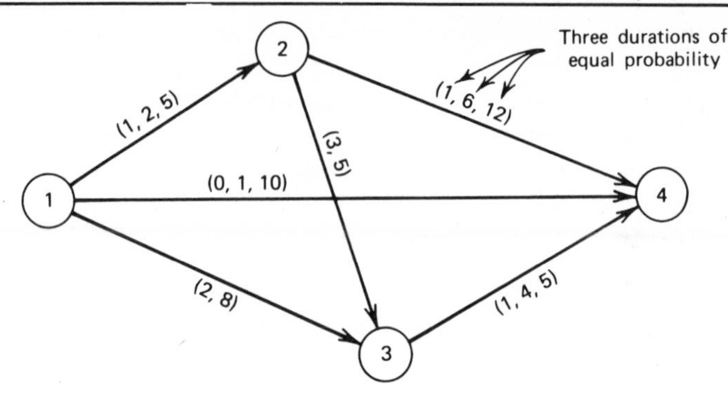

Figure 4.4. Example network with random durations.

prove that it is true for all nodes, assume that it is true for all nodes $j \leqslant j_0$ (the induction hypothesis); that is, assume that for $j = 1, 2, \ldots, j_0$ it is true that $g_j \leqslant f_j$. Now we must show that it is also true for $j_0 + 1$. By definition,

$$f_{j_0+1} = \sum_{y(\mathfrak{B}_{j_0+1})} \max_{1 \leqslant k \leqslant r_{j_0+1}} \{(f_k + y_k)\} p[y(\mathfrak{B}_{j_0+1})]$$

Interchanging the summation and maximization operations, we obtain the inequality

$$f_{j_0+1} \geqslant \max_{1 \leqslant k \leqslant r_{j_0+1}} \left\{ \left(\sum_{y(\mathfrak{B}_{j_0+1})} (f_k + y_k) p[y(\mathfrak{B}_{j_0+1})] \right) \right\}$$

$$= \max_{1 \leqslant k \leqslant r_{j_0+1}} \{ f_k + \bar{y}_k \} \tag{4.11}$$

The inequality asserts that the expected value of the maximum of a finite number of RVs is larger than or equal to the maximum of the expected values of the RVs. This may be established from first principles by the repeated application of the following argument. (The inequality also follows trivially from Jensen's inequality [see Parzen (1960), p. 434]). Let $Z = \max(Y_1, Y_2)$, and let $F(y_1, y_2)$ be the joint PDF of the two RVs Y_1 and Y_2. Then $\mathcal{E}(Z) = \iint_{-\infty}^{\infty} \max(y_1, y_2) dF(y_1, y_2) dy_1 dy_2$. But $\max(y_1, y_2)$ is at least as large as either variable by itself, leading to the inequality:

$$\geqslant \max\left(\iint_{-\infty}^{\infty} y_1 dF(y_1, y_2) dy_1 dy_2; \iint_{-\infty}^{\infty} y_2 dF(y_1, y_2) dy_1 dy_2 \right)$$

$$= \max(\mathcal{E}(Y_1), \mathcal{E}(Y_2)).$$

Finally, the equality in (4.11) follows from the fact that f_{i_k} is a constant and that $\Sigma_{y(\mathcal{B}_{j_0+1})}\bar{y}_k p[y(\mathcal{B}_{j_0+1})] = \bar{y}_k$. Now, by the induction hypothesis, the r.s. of (4.11) is

$$\geqslant \max_{1 \leqslant k \leqslant r_{j_0+1}} (g_k + \bar{y}_k) = g_{j_0+1}$$

which completes the induction and proves that $g_j \leqslant f_j$ for all j. (2) To show that $f_j \leqslant e_j$ for all j. Again, the assertion is true for $j = 1,2$. Suppose it is true for j_0 (the induction hypothesis); we must show that it is true for $j_0 + 1$. By definition,

$$e_{j_0+1} \triangleq \sum_{y(\mathcal{P}_{j_0+1})} p\left[y(\mathcal{P}_{j_0+1})\right] l_{j_0+1}(\mathcal{P}_{j_0+1}) \tag{4.12}$$

By the assumption of independence of the bunch $\mathcal{B}_{j_0+1}$ from all other bunches, we conclude that $\mathcal{B}_{j_0+1}$ is independent of the union of all other bunches, or simply $\cup \mathcal{P}_j$, where the union is over all j, $1 \leqslant j \leqslant j_0$. Hence the probabilities multiply:

$$p\left[y(\mathcal{P}_{j_0+1})\right] = p\left[y(\mathcal{B}_{j_0+1})\right] \times p\left[y(\cup \mathcal{P}_j)\right]$$

Furthermore, the CP to node $j_0 + 1$ for a particular realization of the subnetwork $\mathcal{P}_{j_0+1}$ is given by

$$l_{j_0+1}(\mathcal{P}_{j_0+1}) = \max_{1 \leqslant k \leqslant r_{j_0+1}} \left\{ l_k(\mathcal{P}_{j_0+1}) + y_k \right\}$$

where l_k denotes the CP to node $i_k \in \mathcal{B}_{j_0+1}$. Substituting in Eq. (4.12),

$$\begin{aligned} e_{j_0+1} &= \sum_{y(\mathcal{B}_{j_0+1})} \sum_{y(\cup \mathcal{P}_j)} p\left[y(\mathcal{B}_{j_0+1})\right] p\left[y(\cup \mathcal{P}_j)\right] \times \max_{i \leqslant k \leqslant r_{j_0+1}} \left\{ l_k(\mathcal{P}_{j_0+1}) + y_k \right\} \\ &= \sum_{y(\mathcal{B}_{j_0+1})} p\left[y(\mathcal{B}_{j_0+1})\right] \sum_{y(\cup \mathcal{P}_j)} p\left[y(\cup \mathcal{P}_j)\right] \times \max_{1 \leqslant k \leqslant r_{j_0+1}} \left\{ l_k(\mathcal{P}_{j_0+1}) + y_k \right\} \end{aligned}$$

interchanging the max and summation operators yields

$$\geqslant \sum_{y(\mathcal{B}_{j_0+1})} p\left[y(\mathcal{B}_{j_0+1})\right] \times \max_{1 \leqslant k \leqslant r_{j_0+1}} \left\{ \sum_{y(\cup \mathcal{P}_j)} p\left[y(\cup \mathcal{P}_j)\right] \left(l_k(\mathcal{P}_{j_0+1}) + y_k \right) \right\}$$

Expanding the quantity between the braces we obtain two terms, the first is $\Sigma_{y(\cup \mathcal{P}_j)} p[y(\cup \mathcal{P}_j)] l_k(\mathcal{P}_{j_0+1}) = e_{i_k}$, by definition; and the second is $\Sigma_{y(\cup \mathcal{P}_j)}$

$p[y(\cup \mathcal{P}_j)]y_k = y_k$, by virtue of the independence of activity $(i_k, j_0+1) \in \mathcal{B}_{j_0+1}$ from the subnetwork $\cup_{j=1}^{j_0} \mathcal{P}_j$. Hence,

$$e_{j_0+1} \geqslant \sum_{y(\mathcal{B}_{j_0+1})} p\left[y\left(\mathcal{B}_{j_0+1}\right)\right] \times \max_{1 \leqslant k \leqslant r_{j_0+1}} \{e_k + y_k\}$$

$$\geqslant \sum_{y(\mathcal{B}_{j_0+1})} p\left[y\left(\mathcal{B}_{j_0+1}\right)\right] \times \max_{1 \leqslant k \leqslant r_{j_0+1}} \{f_k + y_k\}, \quad \text{by the induction hypothesis,}$$

$$\triangleq f_{j_0+1}$$

and the induction is complete. This completes the proof that $g_j \leqslant f_j \leqslant e_j \forall j$.

Note that it follows trivially from this theorem that $g_j \leqslant e_j$, as asserted before. Actually, this inequality is always true, whether the bunch $\mathcal{B}_{j_0+1}$ is independent of the subnetwork $\cup \mathcal{P}_j$ or not. This independence is necessary, however, for the double inequality of Theorem 4.1.

Thus f_j is an improvement on g_j. But two further improvements are possible. Both methods approach e_n from below, and the second method has the advantage of possible straightforward generalization to approximate e_n to any desired degree of accuracy. Both methods are rather elementary and, once stated, are easy to prove. They may be of assistance if better approximations of e_n are deemed important. As is almost always the case, the improved result is obtained at a cost, in the form of extra computing effort. However, as we see below, under the assumption of *independence* of arc durations, the total effort is still well below the effort required for the calculation of e_n.

§2.2 The Second Approach

The second approach is based on the following (rather obvious) observation: if all arrows in a directed acyclic network, such as in PERT, are reversed, the average duration of the project e_n remains unchanged. However, intermediate values of f_i, $1 < i \leqslant n$, do not necessarily remain the same. Consequently, if we substitute in the expression of f_i the maximum of these two values obtained from the "as given" and the "reversed" subnetwork $\mathcal{P}_i$, we can only improve f_i (although still approaching e_i from below).

We clarify this notion below. For the moment, we wish to discuss the above statement concerning the invariance of e_i under reversal of the arrows. A proof of this statement, if one is needed, goes roughly as follows. Consider any realization of $\mathcal{P}_i$, say $\mathcal{P}_i(k)$. For this realization, evaluate the length of the CP, say $l_i(k)$. Obviously, we would obtain the same length

$l_i(k)$ if all the arrows of $\mathcal{P}_i$ are reversed. But $e_i \triangleq \mathcal{E}\{T_i\} = \Sigma_k p \cdot [\mathcal{P}_i(k)] l_i(k)$,† which must be invariant to the direction of the arrows because both $p[\mathcal{P}_i(k)]$ and $l_i(k)$ are invariant.

We talk about *forward* movement when we evaluate the various functions in the order of things "as given," from node *1* to node n. On the other hand, *backward* movement from node i means that the arrows of the subnetwork $\mathcal{P}_i$ have been reversed and that we are evaluating functions in the reverse direction, from node i to node *1*, in which case we write $\overleftarrow{\mathcal{P}}_i$.

By virtue of the introduction of the reversed subnetwork $\overleftarrow{\mathcal{P}}_i$ we now need to define three variables u_i, μ_i and s_i as follows:

$$s_1 = u_1 = 0$$

$$u_i = \sum_{y(\mathcal{B}_i)} \max_{k \in \mathcal{B}_i} \{s_i + y_{ki}\}; \quad \text{forward} \tag{4.13}$$

μ_i: evaluated similar to u_i but in the reversed subnetwork $\overleftarrow{\mathcal{P}}_i$

$$s_i = \max\{u_i; \mu_i\}; \quad i = 2, 3, \ldots, n. \tag{4.14}$$

s_n is the new estimate of e_n, and we assert that

$$f_n \leqslant s_n \leqslant e_n. \tag{4.15}$$

In case of *independent* activities, the calculations of Eqs. (4.10) and (4.13) simplify to

$$f_1 = 0$$

$$f_i = \sum_{z=\alpha_i}^{\beta_i} z \left[\prod_k P_k(z - f_k) - \prod_k P_k(z^- - f_k) \right], \text{ backward,} \tag{4.16}$$

and

$$u_i = \sum_{z=\alpha_i}^{\beta_i} z \left[\prod_k P_k(z - s_k) - \prod_k P_k(z^- - s_k) \right], \text{ forward;} \quad i = 2, 3, \ldots, n \tag{4.17}$$

†T_i is a RV denoting the length of the CP to node i. For any realization of the subnetwork $\mathcal{P}_i$, T_i takes on the value l_i.

where,

$$\alpha_i = \begin{bmatrix} \max\limits_{k \in \mathcal{B}_i} \{f_k + a_{ki}\} & \text{for Eq. (4.16)} \\ \max\limits_{k \in \mathcal{B}_i} \{s_k + a_{ki}\} & \text{for Eq. (4.17)} \end{bmatrix}$$

$$\beta_i = \begin{bmatrix} \max\limits_{k \in \mathcal{B}_i} \{f_k + b_{ki}\} & \text{for Eq. (4.16)} \\ \max\limits_{k \in \mathcal{B}_i} \{s_k + b_{ki}\} & \text{for Eq. (4.17)} \end{bmatrix}$$

Similar expressions are written for μ_i in the reversed networks. $P_k(x)$ is the (cumulative) probability DF along arc k, $k \in \mathcal{B}_i$ and $z^- = z - \delta$ for small $\delta > 0$. In Eqs. (4.16) and (4.17) the variable of summation, z, actually represents the CP as given by $\max_k \{f_k + t_{ki}\}$ in the case of Eq. (4.16) or $\max_k \{s_k + t_{ki}\}$ in the case of Eq. (4.17); a_{ki} is the smallest duration of activity (k, i), and b_{ki} is the largest duration.

The proof that the estimate s_i is uniformly better than f_i follows trivially from Eq. (4.14). The proof that $s_i \leqslant e_i$ can be easily established by induction on the nodes and follows identical lines to the proofs given above.

Example 4.1 (continued). Consider, again, the network shown in Fig. 4.4. We evaluate s_4 as an estimate of e_4. Clearly,

$$s_1 = 0$$

$$u_2 = \frac{8}{3}, \; \mu_2 = \frac{8}{3}; \quad \text{whence}$$

$$s_2 = u_2 = \mu_2 = \frac{8}{3}$$

In order to evaluate s_3, we need the value of μ_3, which necessitates the reversal of the subnetwork $\mathcal{P}_3$. This is shown in Fig. 4.5. Also shown in Fig. 4.5 are the values μ_k, $k \in \overleftarrow{\mathcal{P}}_3$, for the reversed network. Consequently, we see that $\mu_3 = 15/2$. Since $u_3 = f_3 = 22/3$, we have that

$$s_3 = \max\left\{\frac{22}{3};\; \frac{15}{2}\right\} = \frac{15}{2}$$

Finally,

$$u_4 = \sum_{\alpha_4}^{\beta_4} z \left[\prod_1^3 P(z - s_k) - \prod_1^3 P(z^- - s_k) \right]$$

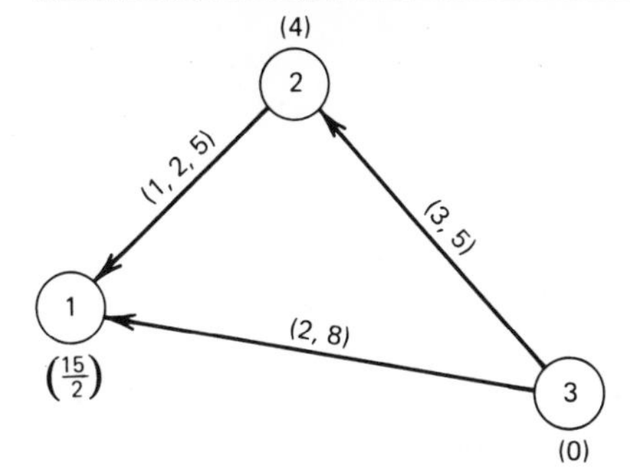

Figure 4.5. The subnetwork $\overleftarrow{\mathcal{P}}_3$.

where

$$\alpha_4 = \max\left\{0+0;\quad \frac{8}{3}+1;\quad \frac{15}{2}+1\right\} = \frac{17}{2}$$

$$\beta_4 = \max\left\{0+10;\quad \frac{8}{3}+12;\quad \frac{15}{2}+5\right\} = \frac{44}{3}.$$

Hence, $u_4 = (991/81) \cong 12.234$, while μ_4 is evaluated from $\overleftarrow{\mathcal{P}}_4$ (see Fig. 4.6)

$$\mu_4 = \frac{962}{81} \cong 11.876.$$

Consequently,

$$s_4 = \max[12.234;\quad 11.876] = 12.234$$

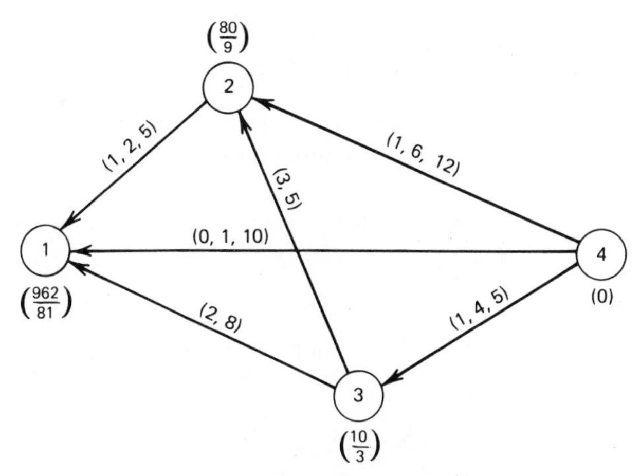

Figure 4.6. The subnetwork $\overleftarrow{\mathcal{P}}_4$.

These results illustrate the inequalities:

$$g_4 = 10 < f_4 = 12.148 < s_4 = 12.234 < e_4 = 12.315 \tag{4.18}$$

§2.3 The Third Approach

The third approach to the estimation of the expected duration of a project, e_n, involves a rather simple generalization of the first approach. In particular, instead of estimating e_i, $i \leqslant n$, with one "point estimate" f_i, we estimate e_i with two "point estimates," $f_i^{(1)}$ and $f_i^{(2)}$, or better still, $s_i^{(1)}$ and $s_i^{(2)}$, with the following two properties:

1. The probability of $s_i^{(r)}$ is $\frac{1}{2}$; $r=1$ or 2. This is always possible except when the space of T_i contains only two points whose probabilities are different from $\frac{1}{2}$.
2. The expected value of the two $s_i^{(r)}$ is equal to the one "point estimate" s_i, as defined by Eq. (4.14), $r=1,2$.

In order to distinguish between the result obtained from this procedure and the previous procedures, we denote the estimate of the expected duration to node i by w_i. Thus,

$$w_i = p_i^{(1)} s_i^{(1)} + p_i^{(2)} s_i^{(2)}; \quad i = 2, 3, \ldots, n$$

where

$$p_i^{(r)} > 0, \quad p_i^{(1)} + p_i^{(2)} = 1, \quad \text{and usually } p_i^{(r)} = \frac{1}{2}$$

We assert that

$$g \leqslant f \leqslant s \leqslant w \leqslant e$$

The proof of this assertion is simple and is left as an exercise to the reader (see Problem 12). We illustrate the implied procedure by application to the same network of Fig. 4.4.

On continuation of the numerical example, it is easy to see that

$$s_1^{(1)} = s_1^{(2)} = 0$$

$$s_2^{(1)} = \frac{4}{3}, \quad \text{with probability } \frac{1}{2}$$

$$s_2^{(2)} = 4, \quad \text{with probability } \frac{1}{2}$$

Consequently, $w_2 = (8/3)(= s_2 = f_2)$.

In order to evaluate w_3, we need $s_3^{(1)}$ and $s_3^{(2)}$, which are, in turn, dependent on $u_3^{(1)}$, $u_3^{(2)}$, $v_3^{(1)}$, and $v_3^{(2)}$, where $v_3^{(r)}$ is defined similar to $u_3^{(r)}$ but relative to subnetwork $\mathcal{P}_3$. Without cluttering this section with long calculations, these four values are given by:

$$u_3^{(1)} = \frac{77}{12}, \quad u_3^{(2)} = \frac{102}{12}, \quad \text{each with probability } \frac{1}{2},$$

$$v_3^{(1)} = \frac{76}{12}, \quad v_3^{(2)} = \frac{104}{12}, \quad \text{each with probability } \frac{1}{2}$$

Since

$$\frac{1}{2}\left(\frac{76}{12} + \frac{104}{12}\right) > \frac{1}{2}\left(\frac{77}{12} + \frac{102}{12}\right),$$

we have

$$s_3^{(1)} = \frac{76}{12} = \frac{19}{3}, \quad s_3^{(2)} = \frac{104}{12} = \frac{26}{3},$$

each with probability $\frac{1}{2}$, and,

$$w_3 = \frac{15}{2} (= s_3)$$

Finally, in order to evaluate w_4 we need $u_4^{(1)}$, $u_4^{(2)}$, $v_4^{(1)}$, and $v_4^{(2)}$, which are given by:

$$u_4^{(1)} = \frac{558}{54}, \quad u_4^{(2)} = \frac{769}{54}, \quad \text{each with probability } \tfrac{1}{2}$$

$$v_3^{(1)} = \frac{6863}{648}, \quad v_3^{(2)} = \frac{9096}{648}, \quad \text{each with probability } \tfrac{1}{2}$$

Since $(1/2)(6863/648 + 9096/648) > (1/2)(558/54 + 769/54)$, we have,

$$w_4 = (1/2)\left[v_4^{(1)} + v_4^{(2)}\right] = 12.310$$

We observe that, noting relation (4.18),

$$g_4 = 10 < f_4 = 12.148 < s_4 = 12.234 < w_4 = 12.310 < e_4 = 12.315$$

We state two final remarks regarding these various estimates of e_n. First, the choice of the estimator is necessarily dictated by the importance of accuracy in estimating the exact mean. Second, our third procedure can be generalized in an obvious manner to three or more point estimates instead of only the two values: $s_i^{(1)}$ and $s_i^{(2)}$. Evidently, as one increases the number of such point estimates, one eventually encompasses the whole DF of T_n.

§2.4 Generalization to Continuous and Independent Distribution Functions

For the special case of *statistically independent* arc durations, a straightforward generalization of the first approach to continuous probability distribution functions is easily obtained [Clingen (1964)]. In particular, the probability element $p[y(\mathcal{B}_j)]$ for the bunch $\mathcal{B}_j$ is given by

$$p[y(\mathcal{B}_j)] = p(y_{i_1})p(y_{i_2})\cdots p(y_{i_r})\partial y_{i_1}\partial y_{i_2}\cdots\partial y_{i_r}$$

Now define the numbers c_j recursively by:

$$c_1 \equiv 0$$

$$c_j = \int_{y_{i_1}}\cdots\int_{y_{i_r}} \max_{1\leqslant k\leqslant r_j}(c_{i_k}+y_{i_k})\prod_{k=1}^{r_j} p(y_{i_k})\partial y_1\cdots\partial y_r$$

$$= \int_{y_{i_1}}\cdots\int_{y_{i_r}} \max_{1\leqslant k\leqslant r_j}(c_{i_k}+y_{i_k})\prod_{k=1}^{r_j} dF_{i_k}(y_{i_k});\quad j=2,3,\ldots,n;$$

where $F_{i_k}(t)=Pr(Y_{i_k}\leqslant t)$; the PDF of the RV Y_{i_k}. Let

$$Z_j = \max_{1\leqslant k\leqslant r_j}(c_{i_k}+y_{i_k})$$

then

$$Pr(Z_j\leqslant t) = \prod_{k=1}^{r_j} F_{i_k}(t-c_{i_k})$$

Consequently,

$$c_{j+1} = \mathcal{E}(Z_{j+1}) = \int_{\alpha}^{\beta} zd\left[\prod_{k=1}^{r_{j+1}} F_{i_k}(z-c_{i_k})\right],$$

where z lies in the interval $[\alpha,\beta]$. Integrating by parts, put $du=dz$, and

$$dv = d\left[\prod_{k=1}^{r_{j+1}} F_{i_k}(z-c_{i_k})\right], \quad \text{to obtain}$$

$$c_{j+1} = z\prod_{k=1}^{r_{j+1}} F_{i_k}(z-c_{i_k})\Big|_{\alpha}^{\beta} - \int_{\alpha}^{\beta}\prod_{k=1}^{r_{j+1}} F_{i_k}(z-c_{i_k})\,dz$$

where $\alpha=\max_{1\leqslant k\leqslant r_{j+1}}(c_{i_k}+a_{i_k})$; a_{i_k} is the shortest duration of arc $(i_k,j+1)$, and $\beta=\max_{1\leqslant k\leqslant r_{j+1}}(c_{i_k}+b_{i_k})$; b_{i_k} is the longest duration of arc $(i_k,j+1)$. Now,

$$z\prod_{k=1}^{r_{j+1}} F_{i_k}(z-c_{i_k})\Big|_{\alpha}^{\beta} = \beta\prod_{k=1}^{r_{j+1}} F_{i_k}(\beta-c_{i_k}) - \alpha\prod_{k=1}^{r_{j+1}} F_{i_k}(\alpha-c_{i_k})$$

But $\beta=\max_{1\leqslant k\leqslant r_{j+1}}(c_{i_k}+b_{i_k})=c_{i_m}+b_{i_m}$, say; therefore, $\beta\geqslant c_{i_k}+b_{i_k}$ for all $1\leqslant k\leqslant r_{j+1}$, and hence $F_{i_k}(\beta-c_{i_k})=1$ for all k, from which we conclude that

$$\beta\prod_{k=1}^{r_{j+1}} F_{i_k}(\beta-c_{i_k})=\beta$$

On the other hand,

$$F_{i_k}(\alpha-c_{i_k})=0 \text{ for some } k\text{, which yields}$$

$$\alpha\prod_{k=1}^{r_{j+1}} F_{i_k}(\alpha-c_{i_k})=0$$

Substituting these results above, we obtain

$$c_{j+1}=\beta-\int_{\alpha}^{\beta}\prod_{k=1}^{r_{j+1}} F_{i_k}(z-c_{i_k})\,dz,$$

which can be evaluated either analytically (if $\prod F_{i_k}$ is a "decent" function of z) or numerically.

§3. ON ESTIMATION OF THE VARIANCE

The PERT model based its probability statements on *estimates* of the mean and the variance of the realization of the last node in the network (i.e., the time to project completion). The previous section gave a detailed discussion of the PERT estimate of the mean; in the present section we discuss its estimate of the variance.

Let $l_n(\mathcal{P}_n)$, or l_n for short, denote, as before, the length of the CP to the terminal node n given any realization of the network. Let T_n denote the duration of the project. T_n takes on the value $l_n(\mathcal{P}_n)$ at any realization. Clearly, T_n is a RV; denote its mean by $\overline{\mathrm{T}}_n$. Then the variance of the duration of the project is given by

$$\operatorname{var}(\mathrm{T}_n) = E\left[\mathrm{T}_n - \overline{\mathrm{T}}_n\right]^2$$

$$= \int_{y(\mathcal{P}_n)} l_n^2(\mathcal{P}_n) p\left[y(\mathcal{P}_n)\right] - \left\{\int_{y(\mathcal{P}_n)} l_n(\mathcal{P}_n) p\left[y(\mathcal{P}_n)\right]\right\}^2$$

The PERT estimate of this variance is based on the longest path (the CP) evaluated when all activity durations, $\{Y_u\}$, are replaced by their expected values, $\{\mathcal{E}(Y_u)\}$. If we denote such estimate by $\sigma_n^2(\mathrm{PERT})$, we have [see Eq. (1.9)]

$$\sigma_n^2(\mathrm{PERT}) = \sum_{u \in \pi_C} \sigma_u^2 \tag{4.19}$$

where σ_u^2 is the variance of the RV Y_u and π_C is the CP evaluated as specified above. Program evaluation and review technique stipulates that each Y_u is distributed as a beta distribution and estimates the individual activity variance, $\sigma_u^2(\mathrm{PERT})$, by

$$\sigma_u^2 = \frac{(b_u - a_u)^2}{36}, \tag{4.20}$$

where a_u and b_u are the "optimistic" and "pessimistic" durations of activity u, respectively, and $b_u - a_u$ is the range of the beta distribution (see Section 2.2 of Chapter 1).

The natural question here is: how good is such an estimate? The discussion of the "goodness" of the PERT procedure in its estimate of the variance usually resolves itself into two distinct, although by no means unrelated, considerations relative to: (a) the individual activity and (b) the total network. We discuss the estimates in that order.

§3.1 The PERT Estimate Relative to a Single Activity

First, suppose we go along with the assumption that each activity duration is beta distributed;

$$f(y)=\frac{1}{(b-a)^{k_1+k_2-1}\beta\cdot(k_1;k_2)}(y-a)^{k_1-1}(b-y)^{k_2-1};a\leqslant y\leqslant b$$

We have that (see Appendix D)

$$\operatorname{var}(Y)=(b-a)^2\frac{k_1k_2}{(k_1+k_2)^2(k_1+k_2+1)} \tag{4.21}$$

so that, on the surface, it appears that the ratio between the PERT estimate of the variance [given by Eq. (4.20)] and the true variance is given by

$$r=\frac{\sigma_u^2(\text{PERT})}{\operatorname{var}(Y_u)}=\frac{(k_1+k_2)^2(k_1+k_2+1)}{36k_1k_2} \tag{4.22}$$

However, a closer look indicates that PERT also assumes that the *mode is given*. This immediately fixes a relationship between the two shape parameters, k_1 and k_2. In particular, since the mode of the beta distribution is given by (see Appendix D)

$$m=\frac{a(k_2-1)+b(k_1-1)}{k_1+k_2-2}=\frac{a\alpha_2+b\alpha_1}{\alpha_1+\alpha_2}$$

where $\alpha_1=k_1-1$ and $\alpha_2=k_2-1$, it immediately follows that we have the relationship

$$\frac{\alpha_1}{\alpha_2}=\frac{m-a}{b-m}$$

For the sake of simplicity, and in order not to obscure the essential point to be made, consider the standardized version of the beta distribution in which $a=0$, $b=1$, and $0\leqslant m\leqslant 1$. Furthermore, assume $\alpha_2=1$, which implies that $f(y)$ is tangential to the axis at $y=b=1$; an assumption mentioned in the literature and considered "reasonable" [Coon (1965) and Donaldson (1965)]. (Results for $\alpha_1=1$ in the interval $0\leqslant m\leqslant 1$ can be obtained by symmetry.) Then we immediately have that

$$\alpha_1=\frac{m}{1-m} \tag{4.23}$$

and that the ratio r of Eq. (4.22) is given by

$$r=(\alpha_1+3)^2\frac{\alpha_1+4}{72(\alpha_1+1)} \tag{4.24}$$

Substituting from Eq. (4.23) into Eq. (4.24), it is easy to construct Table 4.2 for r versus m.

It is obvious that, at least for this particular value of α_2, the PERT estimate is biased for all values of m between 0 and 1 except at one particular point which is dependent on the value of α_2 (in the above example, it occurs at approximately $m \cong 0.78$) at which the PERT estimate is unbiased. Furthermore, note that it is biased on either side of the true value (i.e., it is either higher or lower than the true value) depending on the specific values of m. Similar results can be obtained for other values of α_2, but not with equal ease.

Second, consider the case in which the true PDF is *not* the beta distribution. Then the PERT procedure involves two kinds of errors: the first is that it approximates the true DF with the beta distribution, say *with the same* (or approximately the same) *range and mode*, and the second is that it approximates the variance of the surrogate beta distribution with σ_u^2(PERT).

It is easy to see that the var(Y_u) may vary between two extremes represented by the "almost sure" duration of $Y_u = m$ with probability 1 and the "complete ignorance" case of almost even distribution between a and b. The variance in the first extreme case is 0 and in the second extreme case is approximately $(b-a)^2/12$. Consequently, the worst (magnitude of) error in the PERT variance estimate is

$$\max\left\{\left|\frac{(b-a)^2}{36}-0\right|;\left|\frac{(b-a)^2}{36}-\frac{(b-a)^2}{12}\right|\right\}=\frac{(b-a)^2}{18}.$$

If r represents again the ratio of the PERT estimate to the true value of the variance of the activity, then

$$\frac{1}{2} \leqslant r \leqslant \infty$$

which is the same range of r as that given by Eq. (4.24). This is an illuminating result, because it implies that the range of uncertainty in the estimate of the variance of an activity is not diminished by having assumed that the activity is beta distributed, mainly due to the other assumptions of the PERT Model!

Table 4.2. The Ratio r versus (2) mode m

Mode m	0.05	0.10	0.15	0.20	0.25	0.30	0.35	0.40	0.45	0.50
Ratio r	0.498284	0.497383	0.497493	0.498785	0.5015212	0.506123	0.513001	0.522840	0.536569	0.555
Mode m	0.55	0.60	0.65	0.70	0.75	0.80	0.85	0.90	0.95	1.00
Ratio r	0.581858	0.618750	0.643240	0.750617	0.8750	1.120	1.51265	2.60	7.73055	∞

A discussion of the error in the PERT estimate of the variance of an individual activity is incomplete without mentioning the following important observation: σ_u^2(PERT) is always written as if the parameters a and b were known without error. In fact, the notation $\hat{a}$ and $\hat{b}$ should be used to indicate that these are *subjective estimates* of the "true" a and b, and as such are subject to sampling errors. What is the effect on σ_u^2(PERT) of the assumption that a and b are estimated error-free? It is extremely difficult to obtain an explicit expression of the var(Y) taking into account the assumed variability in these estimates. Undoubtedly, the var(Y) is increased due to such variability—the question is by how much, and what then is the error involved in the PERT estimate.

The simple-minded argument on this point presented in Section 1 (see p. 237), is still valuable in exposing the contradictions inherent in the PERT assumptions and derivations.

§3.2 The PERT Estimate Relative to the Entire Project

The network-based questions may be formulated somewhat as follows. What is the effect of the following factors on the PERT estimate of the variance: the structure of the network? The parameters of the individual activities? The degree of activity subdivision?

Consider the last question first, discussed in Section 1 (see p. 234), where we concluded that the subdivision of an activity into several activities (which constitute a subnetwork), so that the following two conditions are satisfied, results in the variance of the terminal event be necessarily *decreased*:

1. The expected times of the small activities add up to the expected duration of the "parent" activity.
2. The range of the small activities combine into the range of the "parent" activity.

In other words, the PERT estimate of the duration of the project is dependent on the very manner in which the various activities were defined, which is arbitrary and subjective!

Consider next the first question posed above, that concerning the effect of the structure of the network. Since the PERT estimate, σ_n^2(PERT), is determined as the sum of the variances along the CP (the latter evaluated when the expected values of durations are substituted for the duration of each activity), the variance estimate will *increase* with an increase in the number of activities along the CP. Mind you, this is independent of the structure of the network, and irrespective of the closeness of the second "runner up" (i.e., the second CP), or, in general, the closeness of the kth runner up (i.e., the kth CP).

Needless to say, this is contrary to known results, and is true only if the CP is really longer than any other path in the network by a sufficient margin (say at least three σ values). Otherwise, if several paths are of almost the same length, the variance of the project duration is smaller than the variance of the length of any single path, and the variance will *decrease* as the number of competing paths increases.

For instance, suppose all paths of the network are identically distributed as the uniform distribution between 0 and b. Let X_i be the duration of path i and let $Y=\max(X_i)$; $i=1,2,\dots,n$. Then Y has the DF

$$F_Y(y)=(F_X(y))^n$$

where $F_X(x)=x/b$ for $0\leqslant x\leqslant b$; hence its "density" function is $dF_X(x)=1/b$; $0\leqslant x\leqslant b$. Hence,

$$dF_Y(y)=n(F_X(y))^{n-1}\partial F_X(y)=ny^{n-1}/b^n$$

Clearly, $\mathcal{E}(X_i)=b/2$ and $\text{var}(X_i)=b^2/12$. Furthermore, $\mathcal{E}(Y)=nb/(n+1)$ and

$$\text{var}(Y)=\mathcal{E}(Y^2)-(\mathcal{E}(Y))^2=\int_0^b n\frac{y^{n+1}}{b^n}\,\partial y-\left(\frac{nb}{n+1}\right)^2=\frac{nb^2}{(n+1)^2(n+2)}.$$

But $n/(n+1)^2(n+2)$ is $\leqslant 1/12$ for all $n\geqslant 1$ (equality holds at $n=1$); which verifies the assertion that $\text{var}(Y)\leqslant\text{var}(X_i)$ for all i. Furthermore, $\text{var}(Y)\to 0$ as $n\to\infty$.

The same conclusion was arrived at by Tippett (1925) for the case X_i is distributed $N(0,1)$, and again by Clark (1960).

We are thus forced to conclude that the more parallelism is in the network, and the more close are the means of the activities to each other,

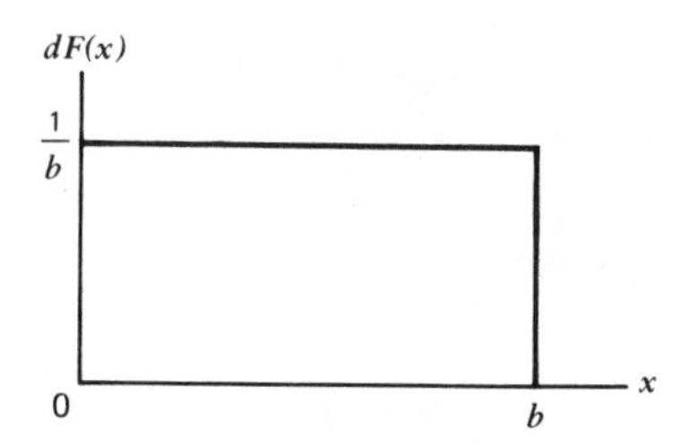

the greater is the error in σ_n^2(PERT). In particular, σ_n^2(PERT) is $>$ or $\gg$var(T_n).

Finally, there is little doubt that the estimates of the parameters of individual activities combine in a complex manner to yield the estimate of the variance of the project duration. Errors in such parameters introduce errors in the final result, whose magnitude and direction remain, to date, largely undetermined.

§4. ESTIMATION OF THE DF OF PROJECT DURATION

Rather than concentrate on the expected value of the RV T_n, as we did in Section 2, suppose we inquire into the complete PDF of T_n: can it be determined, approximated, or at least bounded? One advantage of such determination is that we would then be able to not only determine $\mathcal{E}(T_n)$, but also give precise probability statements concerning the completion time of the project. The reader will recall (see Sections 1 and 3) that the PERT probability statements left a great deal to be desired.

In general, there are two approaches: (a) the analytical approach and (b) the population sampling (Monte Carlo) approach. The first approach applies theoretical constructs to derive the desired PDF, or at least "good" approximations, or bounds, to it. If one insists on the exact PDF, the analytical approach suffers from the almost fatal drawback of being computationally unwieldy to the point of impracticality. On the other hand, if one is content with an approximation, the analytical approaches, unfortunately, yield approximations (or bounds) of unknown "closeness" to the true PDF! The second approach, that of Monte Carlo sampling, suffers from sampling errors, although its generality and practicality more than offset such a drawback. In fact, it forms the "standard" against which other approximations are compared.

Our discussion in this section proceeds as follows. In Section 4.1 we give the analytical approach to the exact determination of the PDF and illustrate it by application to simple networks. In Section 4.2 we briefly discuss two analytical approximations. This should give the reader a glimpse of the mathematical "machinery" needed for such approaches. Finally, Section 4.3 gives a more detailed treatment of the estimation problem through population sampling.

§4.1 The Analytical Approach: Determination of the Exact Probability Distribution Function

The approach involves following four general steps, which are applicable for *any* analytical approach, whether intended for the determination of the exact PDF or its bounds and approximations (treated in the text that follows).†

STEP 1. *Identification.* Identify various subnetworks as special activity configurations whose PDFs are known (the so-called "generic" subnetworks).

STEP 2. *Simplification.* Replace the various configurations of Step 1 and their associated completion time distributions by single equivalent activities and completion time distributions.

STEP 3. *Decomposition.* Decompose the simplified network into several subnetworks by separating subnetworks at each *cut vertex*. A cut vertex is any node such that every path from the origin to the terminal passes through it. Each subnetwork should be a set of parallel paths.

STEP 4. *Synthesis.* Reduce each subnetwork to a single equivalent activity, then combine the set of equivalent activities, which are now in series, in a grand equivalent activity. The result is the completion time of the entire project.

To better understand the approach, we start with a network to which we can apply the last step. Suppose we have a network composed of a set of independent paths in parallel and in series. It is well known that:

1. For two paths in parallel, the PDF of their equivalent activity is the product of the two PDFs.
2. For two paths in series, the PDF of their equivalent activity is the convolution of the PDFs of the durations of the two paths. Consequently, the PDF of the duration of the project, T_n, is accumulated step-by-step by combining paths in parallel, then in series, and so on, until a single equivalent arc is reached.

To be more explicit, consider two arcs in parallel, as shown in Fig. 4.7*a* joining two nodes i and j. Let X_{ij} denote the duration of one arc, and Y_{ij} the duration of the other. If i is realized at time zero, then $\mathrm{T}_j = \max(X_{ij}, Y_{ij})$. Clearly,

$$
\begin{aligned}
Pr[\mathrm{T}_j \leqslant \tau] &= Pr[\max(X_{ij}, Y_{ij}) \leqslant \tau] \\
&= Pr[X_{ij} \leqslant \tau \text{ and } Y_{ij} \leqslant \tau] \\
&= Pr[X_{ij} \leqslant \tau] \times Pr[Y_{ij} \leqslant \tau]; \text{ by independence,}
\end{aligned}
$$

†The basic idea for these steps is due to Sielken, Hartley, and Arseven (1975).

that is,

$$F_j(\tau) = F_X(\tau) F_Y(\tau) \tag{4.25}$$

and the assertion made in (1) above follows.

Next, consider two arcs in series, as shown in Fig. 4.7*b*. Let Y_{ij} denote the duration of the first arc (ij), with PDF $F_{ij}(\tau)$, and define Y_{jk} and $F_{jk}(\tau)$ similarly. If i is realized at time zero, then the time of realization of k is a RV, denoted by T_k, and given by $T_k = Y_{ij} + Y_{jk}$. The PDF of T_k is given by

$$F_k(\tau) \triangleq Pr[T_k \leqslant \tau] = \int_0^{\tau} F_{ij}(\tau - y)\,\partial F_{jk}(y) = F_{ij}(\tau - y) * F_{jk}(y) \tag{4.26}$$

where the asterisk in the rightmost expression denotes convolution. Hence the assertion made in (2) above follows.

It is thus seen that step 4 reduces to the successive application of Eqs. (4.25) and (4.26).

Example 4.2. Consider the network of Fig. 4.8. Assume, for simplicity, that the duration of all arcs are independently and identically distributed as the random variable Y that is exponentially distributed,

$$\partial F_Y(t) = be^{-bt}\,\partial t$$

(This is just about the simplest function to convolve or multiply.) The steps of analysis are summarized in Table 4.3 and are self-explanatory. The important remark to make is that in order to evaluate the PDF of T_5 it was necessary to proceed with caution *in order to conserve the independence of the RVs*. Thus it was necessary to evaluate first the $\max(Y_{3,4} + Y_{4,5}; Y_{3,5})$

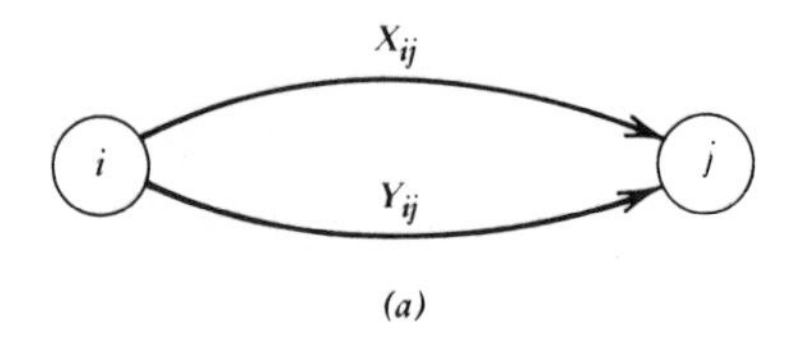

(*a*)

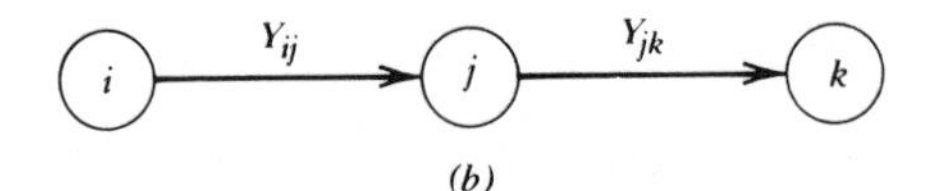

(*b*)

Figure 4.7.

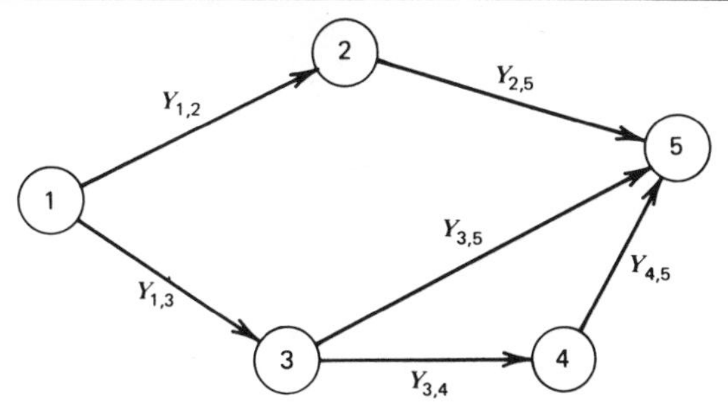

Figure 4.8. Example network.

before evaluating $Y_{1,3} + \max(Y_{3,4} + Y_{4,5}; Y_{3,5})$; which is subsequently compared with the duration of the path *1–2–5*. If another approach were adopted, say, for example, put $Z_1 = Y_{1,3} + Y_{3,4} + Y_{4,5}; Z_2 = Y_{1,3} + Y_{3,5}$; and determine F_{Z_1} and F_{Z_2} [using the arce-in-series Eq. (4.26)], then we could *not* have evaluated $\max(Z_1, Z_2)$ because the RVs Z_1 and Z_2 are *not independent* [the two paths defining them share one arc, arc $(1,3)$].

As can be easily seen, the application of Step 4 is straightforward *if* the network is amenable to such composition of variables, or *if it has been so preconditioned*. (This is basically the function of the first two steps of the general procedure.) Yet it is not difficult to encounter networks that cannot be analyzed according to the simple sequential synthesis of Step 4, in which case one must resort to more elaborate techniques, which we discuss next.

Consider the simple network shown in Fig. 4.9; there is no way in which the three paths from node *1* to node *4* can be treated as independent segments which are then combined into equivalent elements according to Step 4 to yield the PDF of T_4. How can one then proceed?

Table 4.3

Random Variable	Representation	Probability Distribution Function
T_1	0	
T_2	$Y_{1,2}$	$1 - e^{-bt}$
T_3	$Y_{1,3}$	$1 - e^{-bt}$
W_1	$Y_{3,4} + Y_{4,5}$	$1 - e^{-bt}(1 + bt)$
W_2	$Y_{1,2} + Y_{2,5}$	$1 - e^{-bt}(1 + bt)$
W_3	$\max(Y_{3,5}, W_1)$	$(1 - e^{-bt})[1 - e^{-bt}(1 + bt)]$
W_4	$Y_{1,3} + W_3$	$1 - e^{-bt}[-1 + 2b_t + \frac{1}{2}(bt)^2 + e^{-bt}(2 + bt)]$
T_5	$\max(W_2, W_4)$	$[1 - e^{-bt}(1 + bt)][1 - e^{-bt}\{-1 + 2bt + \frac{1}{2}(bt)^2 + e^{-bt}(2 + bt)\}]$

Evidently, the network must be analysed as an entity. The general principle to apply in such a case is to evaluate the PDF of the terminal node *conditioned on the values of some arcs*, and then remove the conditional nature of the result thus obtained through integration over the DF of these common arcs. The final result is the desired (unconditional) PDF.

For example, in the network of Fig. 4.9 it is clear that there are three paths, two of which (viz., paths *1–2–4* and *1–3–4*) can be analyzed in a straightforward fashion because they are independent, but the third path cannot because of the existence of a common arc between it and each of the other two paths. Since

$$T_4 = \max[Y_{1,2} + Y_{2,4}; Y_{1,2} + Y_{2,3} + Y_{3,4}; Y_{1,3} + Y_{3,4}]$$

one may condition the PDF of T_4 on the two activities $Y_{1,2}$ and $Y_{3,4}$. To this end, let

$$W_1 = Y_{1,2} + Y_{2,4}; W_2 = Y_{1,2} + Y_{2,3} + Y_{3,4}; W_3 = Y_{1,3} + Y_{3,4}$$

Then,

$$F_4(\tau|y_{1,2}, y_{3,4}) = F_{W_1}(\tau|y_{1,2}, y_{3,4}) F_{W_2}(\tau|y_{1,2}, y_{3,4}) F_{W_3}(\tau|y_{1,2}, y_{3,4})$$

and

$$F_4(\tau) = \int\int F_4(\tau|y_{1,2}, y_{3,4}) \partial F_{1,2}(y_{1,2}) \partial F_{3,4}(y_{3,4})$$

Naturally, it is essential that one conditions on the least number of arcs in order to minimize the chore of multiple integration. The determination

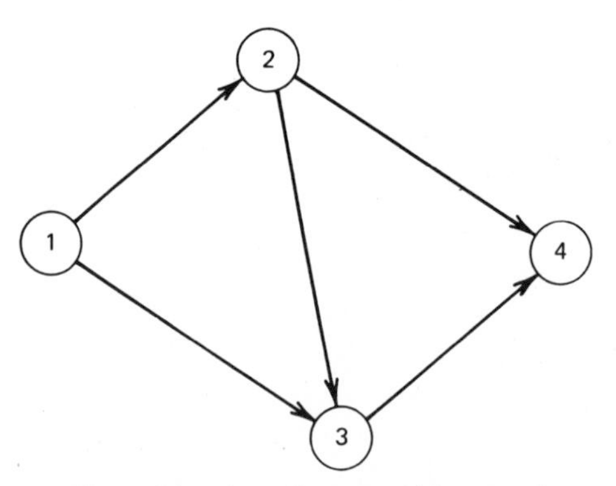

Figure 4.9. A nondecomposable network.

of the minimum number of arcs can be achieved following a procedure due to Burt and Garman (1971). We delay the specification of this procedure to the discussion of conditional sampling in Section 4.4 (in particular, p. 299), where the subject is revisited.

The "wheatstone bridge" of Fig. 4.9 was perhaps the first "generic" type of complex networks recognized as such. Others followed in quick succession—the two forms of the "double wheatstone bridge" and the three forms of the "criss-cross" [see Ringer (1971)]. Undoubtedly, others will continue to be discovered. But their analysis follows the same principle stated above; one first obtains the conditional PDF and then eliminates the conditioning through (possibly multiple) integration. In Problem 9 we ask the reader to express the PDF for some of these generic types.

The above discussion should give full meaning to the first two steps of the general procedure. The "identification" step simply requires the recognition of the various generic types which must be analysed as separate entities because the complexity of the subnetworks precludes the straightforward step-by-step application of Eqs. (4.25) and (4.26). The "simplification" step replaces these generic subnetworks by their equivalent arcs.

To illustrate the general procedure in its entirety, we outline the analysis of a more complex sample network.

Example 4.3. Consider the network of Fig. 4.10a. One immediately recognizes that nodes *7* and *11* are cut vertices; consequently the network is decomposable into three subnetworks as shown in Fig. 4.10b_1, b_2, b_3. The subnetwork of (4.10b_1) is further recognized as decomposable into two subnetworks; the first, $\{(1,2),(1,3),(2,3),(2,7),(3,7)\}$, is a wheatstone bridge and the second, $\{(1,4),(1,5),(1,6),(4,5),(4,7),(5,6),(5,7),(6,7)\}$, is a double wheatstone bridge. Denote their equivalent elements by A_1 and A_2; two arcs in parallel. The subnetwork of Fig. 4.10b_2 is next identified as a criss-cross network. Denote its equivalent element by A_3. Finally, the subnetwork of Fig. 4.10b_3 is a still different type of complex subnetwork from all generic types discussed thus far (including those relegated to Problem 9). Its analysis is posed as Problem 15. Denote its equivalent arc by A_4. The original network is now represented by the set of equivalent arcs shown in Fig. 4.10c. Now we are finally ready to apply Step 4. Activities A_1 and A_2 are in parallel and hence can be replaced, using Eq. (4.25), by an equivalent arc, say B_1. Arcs B_1, A_3, and A_4 are three arcs in series that can be combined, following Eq. (4.26), to yield the desired PDF of T_{15}. Problem 16 asks the reader to detail the sequence of steps of iteration required to determine the duration of a complex project network.

A refinement of the above approach was proposed by Martin (1965), who presented a computational method for the evaluation of the DF of T_n

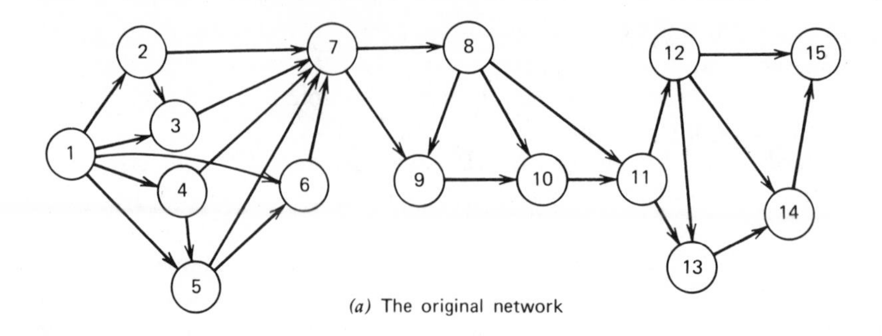

(a) The original network

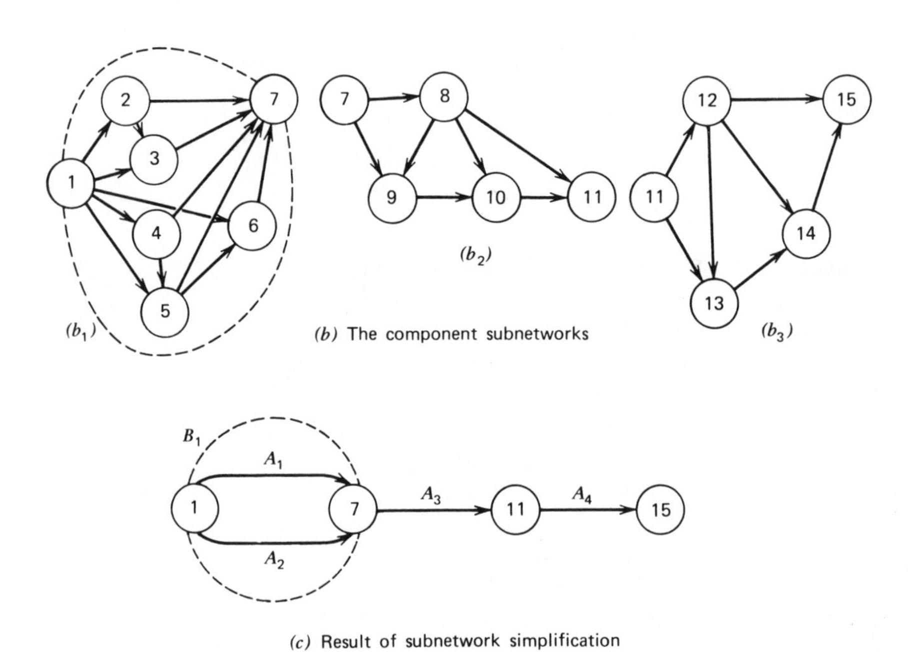

(b) The component subnetworks

(c) Result of subnetwork simplification

Figure 4.10. The decomposition of a network. (*a*) The original network. (*b*) The component subnetworks. (*c*) Result of subnetwork simplification.

under the assumption that the arc duration DFs are polynomials. We do not pursue his approach here since it suffers from the same fatal drawbacks mentioned in the preamble to this section.

§4.2. The Analytic Approach: Approximating the Probability Distribution Function

Admitting that time estimates are very rough images of the manager's subjective feelings, it may seem inordinately sophisticated to even attempt an "exact" determination of the PDF of T_n. One may have the uneasy feeling that such is a misplacement of effort, and that a "reasonable" approximation or bound should be sufficient. This section discusses two approaches that purport to do just that.

Approximating the PDF

Harking back to our discussion of Sections 2.2 and 2.3, the reader will recall that the second and third approaches to the approximation of the expected value of the duration of the project, $\mathcal{E}(T_n)$, relied on approximating the DF of the time of realization of a node by two values, each with probability $\frac{1}{2}$. Can the same approach be used to estimate the complete DF, instead of just the mean?

The answer is, fortunately, yes. In carrying out such estimation, we are also answering, at least partially, the questions raised on p. 234 concerning the best manner of reflecting the manager's feelings. For it seems no less reasonable to inquire about two time estimates than about three time estimates (as demanded by PERT), or about the complete DF of each activity duration. Presumably, the two time estimates represent high and low percentiles (say 0.90 and 0.10 or 0.85 and 0.15, etc.), which are symmetric about the median. One then would be reasonably justified in deducing the midpoint as the expected duration of the activity. The general approach of the previous section (Steps 1–4) is now applied in its entirety with the following two modifications:

1. All activities possess the binomial distribution with probability $\frac{1}{2}$ attached to each time estimate.
2 The DF of individual subnetworks is itself approximated by a binomial DF or any other appropriate discrete DF of the same expected value as the original PDF and of equal probabilities attached to the time estimates.

This second modification may be ignored in the case of "small" networks, especially if one desires the PDF of T_n in as complete form as possible with the data given. Note that with only two time estimates for each activity, the operations of multiplication, convolution, and conditioning are greatly simplified.

The following example illustrates the application of the above ideas to the simple network of Fig. 4.8, whose exact DF was derived in Table 4.3.

Example 4.4. Suppose that $b=1$; that is, assume that all the activities of Fig. 4.8 are exponentially distributed as $dF(t)=e^{-t}dt$. Then Table 4.3 yields that the DF of T_5 is

$$F_5(t)=\left[1-e^{-t}(1+t)\right]\left[1-e^{-t}\left\{-1+2t+\frac{t^2}{2}+e^{-t}(2+t)\right\}\right]$$

This function is plotted in Fig. 4.11 (as the continuous line) for $t\in[0,6]$. The approximation proceeds as shown in Table 4.4, which is based on a two-point approximation to the exponential DF at the 0.85 and 0.15 percentiles; in other words, each activity is assumed binomially distributed

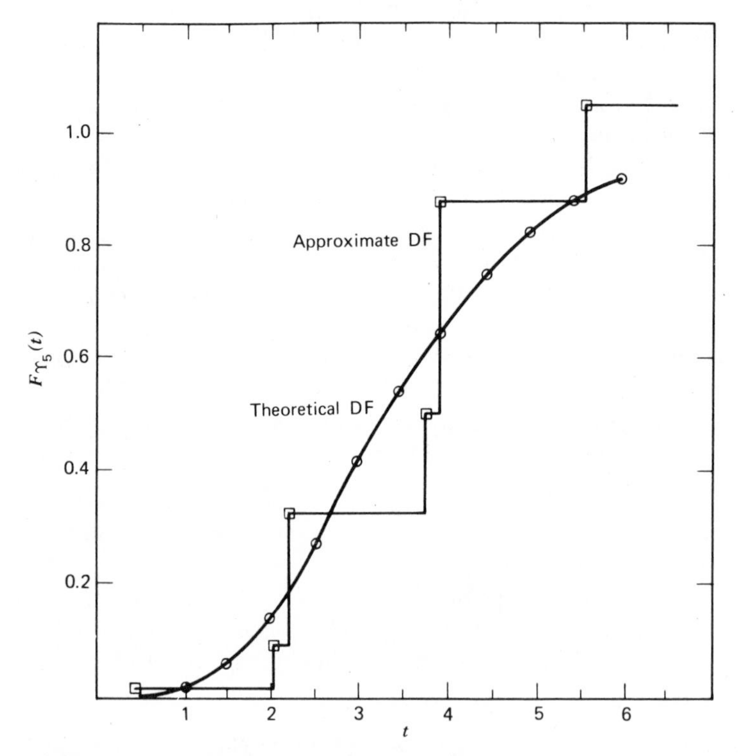

Figure 4.11. Theoretical and approximate distributions of T_5 of Figure 4.8.

Table 4.4. Calculation of the approximate PDF of Fig. 4.8

Random Variable	Representation		Probability Distribution Function					
T_I	0							
T_2	$Y_{I,2}$	t	0.1625	1.8971				
		p	0.5	1.0				
T_3	$Y_{I,3}$	t	0.1625	1.8971				
		p	0.5	1.0				
W_1	$Y_{3,4} + Y_{4,5}$	t	0.325	2.0596	3.7942			
		p	0.25	0.75	1.0			
W_2	$Y_{I,2} + Y_{2,5}$	t	0.325	2.0596	3.7942			
		p	0.25	0.75	1.0			
W_3	$\max(Y_{3,5}, W_1)$	t	0.325	1.8971	2.0596	3.7942		
		p	0.125	0.250	0.750	1.0		
W_4	$Y_{I,3} + W_3$	t	0.4875	2.0596	2.2221	3.7942	3.9567	5.6913
		p	0.0625	0.125	0.4375	0.50	0.875	1.0
T_5	$\max(W_2, W_4)$	t	0.4875	2.0596	2.2221	3.7942	3.9567	5.6913
		p	0.015625	0.09375	0.328125	0.50	0.875	1.0

with equal probability at $y_1=0.1625$, and $y_2=1.8971$. The steps of analysis parallel the steps of Table 4.3. As illustration, consider the entries opposite W_1 and W_3. We have

$$W_1 = Y_{3,5} + Y_{4,5}$$

$$F_{W_1}(\tau) = \sum_y F_{Y_{3,5}}(\tau - y) p_{4,5}(y)$$

$$= \begin{cases} 0 & \text{for} \quad \tau < 0.3250 \\ \frac{1}{4} & \tau \in [0.3250, 2.0596^-] \\ \frac{3}{4} & \tau \in [2.0596, 3.7942^-] \\ 1 & \tau \geqslant 3.7942 \end{cases}$$

and

$$W_3 = \max(Y_{3,5}, W_1)$$

$$F_{W_3}(\tau) = F_{Y_{3,5}}(\tau) \times F_{W_1}(\tau)$$

$$= \begin{cases} 0 & \text{for} \quad \tau < 0.3250 \\ \frac{1}{8} & \tau \in [0.3250, 1.8971^-] \\ \frac{1}{4} & \tau \in [1.8971, 2.0596^-] \\ \frac{3}{4} & \tau \in [2.0596, 3.7942^-] \\ 1 & \tau \geqslant 3.7942 \end{cases}$$

The approximate DF of T_5 is also plotted in Fig. 4.11 (as the stepped line) for the same range of t. The reader is left to judge for himself the "goodness" of the approximation.

In large networks, even this approximate procedure may be too demanding in computing effort, and still additional "shortcuts" may be needed to bring such effort to within reasonable limits. The following observation is appropriate in this respect.

A known fact, to which we alluded several times in our discussions of PANs, is that almost all paths of the network have some chance of being CPs, but that the probability of such eventuality varies considerably among the paths.

Consequently, minor losses in accuracy would be incurred if we ignore the paths of low probability (i.e., ignore their arcs) and concentrate on the (usually few) paths where the bulk of the probability lies. But how can one determine such paths in an efficient manner? For indeed, if the detection of the "core subnetwork" itself requires elaborate and lengthy procedures, nothing would be gained.

This question was addressed by Sielken, Hartley, and Arseven (1975), who introduced the concept of "clusters" whose analysis should be sufficient to determine the desired approximation. We shall not pursue their procedure here since, as of the time of this writing, its computing efficiency has not been demonstrated yet to be superior to, say, Monte Carlo sampling.

§4.3 The Analytical Approach: Bounding the Probability Distribution Function*

Bounds are interesting from at least two points of view: (a) If one can obtain upper and lower bounds on a quantity, and the bounds are quite close to each other, then one has virtually estimated the quantity itself. (b) Bounds also indicate the degree of "spread" of the interval of uncertainty, which is a valuable piece of information in its own right that cannot be gleaned from a point estimate.

Consequently, it is legitimate to inquire into the bounding of the PDF of T_n, especially in view of the complexity of the procedures discussed in the previous two subsections. The following ingenious procedure due to Robillard and Trahan (1977) is interesting from both a theoretical as well as computational points of view.†

Consider an AN with random activity durations; and let π be the index of the paths from node *1* to node n; $\pi = 1,2,\ldots,r$, and T_π its duration. Furthermore, let $\Pi(i)$ denote the set of paths to node $i \in N$. (Unless otherwise stated, we are always interested in the terminal node n and its associated parameters.) Then the PDF $F(y)$ is bounded from below by $\lambda(y)$, which is defined by

$$\lambda(y) \triangleq \max\left\{0; 1 - \sum_{\pi} Pr(T_\pi \geqslant y)\right\} \leqslant F(y)$$

Note that the possible dependence among the paths (which share one or more activities) causes the sum $\Sigma_\pi Pr(T \geqslant y)$ to be $\geqslant Pr(T_n \geqslant y)$; whence the inequality. (When does equality hold?) Since $Pr(T_\pi \geqslant y) = 1 - Pr(T_\pi \leqslant y) = 1 - F_\pi(y)$, the ℓ.b. may be written as

$$\lambda(y) = \max\left\{0; 1 - r + r\sum_{\pi} \frac{1}{r} F_\pi(y)\right\}$$

where r is the number of paths.

*Starred sections may be omitted on first reading.

†Pierre Robillard met his premature death in a tragic automobile accident in 1974; thus ended the brief but luminous career of a brilliant and amiable colleague.

We have argued above [see Eq. (4.26)] that the PDF of the terminal node of a chain of arcs in series is obtained by the convolution of their respective PDFs. The operation of convolution always evokes the use of either the "probability generating function" (sometimes also referred to as the "moment generating function," MGF) or the "characteristic function," with the latter being the more general of the two, because the operation of integration is replaced by the operation of multiplication in the transform domain. The following argument utilizes the characteristic function. (We utilize the MGF in Section 2 of Chapter 5.)

Let $\psi(z)$ denote the characteristic function† of the duration T_π of the path π [i.e., $\psi_\pi(z)=\mathcal{E}(e^{jT_\pi})$]. By the linearity of the characteristic function,

$$\sum_{\pi} \frac{1}{r} F_\pi(y) \leftrightarrow \sum_{\pi} \frac{1}{r} \psi_\pi(z)$$

Define I_y as the "inverse characteristic function" evaluated at y; then we may write

$$\lambda(y)=\max\left\{0; 1-r+rI_y \sum_{\pi} \frac{1}{r} \psi_\pi(z)\right\}$$

Since the activities along one path are assumed independent, $\psi_\pi(z)=\Pi_{u\in\pi}\psi_u(z)$; where Π denotes the product, and hence

$$\sum_{\pi} \psi_\pi(z)=\sum_{\pi} \prod_{u\in\pi} \psi_u(z),$$

denoted by $\phi(z)$. Note that the function ϕ is independent of y. To evaluate the function $\phi(z)$ for many values of z in "one pass" of the network, define

$$\phi(z,i)=\sum_{\pi\in\Pi(i)} \prod_{u\in\pi} \psi_u(z); i\in N; \phi(z,1)\equiv 1;$$

where $\Pi(i)$ denotes the set of paths to node i. Then

$$\phi(z,k)=\sum_{i\in\mathcal{B}(k)} \phi(z,i)\psi_{(ik)}(z); k\neq 1$$

where $\mathcal{B}(k)$ is the set of nodes immediately preceding node k, which

†The characteristic function of a random variable T is defined as $\mathcal{E}(e^{jT})$; $j=\sqrt{-1}$; see the work by Feller (1971) for rigorous mathematical derivations of all related concepts utilized in this discussion.

allows for the computation of $\phi(z,n) \triangleq \phi(z)$ in a sequential fashion such that node i is enumerated before node j if $(ij) \in A$.

The inverse transformation of the characteristic function $\phi(z)$ can be efficiently evaluated by the approximations proposed by Davies (1973). For discrete DFs of activity durations we have

$$\lambda(y) \cong \max\left\{0; 1-\frac{r}{2}-\int_{-\pi}^{\pi} Re\left[\phi(u)\exp\{-ju(y+1)\}/2\pi(1-e^{-ju})\right]du\right\}$$

where $Re[CN]$ is the real part of the complex number $CN; j=\sqrt{-1}$, and π is the constant 3.1428.... For continuous DFs we have

$$\lambda(y)$$

$$\cong \max\left\{0; 1-\frac{r}{2}-\sum_{k=0}^{K} Im\left[\phi(k+0.5)\Delta\exp\{-j(k+0.5)\Delta y\}/\pi(k+0.5)\right]\right\}$$

where $Im[CN]$ is the imaginary part of the complex number CN. The choice of K and Δ is made to ensure that

$$\max\left\{Pr\left[\mathrm{T}_n < y-\frac{2\pi}{\Delta}\right]; Pr\left[\mathrm{T}_n > y+\frac{2\pi}{\Delta}\right]\right\}$$

and the truncation error are both less than half the maximum allowable error.

While the above derivation yields $\lambda(y)$ as a lower bound on the value of $F(y)$, it is also possible to obtain bounds for different moments of the PDF of T_n. From Rao (1971), we know that if T is a nonnegative RV and if the mean of T, $\mathcal{E}(\mathrm{T})$, exists, then

$$\mathcal{E}(\mathrm{T}) = \int_0^\infty [1-F(\tau)]\,d\tau$$

From the lower bound $\lambda(y) \leqslant F(y)$, we have

$$\mathcal{E}(\mathrm{T}) \leqslant \int_0^\infty [1-\lambda(\tau)]\,d\tau \triangleq \overline{\mathcal{E}(\mathrm{T})}$$

thus obtaining an upper bound on $\mathcal{E}(\mathrm{T}_n)$. A lower bound, $\underline{\mathcal{E}(\mathrm{T}_n)}$, for $\mathcal{E}(\mathrm{T}_n)$ may be obtained by any of the four approaches described in Section 2.

The formula for the mean can be generalized for the other moments,

$$\mathcal{E}[\mathrm{T}^k] = k\int_0^\infty \tau^{k-1}[1-F(\tau)]\,d\tau$$

In particular, we have for the second moment,

$$\mathcal{E}\left[\mathrm{T}_n^2\right] \leq 2\int_0^\infty \tau\left[1-\lambda(\tau)\right]d\tau \triangleq \overline{\mathcal{E}\left[\mathrm{T}_n^2\right]}$$

thus obtaining an upper bound on the second moment. By definition of T_n, we have

$$\mathcal{E}(\mathrm{T}_n^2) = \mathcal{E}\left[\max_{\pi \in \Pi_n} T_\pi^2\right]$$

where T_π is the duration of path π, a random variable. Since the function between square brackets is convex, and using the Jensen inequality we can write

$$\mathcal{E}\left[\mathrm{T}_n^2\right] \geq \max_{\pi \in \Pi_n} \mathcal{E}\left[T_\pi^2\right]$$

Recall that

$$\mathcal{E}\left[T_\pi^2\right] = \mathrm{var}(T_\pi) + (\mathcal{E}[T_\pi])^2$$

Denote by C the critical path determined by the classical PERT procedure (see Section 2.2 of Chapter 1). Since, from above, we have just derived $\underline{\mathcal{E}\left[\mathrm{T}_n\right]}$ as a lower bound on $\mathcal{E}[\mathrm{T}_n]$; then clearly

$$\max_{\pi \in \Pi_n} \mathcal{E}\left[T_\pi^2\right] \geq \max_{\pi \in \Pi}\left\{\mathrm{var}(T_n) + \underline{(\mathcal{E}\left[\mathrm{T}_n\right])^2}\right\}$$

$$\geq \sum_{u \in C} \mathrm{var}(Y_u) + \underline{(\mathcal{E}\left[\mathrm{T}_n\right])^2} \triangleq \underline{\mathcal{E}\left[\mathrm{T}_n^2\right]}$$

where Y_u is the duration of activity $u \in C$. Using the relation $\mathrm{var}[T] = \mathcal{E}[T^2] - (\mathcal{E}[T])^2$, we can obtain bounds for $\mathrm{var}(\mathrm{T}_n)$ as follows:

$$\underline{\mathrm{var}(\mathrm{T}_n)} \triangleq \max\left(0, \underline{\mathcal{E}\left[\mathrm{T}_n^2\right]} - \overline{\mathcal{E}\left[\mathrm{T}_n\right]}^2\right)$$

$$\leq \mathrm{var}(\mathrm{T}_n) \leq \overline{\mathcal{E}\left[\mathrm{T}_n^2\right]} - \underline{(\mathcal{E}\left[\mathrm{T}\right])^2} \triangleq \overline{\mathrm{var}(\mathrm{T}_n)}.$$

The approximation method described above was programed in FORTRAN EXTENDED on the CDC CYBER-74 computer. The program was run, for the sake of comparison, on two networks: one proposed by Martin (1965) and the second by Pritsker and Kiviat (1969).

Martin studies the simple network given in Fig. 4.12. The activity duration DFs are uniform with ranges as follows:

Activity	Range of density function
$(1,2)$	[1,2]
$(2,3)$	[2,4]
$(2,4)$	[1,3]
$(3,8)$	[1,5]
$(4,5)$	[2,3]
$(5,6)$	[3,5]
$(5,7)$	[3,7]
$(6,8)$	[1,2]
$(7,8)$	[4,5]
$(8,9)$	[4,6]

This table is to be read as follows: $f_{1,2}(y)=1$ for $1 \leqslant y \leqslant 2$, $f_{2,3}(y)=\frac{1}{2}$ for $2 \leqslant y \leqslant 4$; and so forth.

Robillard and Trahan (1974) evaluated the approximate expression for $\lambda(y)=\max\{0; 1-\Sigma_{\pi} Pr(T_{\pi} \geqslant y)\}$ for various values of y. The results are shown in Table 4.5, together with the values obtained using Martin's approach. The results are remarkably good. However, they leave some doubts on the precision of Martin's results, since the lower bounds for values of $y=21$ and $y \geqslant 23$ are *higher* than Martin's estimates! This is because a small deviation (even at the tenth decimal place) in a polynomial coefficient of Martin's formula may give very different results.

The above approximation of the PDF was obtained in less than 0.11 sec of central processor time with $\Delta=0.1$ and $K=50$. The same DF was obtained for K as large as 1000 and for different values of Δ around 0.1.

The bounds on the mean and the second moment are

$$\underline{\mathcal{E}[\mathrm{T}]}=20.500 \leqslant \hat{\mathcal{E}}[\mathrm{T}]=20.524 \leqslant 20.562=\overline{\mathcal{E}[\mathrm{T}]}$$

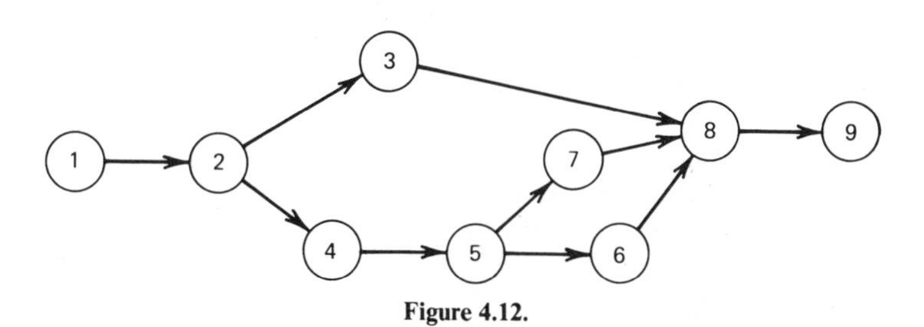

Figure 4.12.

Table 4.5. Comparison Between the Lower Bound $\lambda(y)$ and the "Exact" Values Obtained by Martin for $F(y)$

y	Martin's $F(y)$	Approximate Lower Bound $\lambda(y)$
16	0.00011	0.00000
17	0.00528	0.00000
18	0.04665	0.00000
19	0.17252	0.16111
20	0.38079	0.38003
21	0.61857	0.61979
22	0.82835	0.82830
23	0.94753	0.95330
24	0.98619	0.99470
25	0.99258	0.99991
26	0.99258	1.00000

and

$$\underline{\mathcal{E}[T^2]} = 422.250 \leqslant \hat{\mathcal{E}}[T^2] = 423.139 \leqslant 424.759 = \overline{\mathcal{E}[T^2]},$$

where $\hat{\mathcal{E}}[T]$ and $\hat{\mathcal{E}}[T^2]$ are the values given by Martin (1965).

The bounds on the variance of T are easily seen to be

$$0 \leqslant \hat{\text{var}}(T) = 1.905 \leqslant 4.509$$

where the value $1.905 = 423.139 - (20.524)^2$ is obtained from Martin's values.

The second example, due to Pritsker and Kiviat (1969), was used by those authors to test the Monte Carlo sampling approach on the PERT network depicted in Fig. 4.13. All the activity times were assumed Normally distributed with the following means and standard deviations.

Activity	Mean	Standard Deviation
$(1,2)$	5.5	1.18
$(1,3)$	13.0	3.00
$(2,3)$	7.0	1.00
$(2,4)$	5.2	0.84
$(2,6)$	16.5	2.50
$(3,6)$	14.7	1.34
$(4,5)$	6.0	1.00
$(4,7)$	10.3	1.67
$(5,6)$	20.0	3.32
$(5,8)$	4.0	0.71
$(6,9)$	3.2	0.50
$(7,8)$	3.2	0.84
$(8,9)$	16.5	1.18

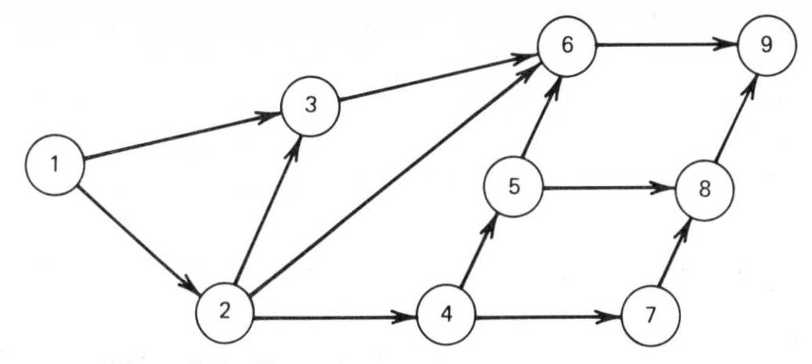

Figure 4.13. Example network of Pritsker and Kiviat.

This is a slightly more complicated network. The estimates of the PDF obtained by Pritsker and Kiviat from 400 "simulations" are compared with the lower bounds in Table 4.6.

The bounds for the mean and second moment are

$$\underline{\mathcal{E}[\mathrm{T}]}=40.700 \leqslant \tilde{\mathcal{E}}[\mathrm{T}]=42.1146 \leqslant 42.457=\overline{\mathcal{E}[\mathrm{T}]}$$

Table 4.6. Comparison Between Monte Carlo Sampling Results and The Lower Bound $\lambda(y)$ For the Network of Fig. 4.13

y	Monte Carlo Sampling, $F(y)$	Lower Bound, $\lambda(y)$
34	0.0000	0.0000
35	0.0050	0.0000
36	0.0100	0.0000
37	0.0375	0.0000
38	0.0725	0.0000
39	0.1475	0.0000
40	0.2400	0.0000
41	0.3550	0.1136
42	0.4950	0.3825
43	0.6550	0.5963
44	0.7525	0.7532
45	0.8300	0.8586
46	0.9050	0.9237
47	0.9575	0.9609
48	0.9675	0.9808
49	0.9850	0.9910
50	0.9950	0.9959
51	0.9975	0.9983
52	1.0000	0.9993
53	1.0000	0.9997
54	1.0000	1.0000

and

$$\underline{\mathcal{E}[T^2]} = 1665.3 \leqslant \tilde{\mathcal{E}}[T^2] = 1782.3 \leqslant 1848.9 = \overline{\mathcal{E}[T^2]}$$

where $\tilde{\mathcal{E}}[T]$ and $\tilde{\mathcal{E}}[T^2]$ are the values given by the Monte Carlo samples.

The lower bound, $\lambda(y)$, is seen to be an excellent approximation to the "true" PDF, especially in the $\pm 1.5\sigma$ range about the mean value. Recognizing its economy in computer time (less than 0.3 sec) and capacity, the utility of the approach is evident.

§4.4 Population-Sampling Approaches: Monte Carlo Methods

As always, whenever analytical approaches fail, or appear to overwhelm one's capacity to obtain numerical answers, one turns to population sampling and statistical techniques. In this instance, Monte Carlo sampling proves to be a most powerful technique for attacking the many thorny problems related to PANs, some of which have been alluded to above. In particular, within any prescribed bounds on the errors, one can determine factors such as: (a) the PDF of project duration; hence its mean and variance; (b) the importance, or lack of it, of any activity or subset of activities as measured by the relative frequency of their appearance along a CP; and (c) the costs involved.

The following is a more detailed discussion of this useful approach. The alert reader will realize that interwoven with the discussion is a detailed presentation of the *theory* of Monte Carlo sampling, which is of much more general applicability than only to PANs.

"Straightforward" Sampling or "Crude" Monte Carlo

As before, by a *realization of the network* we mean the network with a fixed value for each of its arc durations. This value is obtained through a random sample from all possible values for each arc, following the PDF of the arc duration. For a particular realization of the network, we let l_n denote the length of the longest path from the initial to the terminal event (the CP for that particular realization). Clearly, l_n can be viewed as a possible value of a RV, which we denoted by T_n, representing the duration of the CP (or, equivalently, the time of realization of node n). Equally obvious, l_n varies between lower and upper limits, given by

$$(l_n | y_a = L_a) \leqslant l_n \leqslant (l_n | y_a = U_a)$$

where $l_n | y_a = L_a$ is the time of realization of the terminal node n when all arc durations are put at their *lower* bounds and $l_n|y_a = U_a$ is the time of realization of the terminal node n when all arc durations are put at their *upper* bounds.

The actual mechanics of carrying out the sampling are patently simple: apply the CP algorithm to a long series of realizations, each one obtained by assigning a sample value to every activity drawn from its proper PDF. Each realization obtained in this way is considered a sample value of T_n. Given this information, one applies standard statistical methods to estimate the PDF and/or other parameters of interest. We demonstrate below how such estimation is performed within specified confidence levels.

First, however, there is an important conceptual notion which population sampling brings to the forefront in a rather forceful manner, which hitherto did not appear of any significance. Namely, we have always emphasized the *critical path*. But surely, a more relevant concept is that of a *critical activity*, simply because it is evident from the probabilistic structure of PERT that any path (or, at least, any path in a particular subset of paths, namely those that "compete" for longest duration), is "critical" (in the sense that it is of longest duration) with a certain probability. Consequently, it seems meaningful to inquire about individual critical *activities* rather than about whole paths!

Now, a possible definition of a critical activity is an activity that is in one or more paths, each of which may be critical with a sufficiently high probability. A possible *measure of the criticality* of any activity (called the *criticality index* of the activity) is the probability that it will be on the CP.

To make this notion precise, let M denote the number of paths in the network (from node *1* to node n) and designate the paths by $\pi_1, \pi_2, \ldots, \pi_r$. (In any realistic network, r can be very large, indeed.) At any realization of the network, at least one of these r paths is the CP. If some paths are never critical, we simply ignore them completely, and they are not counted in the r paths. Clearly, the total probability measure (of unity) must be divided among these r paths; so let $p(\pi_k)$ denote the probability that path π_k is the CP. Let $a \in \pi_k$ denote a particular arc in path π_k. Then the criticality index of activity a is given by

$$\text{Criticality }(a) = \rho_a \triangleq \sum_{\substack{\pi_k \\ (a \in \pi_k)}} p(\pi_k)$$

In other words, it is the sum of probabilities $p(\pi_k)$ of the paths that contain arc a.

For instance, consider once more the simple network of Fig. 4.4; it has a total of 324 realizations, and Table 4.7 summarizes the data on the CPs, their lengths, and the frequency of their occurrence, where

Path *1–2–4*	is labeled	#1
1–2–3–4	"	#2
1–4	"	#3
1–3–4	"	#4

An entry in the table indicates the frequency of occurrence of a particular path (or combination of paths); for instance, the CP is of length 13 because of the CP lying on path 1, or path 2, or path 4, or the simultaneous occurrence of 1 and 4 or 2 and 4, with frequencies 30, 6, 24, 6, and 6, respectively, a total of 72 times. The bottom row gives the frequency of occurrence of a path as the CP; for instance, path 4 is the CP 64 times; 24 times of which it is of length 13; 24 times it is of length 12, 16 times it is of length 9, and so on.

To deduce the criticality index of the various activities we construct Table 4.8, in which the paths of the network are exhibited in terms of their constituent arcs. For instance, path 4 is composed of activities $(1,3)$ and $(3,4)$. Since path 4 is the CP 64 times out of 324 realizations, the number 64 appears next to its constituent arcs in Table 4.8, and so forth. The row sums give the number of times the activity appears on the CP. For example, activity $(1,2)$ appears 236 times on *some* CP, or 72.84% of the total realizations of the network. Observing the rightmost column and the bottom row of Table 4.8 should give a fairly good idea about the relative criticality of the various activities as well as paths. For instance, it is evident that activity $(1,2)$ is the most critical, followed by activity $(3,4)$; and that path 1 (*1–2–4*) is the most frequent CP followed by path 2 (*1–2–3–4*). Hence any effort at reducing the expected project duration should be directed at activities in these two paths and these two activities in particular.

This example is illuminating from another point of view. While it seems reasonable to measure criticality of an activity by its probability of appearing on the CP, the index should not be applied blindly. Indeed, in this example, while activity $(1,2)$ is the most critical, it is intuitively evident that the range of its duration is much smaller than the range of the duration of activities $(2,4)$. If the range is an indication of the "looseness" of the activity, one would expect success in reducing the duration of activity $(2,4)$ since one has 11 time units to "play with" versus only 4 time units for activity $(1,2)$. Thus, the criticality of an activity may not necessarily imply preference for shortening.

Drawing attention to "critical activities" is also appealing from a different point of view. We recall that the main interest in evaluating the CP is

Table 4.7. Analysis of CPs of Network of Fig. 4.4

	Paths in Network									
Length of CP	1	2	3	4	1 or 2	1 or 4	2 or 3	2 or 4	Frequency of Length	Relative Frequency
17	36								36	0.1111
15		12							12	0.0370
14	36	12							48	0.1481
13	30	6		24		6		6	72	0.2222
12		12		24				6	42	0.1296
11	6	18			6				30	0.0926
10		8	24				4		36	0.1111
9		10		16				2	28	0.0864
8	2	6			2				10	0.0309
7	2	2			2				6	0.0185
6		2							2	0.0062
5		2							2	0.0062
Frequency of Path	112	90	24	64	10	6	4	14	324	

Table 4.8. Criticality of Activities of Fig. 4.4

Activities	Paths								Frequency of Length	Relative Frequency
	1	2	3	4	1 or 2	1 or 4	2 or 3	2 or 4		
$(1,2)$	112	90			10	6	4	14	236	0.7284
$(1,3)$				64		6		14	84	0.2593
$(1,4)$			24				4		28	0.0864
$(2,3)$		90			10		4	14	118	0.3642
$(2,4)$	112				10	6			128	0.3951
$(3,4)$		90		64	10	6	4	14	188	0.5802
Path Frequency	112	90	24	64	10	6	4	14	324/782	
Path Relative Frequency	0.3457	0.2778	0.0741	0.1975	0.0309	0.0185	0.0123	0.0432		

to determine the "bottleneck activities" with the purpose of concentrating attention on such activities and "doing something about them", see the discussion on duration-cost trade-off in Chapter 2. Thus, sooner or later, we have to concern ourselves with individual activities. Therefore, it is logical to focus attention from the start on the critical activities as defined above.

An interesting question at this juncture is in what way did population sampling give prominence to the notion of critical activities, and why such an important concept was not brought up before. The answer is simple; with the help of population sampling one can *answer* the question when it is asked, which was not possible to do with the other (analytical or PERT) approaches.

The reader will recall that in the classical PERT model four assumptions were necessitated in order to be able to make probability statements concerning the duration of the project (see Section 2.2 of Chapter 1). Briefly, we assumed that:

1. The duration PDFs are unimodal—in particular, they are beta distributed.
2. The activities are independent.
3. The CP derived from the PERT calculations, denoted by π_C, is "sufficiently longer" than any other path so that the probability of a network realization having a different CP is negligible.
4. If T_n represents the time at which node n obtains, then T_n is approximately Normally distributed with a mean:

$$\mathcal{E}(\mathrm{T}_n) = \sum_{a \in \pi_C} \mathcal{E}(Y_a)$$

and a variance

$$\mathrm{var}(\mathrm{T}_n) = \sum_{a \in \pi_C} \mathrm{var}(Y_a)$$

where π_C is the CP obtained from the PERT calculations.

It is interesting to remark that the population sampling approach requires none of these assumptions to reach its probability statements. Assumption (2), that of independence among activities is, however, usually invoked to simplify the sampling procedure rather than as a necessary condition for arriving at any conclusion.

In fact, whether the activities are independent or not, the estimates of the mean and variance of T_n are *unbiased*. The reader should compare this with our conclusions in Sections 2 and 3 of this chapter.

Finally, we address ourselves to the question of the *precision* of the sampling procedure, and the *confidence* that can be attached to any statement based on the sample taken. For brevity, denote the mean of T_n by $\mu_n \triangleq \mathcal{E}(T_n)$; and let σ_n^2 denote the variance of T_n; $\sigma_n^2 \triangleq \text{var}(T_n)$; and $F_n(\tau)$ the PDF of T_n. We assume given: (a) the structure of the network and (b) the PDF $F_a(y_a)$ for each activity $a \in A$.

At the outset, all three statistics μ_n, σ_n^2, and $F_n(\tau)$ are unknown. Our desire is to determine all three of them with the "utmost precision," which, translated into concrete probability statements, we take to mean that the probability of any statement made is 0.99 or more.

Consider first the *variance*, σ_n^2. Clearly, its "best" unbiased estimate is given by the sample mean-squared-deviation s_n^2 given by

$$s_n^2 = \sum_{k=1}^{K} \frac{\left[l_n(k) - \bar{l}_n\right]^2}{K},^{\dagger}$$

where $l_n(k)$ is the "length" of the CP on the kth sample realization; $\bar{l}_n$ is the sample mean duration and K is the total number of samples taken. In Section 1 we argued that T_n is *not* normally distributed but that, *as a first order of approximation, it can be considered as such.* With this assumption we know that Ks_n^2/σ_n^2 is distributed as the χ^2 distribution with $K-1$ degrees of freedom, and that for large K it is approximately distributed as $N(K-1; 2(K-1))$, that is, as a Normally distributed RV with mean equal to $K-1$ and variance equal to $2(K-1)$. Hence, if we desire to estimate σ_n^2 within 2% of its true value with probability (i.e., confidence) of 0.99, we require that

$$Pr\left\{0.98\sigma_n^2 \leqslant s_n^2 \leqslant 1.02\sigma_n^2\right\} \geqslant 0.99$$

But the above statement is equivalent to

$$0.99 \leqslant Pr\left\{0.98K \leqslant \left(Ks_n^2\right)/\left(\sigma_n^2\right) \leqslant 1.02K\right\}$$

†Strictly speaking, in order to obtain an unbiased estimate of σ_n^2 the denominator should be $(K-1)$, but since K is of the order of several thousands, the error in using this simpler expression is negligible.

$$= Pr\left\{\frac{1-0.02K}{\sqrt{2(K-1)}} \leqslant \frac{Ks_n^2/\sigma_n^2-(K-1)}{\sqrt{2(K-1)}} \leqslant \frac{1+0.02K}{\sqrt{2(K-1)}}\right\}$$

$$= 1-2\Phi\left(\frac{1-0.02K}{\sqrt{2(K-1)}}\right)$$

where $\Phi(z)$ is the PDF of the standard Normal deviate; from which we deduce that

$$\Phi\left(\frac{1-0.02K}{\sqrt{2(K-1)}}\right) \leqslant 0.005;$$

hence

$$\frac{1-0.02K}{\sqrt{2(K-1)}} \leqslant -2.58$$

or K, the sample size is $\geqslant 33{,}295$; say 33,300 samples.

Next, consider the mean duration, μ_n. Since the variance of the RV T_n is unknown one should, strictly speaking, adopt a two-sample procedure in which the first is used to estimate the variance, then based on such estimate one constructs the size of the second sample to yield the desired degree of confidence [see e.g., Hald (1952), p. 496]. The procedure is based on the assumption that the RV T_n is Normal, and, as is to be expected, it utilizes the t-distribution (which is free of any unknown population parameters). However, since the t-distribution converges to the Normal as the sample size increases, and considering the very large sample sizes taken in this approach, it seems appropriate to utilize the Normal theory directly, especially in view of the rather high precision with which the variance, σ_n^2, is estimated.

For instance, suppose it is desired to estimate the mean, μ_n, to within $0.01\sigma_n$ of its true value with probability 0.99; in other words,

$$Pr\{\mu_n - 0.01\sigma_n \leqslant \bar{l}_n \leqslant \mu_n + 0.01\sigma_n\} \geqslant 0.99$$

which is equivalent to

$$Pr\left\{\frac{-0.01\sigma_n}{\sigma_n/\sqrt{K}} \leqslant \frac{l_n-\mu_n}{\sigma_n/\sqrt{K}} \leqslant \frac{0.001\sigma_n}{\sigma_n/\sqrt{K}}\right\} \geqslant 0.99;$$

that is, $\Phi(-0.01\sqrt{K}) \leqslant 0.005$, which yields that

$$0.01\sqrt{K} \geqslant 2.58, \qquad \text{or } K \geqslant 66{,}564 \text{ samples.}$$

Thirdly, consider the estimation of the PDF of T_n, denoted by F_n. If we are willing to assume that F_n is continuous, we can make statements about the greatest absolute difference between the sample cumulative DF and the true PDF *independent* of the distribution itself. The asymptotic results are due to Kolmogorov [see Wilks (1962), p. 341, and the references cited therein] and have been tabulated [see, e.g., Hoel (1954), p. 411].

Let $G_{nK}(\tau)$ denote the cumulative DF derived from a sample of size K; and $D_K = \sup_\tau |G_{nK}(\tau) - F_n(\tau)|$. Then the probability that D_K is less than some specified $d/\sqrt{K}$ is asymptotically given by

$$\lim_{K\to\infty} Pr\left(D_K \leqslant d/\sqrt{K}\right) = \sum_{i=-\infty}^{\infty} (-1)(-1)^i e^{-2i^2 d^2}$$

In the tables we find d and the asymptotic probability of $\sqrt{K}\, D_K \leqslant d$. For example, suppose we wish to estimate F_n by G_{nK} such that the maximum deviation between the two functions does not exceed 0.01 in absolute value more than 1% of the time. We find that for the asymptotic probability $Pr(\sqrt{K}\, D_K \leqslant d) = 0.99$ we have $d = 1.63$. Since $D_K = 0.01$, we solve for $\sqrt{K}\,(0.01) \geqslant 1.63$, to obtain $K \geqslant 26{,}570$.

Of course, knowing the PDF (or the closest thing to it, the sample cumulative DF) enables one to make probability statements concerning the completion time of the project. Unlike such probability statements made with the classical PERT calculations, these statements are accurate within the limit specified (1% in our example).

Finally, consider the measure of criticality of one or more activities. Here, one may be interested in one activity only, or in several activities.

Consider the one activity case first [Van Slyke (1963)]. In any realization of the network, the activity a is either on the CP or it is not. The probability of it being on the CP is given by its criticality index, which was defined above to be equal to

$$\rho_a = \sum_{\substack{\pi_k \\ a \in \pi_k}} p(\pi_k)$$

The dichotomy leads immediately to the binomial DF, which is approximated very well by the Normal DF for the sample sizes used in such experimentations. Consequently, if ρ_a is estimated from a sample of size K

by r_a, we know that

$$Pr\{r_a - 0.01 \leqslant \rho_a \leqslant r_a + 0.01\} = 1 - 2\Phi\left(-0.01/\sqrt{\rho_a(1-\rho_a)/K}\right)$$

Of course, ρ_a is unknown, and r_a is substituted in the r.s. of the equation in place of ρ_a or, better still, we can be conservative and substitute the worst possible value, $\rho_a = 0.5$. The same relationship can be used in the reverse direction to determine the required sample size K to guarantee, with a specified probability 0.99, say, that a particular interval about the sample value r_a contains the true value ρ_a.

The multiple activity case is more complicated. Suppose that attention is focused on a set of activities $A_m = \{a_1, a_2, \dots, a_m\}$. It seems meaningless to inquire whether the set A_m is on the CP in any realization or not because, first, one usually chooses the set A_m with an eye on it being almost always on the CP, and, second, the arcs may not be independent in their behavior any longer. In fact, it is relatively easy to construct examples in which two arcs a_i and a_j are perfectly correlated, in the sense that either arc will be on the CP if and only if the other arc is on the CP. More interesting questions, for instance, are the following:

1. Are all arcs equally critical?
2. Suppose the hypothesis is put forward that the criticality indices of arcs $a_1, \dots, a_m$ are given by $\rho_1, \dots, \rho_m$, respectively. Can one test the truth of such an hypothesis?
3. In the set A_m, which is the most critical arc, which is the second most critical, and so on, and which is the least critical arc?

To the best of our knowledge such questions still remain unanswered to date.

Improved Techniques

A great deal of improvement over the "crude" Monte Carlo sampling scheme described above is possible, due mainly to the substantial recent advances in developing sophisticated procedures for improving the statistical qualities of the sampling procedures. For our purposes, "improvement" is taken to imply smaller variance for the same amount of computing effort, or the expenditure of less amount of effort to achieve the same precision.

The major contributions to the theory of Monte Carlo sampling are due to the distinguished work of several researchers such as Tukey, Feller, Hartley, Hammersley, Handscomb, Shreider, and others; see the book by Shrieder (1966) and the references to Chapter 5 of the book by

Hammersley and Handscomb (1967). The pioneering effort in the application of these general principles to ANs is due to Burt, Gaver, and Perlas (1970), Burt and Garman (1971), and Garman (1972). We take this opportunity to alert the reader to this fertile field of research and to the significant rewards (in the form of computing efficiency) one reaps.

We briefly discuss four approaches: (a) antithetic variates, (b) stratification, (c) control variates, and (d) conditional sampling. We do not discuss other approaches such as importance sampling, regression methods, and the use of orthonormal functions. The interested reader should refer to the literature mentioned above, especially the work by Hammersley and Handscomb (1967) and the references cited therein.

In all attempts at improving the precision of the estimate a price has to be paid in the form of additional computing. Therefore, it seems logical to measure the relative improvement of a procedure over another by the following ratio:

$$\frac{(\text{labor})_1}{(\text{labor})_2} \times \frac{(\text{variance})_1}{(\text{variance})_2}$$

The ratio of the variances is a logical candidate for judging improved precision. (The ratio of the standard deviations is another logical candidate.) However, it must be tempered by the labor expenditure in each procedure. Sometimes, the sample size is itself a good measure of the effort expended; but often it is not a measure at all, especially when the same sample size is used in both procedures. In such a case, the analyst must turn to another measure such as the computer time consumed (exclusive of input–output) or the size of memory used.

Antithetic Variates

The basic idea of this approach is as follows. Suppose we have an unknown parameter θ, which we wish to estimate by a statistic t. For instance, in the context of ANs, θ may be T_n, which we are estimating by the sample duration. We seek another estimation t' having the same expectation as t and a strong *negative* correlation with t. Then $\frac{1}{2}(t+t')$ will be unbiased in θ, but its sampling variance is given by

$$\begin{aligned}\operatorname{var}\left[\tfrac{1}{2}(t+t')\right] &= \tfrac{1}{4}\operatorname{var}(t)+\tfrac{1}{4}\operatorname{var}(t')+\tfrac{1}{2}\operatorname{cov}(t,t')\\ &< \tfrac{1}{2}\operatorname{var}(t), \text{ because } \operatorname{cov}(t,t')<0\end{aligned}$$

A theorem due to Hammersley and Mauldon (1956) asserts that whenever we have an estimation consisting of a sum of random variables, it is possible to arrange for there to be a strict functional relationship between them such that the estimation remains unbiased, while its variance comes arbitrarily close to the smallest that can be attained with these variables. The basic idea is that we "rearrange" the RVs by permuting finite subintervals, in order to make the sum of the rearranged functions as nearly constant as possible; hence their variance is made as small as possible. (If the individual subinterval sums are exactly a constant, their variance is zero!)

Perhaps the best way to introduce the concept of antithetic variables is to refer to the simple network of Fig. 4.14*a*, which is composed of two activities in series. We wish to estimate the PDF of the time of realization of node *3*, denoted by T_3, or at least some of its statistics (e.g., its mean and variance).

The "straightforward" procedure is to select two random numbers (from a uniform distribution between 0 and 1), say r_a and r_b; transform to (realizations of) y_a and y_b by use of their respective DFs, say

$$y_x = F_x^{-1}(r_x); \quad x = a \text{ or } b$$

where F_x is the DF of Y_x and F_x^{-1} is its inverse. Then if we denote a realization of T_3 by τ, and the kth realization by $\tau(k)$, then the first realization is given by

$$\tau(1) = y_a(1) + y_b(1)$$

We could tabulate K such realizations and average to obtain the estimate of the mean

$$\overline{T} = \sum_{k=1}^{K} \frac{\tau(k)}{K} = \sum_{k=1}^{K} \frac{[y_a(k) + y_b(k)]}{K}$$

All random numbers are independent, and the K realizations are generated from their appropriate DFs; so $\mathcal{E}(\overline{T}_3) = \mathcal{E}(Y_a) + \mathcal{E}(Y_b)$ and $\text{var}(\overline{T}_3) = [\text{var}(Y_a) + \text{var}(Y_b)]/K$. Hence, the straightforward procedure gives an unbiased estimate of $\mathcal{E}(T_3)$ whose variance decreases as $1/K$. Unfortunately, if there is a long chain of activities in series, then the sum of the variances in

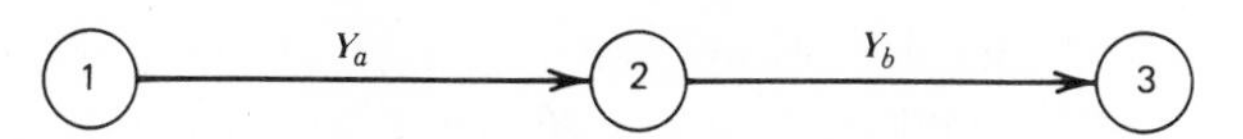

Figure 4.14*a*. Two activities in series.

the numerator becomes large, and a large sample size is required in order to determine $\mathcal{E}(T_3)$ precisely.

Notice first that in order to estimate $\mathcal{E}(T_3)$ two realizations, $\tau(k)$ and $\tau(k+1)$, need not be independent, so long as they have the correct marginal distributions. Therefore, if $\tau(k)$ is large and one "forces" $\tau(k+1)$ to be small, then the average will tend to be closer to the true value than in the case of purely independent samples!

This is accomplished in the following fashion. Suppose we construct *two* realizations of T_3 using the *one* realization of the pair r_a, r_b as follows: first obtain $y_a(1)$ and $y_b(1)$ in the usual fashion as explained above; next, obtain the pair

$$r'_a = 1 - r_a \quad \text{and} \quad r'_b = 1 - r_b$$

and from these the values $y'_a(1)$ and $y'_b(1)$; and finally the second realization $\tau'(1) = y'_a(1) + y'_b(1)$. The experiment is then repeated K times. Now consider the average

$$\overline{T}_3^A = \sum_1^K \frac{\left[y_a(k) + y_b(k) + y'_a(k) + y'_b(k)\right]}{2K}$$

$$= \frac{1}{2}\left\{\sum_1^K \frac{\left[y_a(k) + y_b(k)\right]}{K} + \sum \frac{\left[y'_a(k) + y'_b(k)\right]}{K}\right\}$$

$$= \tfrac{1}{2}\left(\overline{T}_3 + \overline{T}'_3\right); \text{ the superscript } A \text{ is for "antithetical."}$$

By construction, $\mathcal{E}(\overline{T}_3) = \mathcal{E}(\overline{T}'_3) = \mathcal{E}(T_3)$, so the estimate $\overline{T}_3^A$ is, indeed, unbiased. Furthermore, clearly $\text{var}(\overline{T}_3) = \text{var}(\overline{T}'_3) = [\text{var}(Y_a) + \text{var}(Y_b)]/K$. However, it is apparent from the manner of selection of the antithetic variables r'_x, (and, in any case, can be easily proven) that $\overline{T}_3$ and $\overline{T}'_3$ are negatively correlated; that is, $\text{cov}[\overline{T}_3, \overline{T}'_3] < 0$. Since,

$$\text{var}\left(\overline{T}_3^A\right) = \tfrac{1}{4}\left[\text{var}\left(\overline{T}_3\right) + \text{var}\left(\overline{T}'_3\right)\right] + \tfrac{1}{2}\text{cov}\left[\overline{T}_3, \overline{T}'_3\right]$$

$$= \tfrac{1}{2}\text{var}\left(\overline{T}_3\right) + \tfrac{1}{2}\text{cov}\left[\overline{T}_3, \overline{T}'_3\right] < \tfrac{1}{2}\text{var}\left(\overline{T}_3\right)$$

we conclude that the use of antithetic variables is more efficient than doubling the total number of independent samples taken. Or, to put it differently, with the *same* sample size, the use of antithetic variables would result in a smaller variance of the estimate of the mean, $\overline{\mathrm{T}}_3$, and hence a more precise one. The exact value of $\mathrm{Var}(\overline{\mathrm{T}}_3^A)$ may be easily estimated by noting the sample variance of the K independent averages $[y_a(k)+y_b(k)+y_a'(k)+y_b'(k)]/2$; $k=1,2,\ldots,K$. Since $\overline{\mathrm{T}}_3^A$ is the average of K independent terms, approximate confidence limits may be placed on the "true" value of $\mathcal{E}(\mathrm{T}_3)$ using the t-distribution, with the approximation improving with the increase in the sample size K.

The dramatic effect of antitheticizing the sample is most apparent when the DF of Y_x is symmetric for all activities x (e.g., Y_x is uniformly or Normally distributed). In this case, it is easy to see that the estimate $\overline{\mathrm{T}}_3^A$ is of *zero variance*! The reason is that the antithetic approach makes double use of each random number generated; hence, y_x and y_x' tend systematically to appear on opposite tails of the empirical distribution being developed.

It should also be apparent that the use of antithetic variables is helpful in estimating the DF of the random variable T_3. From the first set of observations one determines the empirical cumulative DF $\hat{F}(\tau)$, and from the antithetic set of observations one determines the second empirical cumulative DF $\hat{F}'(\tau)$, the antithetic variate estimate of the "true" DF would then be given by $\hat{F}^A(\tau)=[\hat{F}(\tau)+\hat{F}'(\tau)]/2$, all τ. Evidently, $\hat{F}'(\tau)$ compensates for any deviation of $\hat{F}(\tau)$ from the true DF $F(\tau)$.

As a second illustration of the antithetic approach, consider the simple parallel network of Fig. 4.14*b*. The objective is to determine the PDF of $\mathrm{T}_2=\max(Y_a,Y_b)$. For simplicity, assume both durations to have the same (standardized) rectangular distribution centered about zero. This is no restriction in fact, for even if the Y's were arbitrarily distributed, we have the identity

$$Pr[Y\leqslant y]=Pr[F_Y(y)\leqslant y]=Pr[R\leqslant y]$$

where R is a random variable uniformly distributed over the interval $[0,1]$. A simple change of variables yields the indicated DF.

The crude approach would take K independent samples $\{r_a,r_b\}$ from

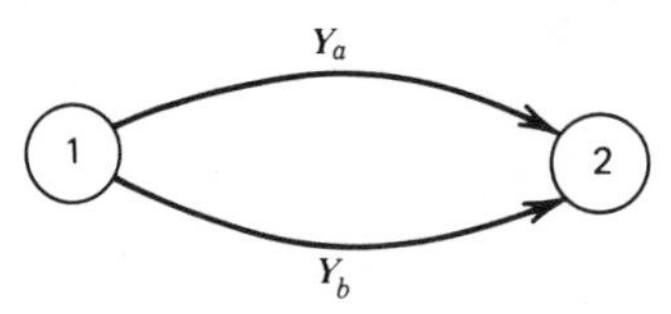

Figure 4.14*b*. Two activities in parallel: $F_Y(y)=0.5+y$; $y\in[-0.5,0.5]$.

each activity. Consider a fixed value of τ in the interval $[-0.5, 0]$, and let

$$\delta_\tau(k) = \begin{bmatrix} 1 & \text{if } r_a, r_b \leqslant \tau + 0.5;\ r_a, r_b \in [0,1] \text{ and } \tau \in [-0.5, 0] \\ 0 & \text{otherwise} \end{bmatrix}$$

on the kth realization of the network; $k = 1, 2, \ldots, K$. Notice that

$$\mathcal{E}[\delta_\tau(k)] = (\tau + 0.5)^2$$

hence $\delta_\tau(k)$ is unbiased in $F_{T_2}(\tau)$, and

$$\text{var}[\delta_\tau(k)] = (\tau + 0.5)^2 [1 - (\tau + 0.5)^2]$$

If K independent samples are taken, let

$$\bar{\delta}_\tau = \sum_{k=1}^{K} \frac{\delta_\tau(k)}{K}$$

It is easy to see that $\bar{\delta}_\tau$ is unbiased in $F_{T_2}(\tau)$. Furthermore, its variance is given by $\text{var}(\bar{\delta}_\tau) = (\tau + .5)^2 [1 - (\tau + .5)^2]/K$. With more than two paths in parallel, say s paths, the expression for the variance becomes $(\tau + .5)^s \times [1 - (\tau + .5)^s]/K$ which decreases as $1/K$.

An antithetic approach may proceed as follows. With each δ_τ defined above associate the antithetic variables δ'_τ defined as follows:

$$\delta'_\tau = \begin{bmatrix} 1 & \text{if } r_a, r_b \geqslant 0.5 - \tau; \quad \tau \in [-0.5, 0] \\ 0 & \text{otherwise} \end{bmatrix}$$

Notice that δ_τ and δ'_τ cannot be equal to 1 simultaneously in the specified range of τ. Furthermore,

$$\mathcal{E}(\delta'_\tau) = [1 - (0.5 - \tau)]^2 = (\tau + 0.5)^2$$

Hence, δ'_τ is also unbiased in $F_{T_2}(\tau)$. Its variance is identical to that of δ_τ. Define $\delta'_\tau(k)$ similar to $\delta_\tau(k)$ and let $\bar{\delta}^A_\tau(k) = [\delta_\tau(k) + \delta'_\tau(k)]/2$. Form the antithetic estimate

$$\delta^A_\tau = \sum_{k=1}^{K} \frac{\bar{\delta}^A_\tau(k)}{K}$$

Since the sum $\delta_\tau(k) + \delta'_\tau(k) = 1$ if either variable equals 1, then $\bar{\delta}^A_\tau(k) = 1/2$

with probability $2(\tau+0.5)^2$, and $\bar{\delta}_\tau^A(k)=0$ with probability $1-2(\tau+0.5)^2$. Hence,

$$\mathcal{E}\left[\bar{\delta}_\tau^A(k)\right]=(\tau+0.5)^2$$

and the estimate is indeed unbiased in $F_{\mathrm{T}_2}(\tau)$. Furthermore,

$$\begin{aligned}\operatorname{var}\left[\bar{\delta}_\tau^A(k)\right]&=1/4\times 2(\tau+0.5)^2-\left(\mathcal{E}\left[\bar{\delta}_\tau^A(k)\right]\right)^2\\&=(\tau+0.5)^2\left[1/2-(\tau+0.5)^2\right]\end{aligned}$$

Consequently,

$$\operatorname{var}(\delta_\tau^A)=\frac{1}{K^2}\sum_{k=1}^{K}\operatorname{var}\left[\bar{\delta}_\tau^A(k)\right]=(\tau+0.5)^2\left[1-2(\tau+0.5)^2\right]/2K$$

The ratio

$$\frac{\operatorname{var}(\bar{\delta}_\tau^A)}{\operatorname{var}(\bar{\delta}_\tau)}=\frac{\left[1-2(\tau+0.5)^2\right]}{2\left[1-(\tau+0.5)^2\right]},\quad \tau\in\left[-0.5,0\right]$$

which varies between $\frac{1}{3}$ and $\frac{1}{2}$ as τ varies from 0 to -0.5. Hence, antitheticizing the variables improved the precision of the estimate by more than twice the original effort, a gain in computing efficiency.

The above construction yields the PDF of T_2 for $\tau\in[-0.5,0]$. We leave it as an exercise to the reader to construct two antithetical variables for τ in the interval $[0,0.5]$; see Problem 13.

As to computing experience with this approach, Burt, Gaver, and Perlas (1970) studied the three simple networks shown in Fig. 4.15 to estimate the expected completion times. The distribution of all arc durations was assumed exponential of the form $F(y)=1-e^{-0.1y}$; $0\leqslant y$. All three examples may be analyzed analytically, which has been done and was used as a check on the Monte Carlo results. These are summarized in Table 4.9. As can be seen, the variance reduction obtained by utilizing the antithetic procedure is greater than that from doubling the sample size and sampling independently. It is worth noting that the estimate of the variance of completion time obtained by simple averaging of the variance estimates from the antithetic experiments (i.e., the estimate in column 5 of Table 4.9) is likely to be biased. Agreement with the analytically calculated values seems, however, to be quite satisfactory for the present example.

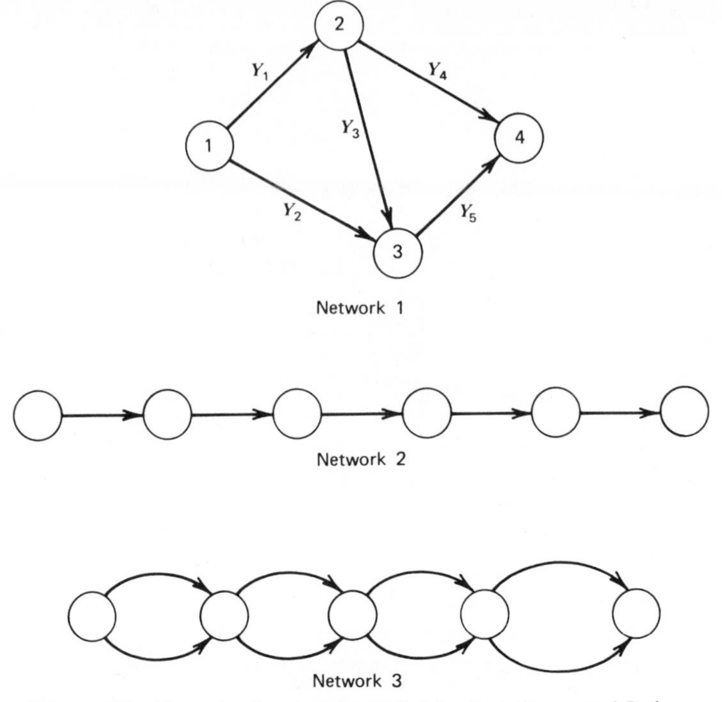

Figure 4.15. Three simple networks studied by Burt, Gaver, and Perlas.

Control Variates (CVTs)

The basic idea underlying antithetic variates was the generation of a twin RV that is *negatively correlated* with the original RV. An alternative approach is to generate a RV that is *positively correlated* with the original RV. Let us call this new RV the *control variate* and assume that it is also endowed with the property of being easily manipulated mathematically. (We can always ensure this property since we choose the CVT). Now, if we wished to estimate some parameter, say θ, of the RV X, say $\theta = g(X)$, where the function g may be expectation, or variance, etc., we let $\phi(X)$ denote the CVT and we can write the identity $\theta = \phi(X) + [g(X) - \phi(X)]$. Since ϕ is positively correlated with g, it "mimics" it in some fashion and the difference, $g(X) - \phi(X)$, should be small. We are now left with θ to be estimated by $\phi(X)$—which can be done analytically—except for the error term $g(X) - \phi(X)$. It is this latter term that is estimated through Monte Carlo sampling, the advantage being that, in all probability, it can be estimated with a great deal of precision.

Table 4.9. Antithetic Variates: Variance Reduction (T^A = Antithetic Realization = $(t+t')/2$)

Network	Sample Size K	$\frac{\hat{\sigma}^2(\mathrm{T})}{K}$	$\frac{\hat{\sigma}^2(\mathrm{T}')}{K}$	Average Columns 3+4	$\frac{\sigma^2(\mathrm{T})}{K}$	$\frac{\hat{\sigma}^2(\mathrm{T}^A)}{K}$	v
(1)	(2)	(3)[a]	(4)	(5)	(6)[b]	(7)	(8)[c]
1	50	5.341	5.986	5.664	5.774	1.849	3.12
1	100	2.830	2.650	2.740	2.887	0.852	3.39
1	150	1.526	1.443	1.485	1.443	0.428	3.37
1	200	0.760	0.691	0.726	0.722	0.227	3.18
2	50	10.110	10.724	10.417	10.000	1.814	5.51
2	100	5.133	4.781	4.957	5.000	0.905	5.53
2	150	2.375	2.531	2.453	2.500	0.444	5.63
2	200	1.233	1.205	1.219	1.250	0.225	5.56
3	50	10.486	9.533	10.011	10.000	3.644	2.74
3	100	4.685	5.553	5.119	5.000	1.855	2.70
3	150	2.601	2.388	2.495	2.500	0.859	3.02
3	200	1.319	1.243	1.281	1.250	0.422	2.96

[a] $\hat{\sigma}^2$ is the estimate of the variance;
[b] Column 6 is $\mathrm{var}(\mathrm{T})/K = \mathrm{var}(\mathrm{T}')/K$, computed analytically.
[c] v is the ratio of column 6 to column 7.

To render these notions concrete, suppose we are interested in the expected duration of the project shown in Fig. 4.16*a*. Suppose, furthermore, that all durations of activities are RVs whose distribution functions are amenable to mathematical manipulation except Y_7. What is $\mathcal{E}(\mathrm{T}_5)$, where T_5 denotes the duration of the project?

We may emulate the network in Fig. 4.16*a* by that in Fig. 4.16*b*, which is much simpler† and whose $\mathcal{E}(\hat{\mathrm{T}}_5)$ can be determined *analytically*, where

$$Y_A = Y_1 + Y_4$$

$$Y_B = Y_1 + Y_3$$

Y_C = a RV highly correlated with Y_7 but much simpler.

Notice that from the identity

$$\mathcal{E}(\mathrm{T}_5) = \mathcal{E}(\hat{\mathrm{T}}_5) + \mathcal{E}\left[\mathrm{T}_5 - \hat{\mathrm{T}}_5\right]$$

†Note that Fig. 4.16*b* ignores the path containing activity $(2,4)$ of duration Y_5.

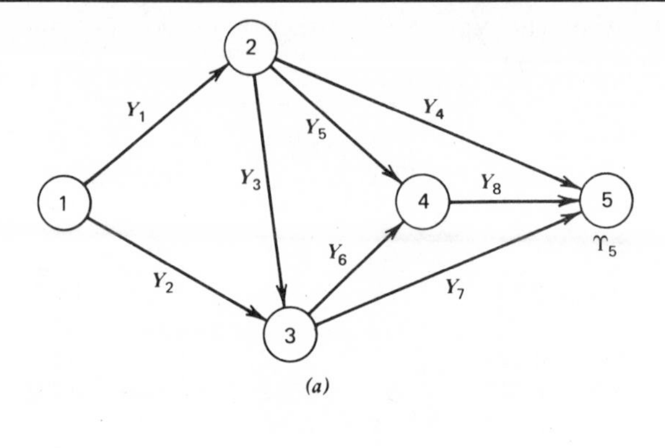

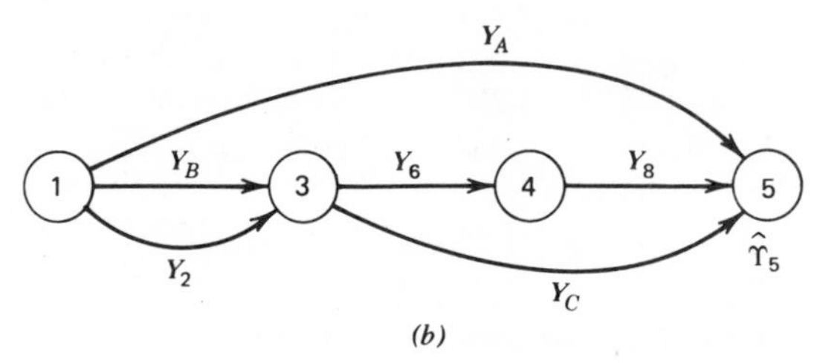

Figure 4.16. Original network and its CVT network.

we may estimate $\mathcal{E}(\mathrm{T}_5)$ from the (analytically determined) value of $\mathcal{E}(\hat{\mathrm{T}}_5)$ and the (sampled) estimate of the difference $\mathcal{E}[\mathrm{T}_5-\hat{\mathrm{T}}_5]$. The latter step involves obtaining K realizations of the eight RVs $Y_1, Y_2, \ldots, Y_8$, plus the surrogate RV Y_C. These, in turn, will yield the values $T(k)$ and $\hat{T}(k)$, $k=1,2,\ldots,K$, where the T values are the sample realizations of the T values. Averaging, we obtain the estimate of $\mathcal{E}[\mathrm{T}_5-\hat{\mathrm{T}}_5]$ as given by

$$\sum_{1}^{K} \frac{\left[T(k)-\hat{T}(k)\right]}{K}$$

Since the arc durations in the two networks share the same random numbers, T_5 and $\hat{\mathrm{T}}_5$ are positively correlated. Thus

$$\operatorname{var}\left(\overline{\mathrm{T}}_5-\overline{\hat{\mathrm{T}}}_5\right)=\frac{1}{K}\left\{\operatorname{var}(\mathrm{T}_5)+\operatorname{var}(\hat{\mathrm{T}}_5)-2\operatorname{cov}(\mathrm{T}_5,\hat{\mathrm{T}}_5)\right\}$$

Hence, it follows that if

$$\operatorname{var}(\hat{\mathrm{T}}_5)<2\operatorname{cov}(\mathrm{T}_5,\hat{\mathrm{T}}_5)$$

a reduction in variance over crude sampling is achieved.

To illustrate the power of this technique, consider the project depicted by network 1 of Fig. 4.15, and suppose we choose the control network of Fig. 4.17. Again, assume all arc durations in both networks to be identically distributed as $F(y)=1-e^{-0.1y}$; $0 \leqslant y$. Let T_4 represent the duration of the network 1 of Fig. 4.15, and let $\hat{T}_4$ denote the duration of the CVT network of Fig. 4.17. Due to its simplicity, the DF of $\hat{T}_4$ can be easily determined (see Section 4.1), and we immediately obtain

$$\mathcal{E}(\hat{T}_4)=32.5; \quad \text{var}(\hat{T}_4)=294$$

Applying crude Monte Carlo to network 1 of Fig. 4.15, we have that

$$T_4=\max\{Y_1+Y_4;\, Y_1+Y_3+Y_5;\, Y_2+Y_5\}$$

Furthermore, the "controlled" estimate of the expected duration of the network† is given by

$$\overline{T}_4^C=\mathcal{E}(\hat{T}_4)+\mathcal{E}(T_4-\hat{T}_4).$$

in which $\mathcal{E}(T-\hat{T})$ is estimated by $\overline{T}_4-\overline{\hat{T}}_4$ obtained from the Monte Carlo samples. The above relation is better written as $\overline{T}_4^C=\mathcal{E}(T_4)+[\overline{T}_4-\overline{\hat{T}}_4]$. Table 4.10 summarizes the experimental findings.

Interestingly enough, an analytic solution can be obtained for the duration of network 1. In particular, the true value of the mean and variance of T_4 are

$$\mathcal{E}(T_4)=34.58; \quad \text{var}(T_4)=289$$

It is instructive to witness the closeness of $\overline{T}_4^C$ to $\mathcal{E}(T_4)$, as well as the drastic gain in precision due to the use of the CVT.

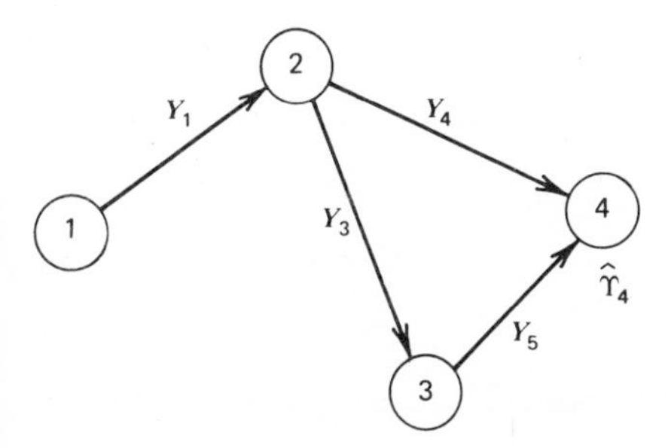

Figure 4.17. The CVT network of Network 1 of Fig. 4.15.

†That is, the estimate of the expected duration using the CVT technique, and is denoted by $\overline{T}^C$.

Table 4.10. Effects of Control Variates

	Number of Realizations, K			
	50	100	200	400
$\bar{T}_4$	37.74	33.84	34.01	34.84
$\text{var}(\bar{T}_4)/K$	6.23	2.35	1.47	0.75
$\mathcal{E}(\hat{T}_4)$	34.84	31.35	32.04	32.70
$\bar{T}_4^C$	35.40	34.99	34.47	34.64
$\text{var}(\bar{T}_4^C)/K$	0.82	0.46	0.25	0.15
var Reduction[a]	7.59:1	5.11:1	5.88:1	5.00:1

[a]Obtained by dividing the entries in the second row by the fifth row.

The success of the CVT technique depends on how closely the control variate $\hat{T}_4$ mimics the realization of T_4 in the given network. The control network must be sufficiently simple so that we can calculate $\hat{F}_4(\tau)$ exactly; however, we must be sure that realizations of the control network are highly correlated with those of the original network. This calls to the foreground the concept of "criticality" mentioned on p. 277. In brief, the criticality of an activity is measured by the probability that it will lie on the CP. If an activity has a high criticality, it should be included in the control network.

There is no necessity to restrict the use of the CVT technique to a simple CVT. If two different control networks are constructed, we may label their durations $\hat{T}_1$ and $\hat{T}_2$; then a good estimate of the desired mean is

$$\bar{T}^C = w_1\left[\frac{1}{K}\sum_1^K (T(k) - \hat{T}_1(k)) + E(\hat{T}_1)\right] + w_2\left[\frac{1}{K}\sum^K (T(k) - \hat{T}_2(k)) + E(\hat{T}_2)\right]$$

The weights w_1 and w_2 could be set at $\frac{1}{2}$ or computed by some method (e.g., regression).

Conditional Sampling

The reader will recall that the major reason for the breakdown of the analytic approach offered in Section 4.1 was the interdependence of paths due to common arcs. Clearly, if such correlation among paths is eliminated, the result would be a set of independent paths, and the DF of the completion time would simply be the product of the DFs of the paths. Conditional sampling is an approach toward such independence. It is best illustrated by an example.

Consider once more the simple network of Fig. 4.9 (or Fig. 4.15), and let T_4 denote the completion time. Our previous analysis (see p.262) indicates that if the values of the RVs $Y_{1,2}$ and $Y_{3,4}$ were fixed then the three paths of the network would be independent and that the *unconditional* PDF is given by

$$F_4(\tau) = \int_{y_{1,2}} \int_{y_{3,4}} F_4 | y_{1,2}, y_{3,4} \partial F_{Y_{1,2}} \partial F_{Y_{3,4}} \tag{4.27}$$

The Monte Carlo approach enters the picture at this juncture; instead of executing (4.27) analytically, one would sample the values $y_{1,2}$ and $y_{3,4}$ and determine the estimate, $\psi_4(\tau)$, of the DF of T_4.

Evidently, for any pair of values $(y_{1,2}, y_{3,4})$, the three paths of the network are independent and in parallel. Hence, one can determine the probability that the total duration is $\leqslant \tau$, for any given $\tau \geqslant y_{1,2} + y_{3,4}$. In particular, it is the product of the three probabilities: $Pr[Y_{2,4} \leqslant \tau - y_{1,2}] \times Pr[Y_{1,3} \leqslant \tau - y_{3,4}] \times Pr[Y_{2,3} \leqslant \tau - y_{1,2} - y_{3,4}]$, since each probability is the probability of one of the three paths being of duration $\leqslant \tau$. Let the above product be denoted by $F_4(\tau | y_{1,2}, y_{3,4})$. Then the desired estimate after K samples of the pair $(y_{1,2}, y_{3,4})$ is given by

$$\psi_4(\tau) = \sum_{k=1}^{K} \frac{F_4(\tau | y_{1,2}, y_{3,4})}{K} ; y_{1,2} + y_{3,4} \leqslant \tau \tag{4.28}$$

One repeats the random sampling K times for each value τ and then averages the values of ψ obtained according to Eq. (4.28). Of course, for any realization of the two samples, one may determine $\psi_4(\tau)$ for all "permissible" values of τ (which depend on the sampled values of $y_{1,2}$ and $y_{3,4}$). However, caution should then be exercised in making any confidence statements since the estimates are correlated.

In the conduct of the sampling experiments, two factors must be considered: (a) the choice of the sample points $\{\tau\}$ at which the DF is estimated and (b) the size of the sample K to be taken at the various sample points.

There is a tendency among experimenters to place the values of τ at equal intervals from the origin, until a "sufficiently high" value is reached beyond which further sampling is considered wasteful. We suggest non-equal spacing of the sample points as more appropriate, since the "rise" in the DF is usually not uniform over its range. In particular, more sample points should be located in the region (or regions) of the steepest rise of the DF. This necessitates that the analyst have some a priori concept about the approximate shape of the DF. Perhaps a small pilot study is appropriate in this respect.

Concerning the question of the size of the random samples (of $Y_{1,2}$ and $Y_{3,4}$) to be determined at each time value, the tendency is again to take the same size sample at all points, say $K=1000$. This may be simultaneously wasteful and insufficient; it is wasteful at the "extremities" and insufficient in the "middle." Consider a small value of τ_1 such that $F_4(\tau_1) \leqslant 0.1$. The variance of the estimate of p is $0.09/K$. On the other hand, for a value of τ_2 close to the median we have that $F_4(\tau_2) \cong 0.5$, in which case the variance of the estimate is approximately $0.25/K$. Clearly, if we utilize the same sample size K, the first value will be estimated more precisely than the second, the improvement in precision is by a factor of $\sqrt{0.25/0.09} = 1.67$. On the other hand, if K is sufficiently large to estimate $F_4(\tau_2)$ with a high degree of precision, it will be excessively wasteful at the "tail ends" of the DF.

Generalizing from the above discussion in a different direction, it is well to recall that we conditioned our estimate on the two values $Y_{1,2}$ and $Y_{3,4}$ because if these RVs were fixed the remaining paths would be independent. In general, we refer to such RVs as *common activities* because they are shared by more than one path. Alternatively, we define a *unique activity* as one that lies on one path only. Evidently, the utility of the approach of conditional sampling is crucially dependent on the ratio of unique activities to common activities; the higher the ratio, the more useful is the approach. For example, in the network of Fig. 4.9, each sample consists of only two values ($y_{1,2}$ and $y_{3,4}$) instead of five; this is a saving of 60% of the sampling effort. However, more effort is involved in performing the multiplication of probabilities for each sample. Hence, the net reduction in effort may be only 50% or 45%.

On the other side of the ledger, the process of conditioning enables us to apply an analytical approach to the remainder of the network. This, in turn, means that we are able to use all the information available on the DFs of the activity durations, rather than just the sample values as in crude Monte Carlo. Consequently, the variability of the estimate $\psi_4(\tau)$ from its true value is greatly reduced. This is a positive and important gain.

To sum up the discussion thus far, the proposed conditional Monte Carlo sampling should contribute toward a more efficient procedure in two respects: (a) by reducing the sampling effort itself because one need only sample from the nonunique activities and (b) by greatly reducing the variability of the probability estimate because of our ability to utilize *all* the information on the DF of the unique activities. The price to be paid for these two advantages is the added burden of multiplying out the conditional probabilities.

There remains one final issue, namely, how to determine in a systematic fashion the unique activities in large networks.† Burt and Garman (1971) suggest the following labeling procedure (for A-on-A representation):

STEP 1. Label the start node with (√).

STEP 2. Select any unlabeled node, all of whose predecessor nodes are labeled. This is the "currently scanned node" (CSN).

STEP 3. If the CSN has two or more immediate predecessor nodes or any immediate predecessor activities marked N, then mark all activities egressing from the CSN with the symbol N. Label (√) the CSN.

STEP 4. If all nodes except the finish node are labeled (√), continue; otherwise, return to Step 2.

STEP 5. Label the finish node with (√).

STEP 6. Select any remaining unlabeled (√) node, all of whose successor nodes are labeled (√). This is the CSN.

STEP 7. If the CSN has two or more immediate successor nodes or any successor activity marked n, then mark all activities immediately preceding the CSN with an n. Label (√) the CSN.

STEP 8. If all nodes except the start node are labeled (√), then stop; otherwise return to Step 6.

STEP 9. Stop: (a) All activities not marked N or n are unique. (b) All activities marked N or n are common.

As an example of the application of this algorithm, consider the network of Fig. 4.18: activities 1, 2, 7, and 8 are common, while activities 3, 4, 5, and 6 are unique.

If the reader is endowed with more patience regarding the concept of conditional sampling, the following more sophisticated approach, suggested by Garman (1972), will result in fewer samples at the expense of more computing because of the need to calculate the convolutions of functions.

Consider again for a moment the network of Fig. 4.9. The conditional Monte Carlo approach described above identified activities $(1,2)$ and $(3,4)$ as the nonunique activities. This led to the sampling of $Y_{1,2}$ and $Y_{3,4}$.

†This question had been raised previously in the discussion of the analytical approach to the determination of the exact PDF (see p. 263), and its answer was deferred to the present section.

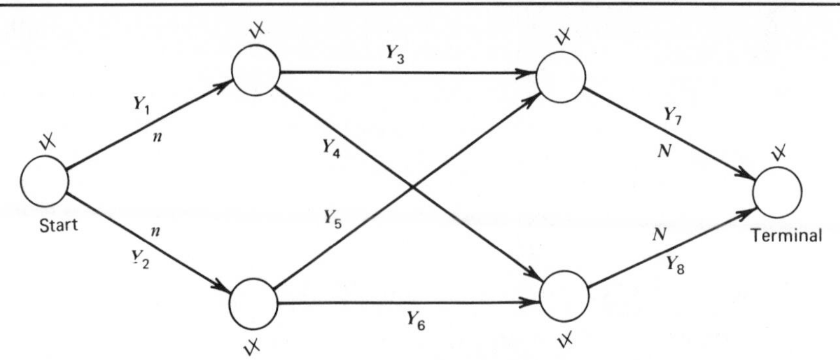

Figure 4.18. Application of procedure to determine the least number of common activities

As an alternative, suppose we consider only $Y_{1,2}$ constant at $y_{1,2}$. Then all paths to node *4* will not be independent. But now there are two paths in parallel from *1* to *3* (see Fig. 4.19), and series-parallel reduction methods to the network of Fig. 4.19 yields the estimate

$$F_4(\tau)|y_{1,2}=\{[F_{1,3}(\tau)F_{2,3}(\tau-y_{1,2})]*F_{3,4}(\tau)\}F_{2,4}(\tau-y_{1,2}) \quad (4.29)$$

where the asterisk in (4.29) denotes the convolution operator, as customary.

Thus we have reduced the number of activities sampled from two to only one—at the price of computing the convolution in (4.29).

It is not difficult to see that some ingenuity, and systematization, are required in order to arrive at the smallest number of samples to be drawn [or, alternatively, to arrive at the maximum confidence (minimum variance) for a given sample size]. In fact, there may be a sequence of "nested samples" before the final convolution is undertaken (if at all). To illustrate, consider the project network shown in Fig. 4.20. Suppose we sample activity (*1*,*2*) and then we obtain the reduced network of Fig. 4.21, which

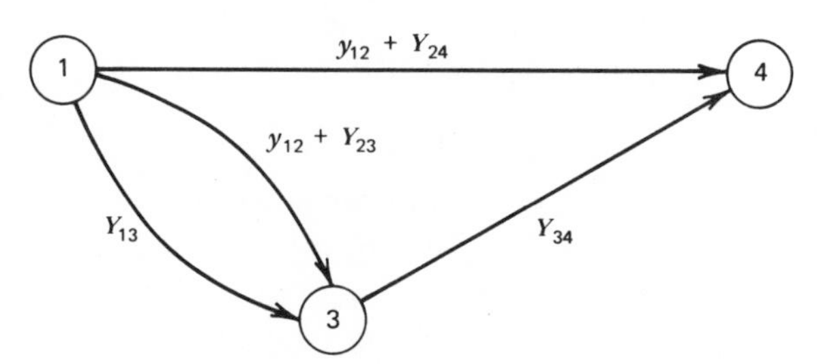

Figure 4.19. Conditioning on $Y_{1,2}$ for network of Figure 4.9.

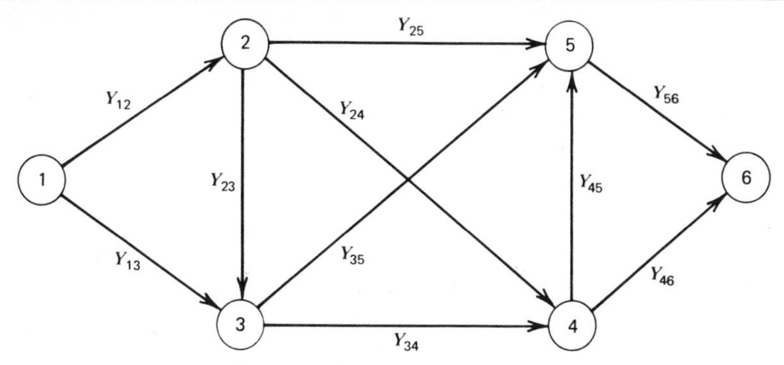

Figure 4.20. Example of nested sampling: Sample $Y_{1,2}$.

is subject to further series-parallel reduction if we sample from the DF of the new RV Z_1 *after* we have sampled from $Y_{1,2}$ (hence the name "nested sampling"). The result is shown in Fig. 4.22, which is really the now too-familiar network 1 of Fig. 4.15 in disguised form. If we now sample from Z_3 (notice that this would constitute the third level of "nesting"), the network reduces, through series-parallel reduction, to the network shown in Fig. 4.23. Here, the analyst has the choice of either resorting to sampling for $Y_{5,6}$, in which case convolution would not be needed, or resort to the convolution of the DF of $\max\{Z_2, z_3 + Y_{4,5}\}$ with the DF of $Y_{5,6}$.

The "systematization" of this procedure runs as follows:

STEP 1. Perform series-parallel reduction on the given network.

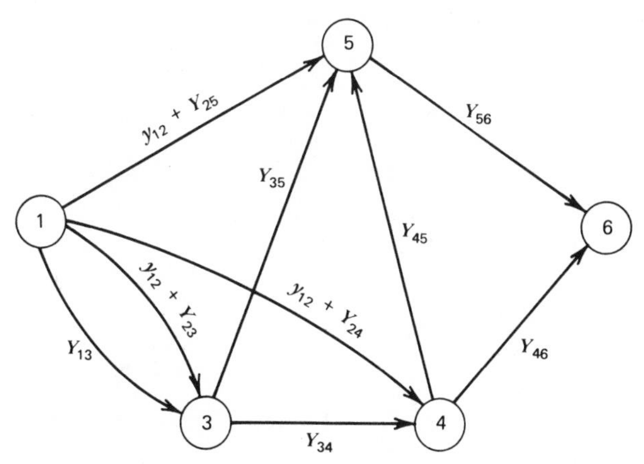

Figure 4.21. Reduced network: Sample Z_1: $Z_1 = \max(Y_{13}, y_{12} + Y_{23})$; $F(z_1) = F_{13}(z_1)F_{23}(z_1 - y_{12})$.

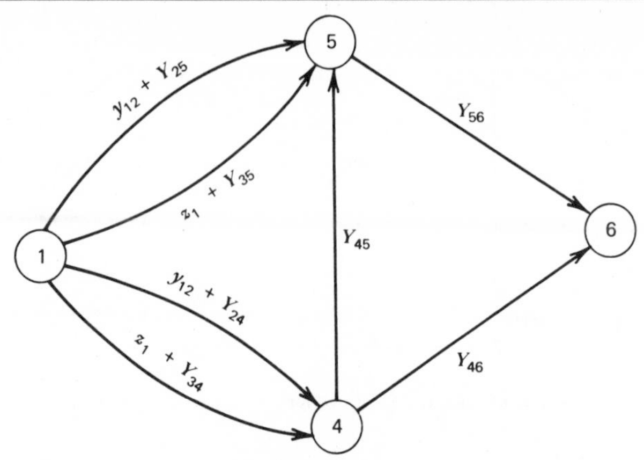

Figure 4.22. Further reduced network: Sample Z_3: $Z_2=\max(y_{12} + Y_{25}; z_1 + Y_{35})$, $F(z_2)=F_{25}(z_2-y_{12})F_{35}(z_2-z_1)$; $Z_3=\max(y_{12}+Y_{24}; z_1+Y_{34})$, $F(z_3)=F_{24}(z_3-y_{12})F_{34}(z_3-z_1)$.

STEP 2. If the reduced network is trivial (i.e., is composed of only one arc), stop; the corresponding product-convolution formula is the estimator.

STEP 3. If the network is nontrivial, choose either: (a) an activity Y with the property that Y has more than one successor while each of its successors has only Y as a predecessor or (b) an activity Y' with the property that Y' has one predecessor while each of its predecessors has only Y as successor.

STEP 4. "Condition" by setting the chosen activity Y (or Y') to a sample y (or y'). Delete Y (or Y'), add y (or y') to the successors of Y (predecessors of Y'), and maintain the implied precedence of activities in the "conditioned" network. Return to Step 1.

Since every iteration of Steps 1–4 must reduce the given network by at least one activity, a trivial network must eventually be reached.

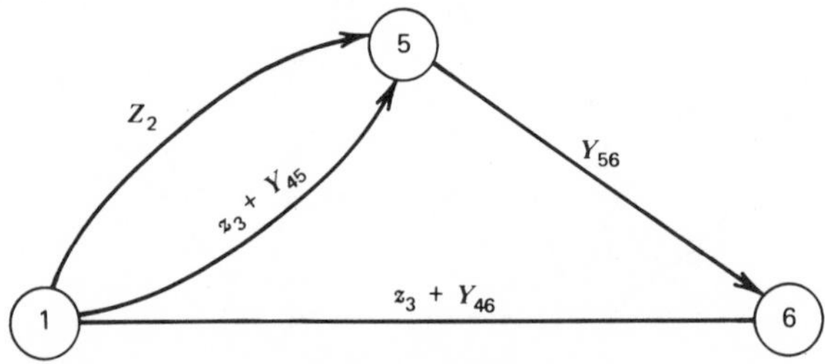

Figure 4.23. Final reduced network: $Z_4=\max(Z_2, z_3+Y_{45})$, $F(z_4)=F_{Z_2}(z_4)F_{45}(z_4-z_3)$; $T_6=\max\{Z_4+Y_{46}, z_3+Y_{46}\}$.

Stratified Sampling

In stratified sampling we break the range of the duration of each activity into several intervals, say

$$\alpha_{j-1} \leqslant Y_j \leqslant \alpha_j; \quad j=1,\ldots,K$$

where $\alpha_0 < \alpha_1 < \cdots < \alpha_K = M$, and M is the maximum duration of Y (or its truncated version). A random value is chosen from each interval and the analysis proceeds on the basis of the samples drawn.

The basic concepts in stratified sampling to yield good results are few and simple. First, consider the α values; they should be chosen so that the differences between the mean values of the DF in the various intervals are greater than the variations within the intervals. Alternatively, the α values should be chosen so that the variation of the estimate is the same in each interval.

Second, consider the number of samples drawn in each interval. These should be proportional to the variance of the estimate in the interval. If, as suggested above, the intervals are chosen to be of approximately equal variance, then the sample size should be the same for all intervals.

Third, there is the operational question of how to draw a random sample from a given interval. We discuss only the case of equal sample size for all intervals—the case of different sample sizes follows similar reasoning.

The K random numbers may be derived from one random number in several ways. We illustrate two such ways:

First, generate the random number r, $0<r<1$, then calculate the values $x_k = \{[10Kr]+k\}(\text{mod}\, K)$, for $k=1,2,\ldots,K$, where $[\cdot]$ denotes the "largest integer less than or equal to the quantity." Then the desired random number in the kth subinterval is given by the fraction in

$$r(k) = \frac{(r+x_k)}{K}; \quad k=1,\ldots,K$$

For instance, suppose that $K=4$ and the intervals are: (0, 0.25), (0.25, 0.5), (0.5, 0.75), and (0.75, 1.0). Let the random number generated be $r=0.308$. Then

$$x_1 = 1 \Rightarrow r(1) = 0.327$$

$$x_2 = 2 \Rightarrow r(2) = 0.577$$

$$x_3 = 3 \Rightarrow r(3) = 0.827$$

$$x_4 = 0 \Rightarrow r(4) = 0.077$$

The advantage of generating the RVs based on the random numbers generated by this scheme is that they are distributed over the various intervals in a random fashion. In other words, the algebra leads to the generation of random numbers that fall in the various segments randomly from experiment to experiment, although for any single experiment they fall sequentially in the various intervals, with the starting interval dependent on the random number r.

Second, generate the random number r, $0<r<1$; then calculate the values:

$$r(k)=\frac{k-1+r}{K}; \quad k=1,2,\ldots,K$$

It is easy to see that in this method the random numbers will always be generated in the same sequence, first $r(1)$, followed by $r(2)$, then $r(3),\ldots,r(K)$. This may be objectionable because of the strong linear correlation among the RVs thus generated in any activity. However, this can be easily remedied by the subsequent random matching of the RVs generated from various experiments before calculating the final result, $\tau(n)$, the time of realization of the final node on the nth experiment.

For instance, consider two activities in tandum with durations Y_1 and Y_2 (see Fig. 4.24). Suppose each activity is divided into three segments and that we have generated 10 RVs in each segment (the r values refer to activity 1 and the ρ values, to activity 2; $r_n(k)$ is the RV generated on the nth draw in segment k; and $\rho_n(k)$ is defined similarly). Now we randomize the matching of the r and the ρ values and may end up with the following realizations:

$$\begin{array}{lll}
r_1(1),\ r_1(2),\ r_1(3) & \qquad & \rho_1(1),\ \rho_1(2),\ \rho_1(3) \\
r_2(1),\ r_2(2),\ r_2(3) & & \rho_2(1),\ \rho_2(2),\ \rho_2(3) \\
\vdots \quad \vdots \quad \vdots & & \vdots \quad \vdots \quad \vdots \\
r_{10}(1), r_{10}(2), r_{10}(3) & & \rho_{10}(1), \rho_{10}(2), \rho_{10}(3)
\end{array}$$

$$r_1(1)+r_5(2)+r_3(3)+\rho_6(1)+\rho_{10}(2)+\rho_8(3) \Rightarrow \tau(1)$$
$$r_2(1)+r_3(2)+r_9(3)+\rho_4(1)+\rho_5(2)+\rho_1(3) \Rightarrow \tau(2) \text{ (etc.)}$$

The estimate of the DF of T will be based on the sample values $\tau(1),\tau(2),\ldots,\tau(N)$.

As may be well imagined, the advantage of the approach of stratified sampling lies in the reduction in the variability of the final estimate, due to the reduction in the variability of the estimates in individual segments. For instance, if the variance of $\tau(k)$ is $1/10$ the variance of Y in the (simple) example of Fig. 4.24, we would obtain a variance of T that is approximately 0.3 its original value without stratification. Of course, this advantage is somewhat diluted by the added expenditure of effort in structuring the subintervals and calculating the individual stratified random variables.

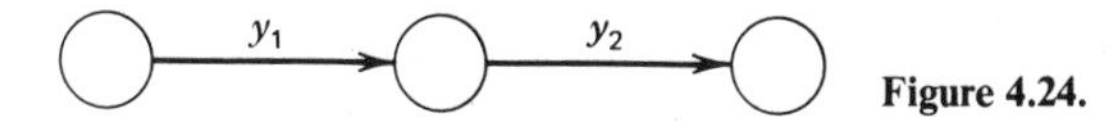

Figure 4.24.

Hybrid Approaches and a Sample Problem

In the above discussion we have elaborated on each individual approach independently. But the alert reader will realize that some of these approaches may be combined to advantage. For instance, the antithetic approach may be combined with any other approach to yield improved results, and so on.

As a second attempt to illustrate the relative merits of the various techniques, consider the "modified Wheatstone bridge" of Fig. 4.25 and assume the following DFs:

$$\text{For activities } 1,4,5, F_Y(y) = 1 - e^{-0.2y}; 0 \leqslant y$$

$$\text{For activities } 2,3,6, F_Y(y) = 1 - e^{-0.1y}; 0 \leqslant y$$

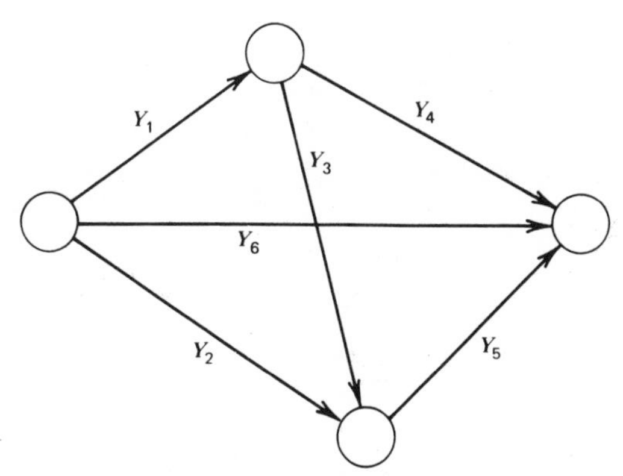

Figure 4.25. Example problem.

Table 4.11. Monte Carlo Simulation for 50 Samples at Each Time Value

Time	True $F(T)$ Value	Crude Estimate	Antithetic Estimate	($K=4$) Stratification Estimate	Conditional Estimate	Antithetic Conditional Estimate	Stratification Conditional Estimate
0	0.0000	0.0000	0.0000	0.0000	0.0000	0.0000	0.0000
5	0.0029	0.0000	0.0100	0.0000	0.0029	0.0032	0.0035
10	0.0540	0.0200	0.0600	0.0400	0.0520	0.0509	0.0518
15	0.1953	0.2200	0.2300	0.1850	0.2529	0.2007	0.2054
20	0.3849	0.3400	0.4100	0.3550	0.4121	0.3802	0.3934
25	0.5655	0.5200	0.5100	0.5300	0.5860	0.5665	0.5715
30	0.7091	0.6400	0.6400	0.7250	0.7436	0.7028	0.7212
35	0.8119	0.8600	0.8300	0.8050	0.8336	0.7990	0.8185
40	0.8810	0.8400	0.8600	0.9000	0.8985	0.8890	0.8891
45	0.9259	0.8800	0.8800	0.9300	0.9092	0.9256	0.9167
50	0.9542	0.9200	0.9200	0.9650	0.9575	0.9580	0.9562
55	0.9719	0.9600	0.9800	0.9600	0.9746	0.9725	0.9722
60	0.9828	0.9600	0.9500	0.9750	0.9842	0.9826	0.9835
65	0.9895	1.0000	1.0000	0.9850	0.9915	0.9911	0.9907
70	0.9936	0.9600	0.9800	0.9850	0.9940	0.9941	0.9939
75	0.9961	1.0000	0.9900	1.0000	0.9966	0.9966	0.9963
80	0.9976	1.0000	1.0000	1.0000	0.9977	0.9978	0.9975
Standard deviation		0.031514	0.027683	0.013389	0.018388	0.004165	0.005313
χ^2 deviation		0.053371	0.043426	0.012984	0.022708	0.000758	0.001406
Maximum absolute deviation		0.0691	0.0555	0.0355	0.0576	0.0129	0.0121

Table 4.12. The Absolute Deviation Between the Theoretical DF and Each of Empirical DFs

Sample Size	5			10			60		
Time	Crude	Conditional	Antithetic–Conditional	Crude	Conditional	Antithetic–Conditional	Crude	Conditional	Antithetic–Conditional
5	0.00295	0.00295	0.00301	0.00295	0.00052	0.00211	0.00295	0.00103	0.00062
10	0.05402	0.02525	0.02616	0.05402	0.01051	0.01426	0.00402	0.00968	0.00834
15	0.19531	0.04303	0.01195	0.00469	0.04544	0.00177	0.00469	0.02025	0.00598
20	0.18491	0.01507	0.03069	0.08491	0.04620	0.02560	0.06824	0.00231	0.00792
25	0.23442	0.00753	0.02882	0.23442	0.04790	0.00773	0.05108	0.01743	0.00487
30	0.29088	0.05678	0.07826	0.19088	0.02440	0.00996	0.05755	0.00071	0.00253
35	0.21190	0.07785	0.02766	0.01190	0.01414	0.00467	0.02143	0.02117	0.01991
40	0.11891	0.00626	0.02108	0.01891	0.01855	0.00570	0.03558	0.00386	0.00224
45	0.07408	0.00953	0.01921	0.02592	0.01366	0.01361	0.00742	0.00581	0.00135
50	0.04572	0.00398	0.00659	0.04572	0.00911	0.00907	0.00428	0.00797	0.00861
55	0.02804	0.00042	0.00358	0.02804	0.00006	0.00322	0.02804	0.00365	0.00006
60	0.01713	0.00695	0.00141	0.01713	0.00450	0.00078	0.01713	0.00466	0.00169
65	0.01044	0.00184	0.00081	0.01044	0.00061	0.00029	0.01044	0.00340	0.00142
70	0.00635	0.00018	0.00116	0.00635	0.00098	0.00156	0.00635	0.00004	0.00024
75	0.00386	0.00125	0.00086	0.00386	0.00113	0.00089	0.00386	0.00050	0.00022
80	0.00234	0.00094	0.00061	0.00234	0.00085	0.00006	0.00234	0.00046	0.00026

It is easy to verify that the theoretical DF of this project is given by

$$F_T(\tau) = 1 - 7e^{-0.1\tau} + e^{-0.2\tau}(12 + 0.4\tau) - 16e^{-0.3\tau}$$
$$+ 19e^{-0.4\tau} - e^{-0.5\tau}(9 + 0.4\tau)$$

This provides a check on the sampling estimates. A sample experiment is shown in Table 4.11, where the first five columns give the "pure" approaches, while the last two columns give "hybrid" approaches, in this case they are the "antithetic-conditional" and the "stratified-conditional."

It is instructive to note the close approximation of all techniques—the maximum deviation being of the order of less than 6% for all but the crude estimate.

In a separate experiment in which the sample size (at each time value) was put at 5, 10, and 60, comparison was made between the crude estimate, the conditional estimate, and the antithetic–conditional estimate only. A sample result is shown in Table 4.12.

The entire experiment was repeated 10 times with different sets of RVs. The results are summarized in Table 4.13, which should be read as follows. For the sample size 5, crude Monte Carlo yields an average absolute deviation (from the true DF) of 0.26329, with variance of 0.011967, while the conditional Monte Carlo gives an average absolute deviation of 0.10725 with variance 0.000693, and so on.

A final remark is in order. While Monte Carlo sampling techniques have been sharpened, and will continue to be sharpened to reduce the variability of the estimates, the reader must realize that at no time should "muscle power" in the form of computer simulation be substituted for "brain power" in the form of analytical approaches. Every effort should be directed at each step of the analysis to resolve the issues analytically, since that usually implies the full utilization of all the information available.

Table 4.13. Summary of Mean and Variance of Absolute Deviations (Based on 10 Experiments)

	5		10		60	
Sample Size	Mean	Variance	Mean	Variance	Mean	Variance
Crude	0.26329	0.011967	0.200883	0.007024	0.079211	0.000765
Conditional	0.10725	0.000693	0.080217	0.000696	0.032512	0.000063
Antithetic–Conditional	0.06237	0.001047	0.035335	0.000211	0.016748	0.000006

APPENDIX D SOME STATISTICAL PROPERTIES OF THE BETA DISTRIBUTION FUNCTION

§A.1 The γ Density Function

Consider the γ-function $\Gamma(k)$ for real positive k defined by

$$\Gamma(k)=\int_0^{\infty} x^{k-1}e^{-x}\,dx \tag{D.1}$$

To evaluate this integral, let $u=x^{k-1}$ and $\partial v=e^{-x}\partial x$; then $\partial u=(k-1)x^{k-2}\partial x$ and $v=-e^{-x}$; hence

$$\begin{aligned}\Gamma(k)&=-x^{k-1}e^{-x}\Big|_0^{\infty}+\int_0^{\infty}(k-1)x^{k-2}e^{-x}\,\partial x\\ &=(k-1)\int_0^{\infty}x^{k-2}e^{-x}\,\partial x\\ &=(k-1)\Gamma(k-1)\end{aligned}$$

From which we deduce that for k positive integer,

$$\Gamma(k)=(k-1)!=(k-1)(k-2)\cdots 2\times 1 \tag{D.2}$$

Since $x^{k-1}e^{-x}$ is $\geqslant 0$ in the range $0\leqslant x\leqslant\infty$, and since $\int_0^{\infty}x^{k-1}e^{-x}\partial x=\Gamma(k)$, we conclude that

$$g(x)\triangleq\frac{1}{\Gamma(k)}x^{k-1}e^{-x} \tag{D.3}$$

may serve as a (continuous) probability "density function," usually referred to as the *Gamma density function*.

§A.2 The Standardized Beta Density Function

Consider next the function

$$f(x)=\frac{\Gamma(k_1+k_2)}{\Gamma(k_1)\Gamma(k_2)}x^{k_1-1}(1-x)^{k_2-1}\quad\text{for } 0\leqslant x\leqslant 1$$

To verify that this is a probability density function we must show that $f(x) \geqslant 0$ for all x in the specified range (which is obvious) and that $\int_0^1 f(x)\,\partial x = 1$. From the definition of $\Gamma(k)$ given in Eq. (D.1) we have

$$\Gamma(k_1)\Gamma(k_2) = \int_0^\infty \partial x_1 \int_0^\infty x_1^{k_1-1} x_2^{k_2-1} e^{-x_1-x_2}\,\partial x_2.$$

The integrals are evaluated by change of variable into polar coordinates; let

$$x_1 = r^2\cos^2\theta \quad \text{and} \quad x_2 = r^2\sin^2\theta;$$

whence,

$$r = \sqrt{x_1 + x_2}\ ; \quad 0 \leqslant r \leqslant \infty;$$

and

$$\theta = \tan^{-1}\sqrt{x_2/x_1}\ ; \quad 0 \leqslant \theta \leqslant \frac{\pi}{2}$$

Therefore, the Jacobian $\partial(x_1,x_2)/\partial(r,\theta)$ is given by

$$\frac{\partial(x_1,x_2)}{\partial(r,\theta)} = \begin{vmatrix} 2r\cos^2\theta & 2r\sin^2\theta \\ -2r^2\sin\theta\cos\theta & 2r^2\sin\theta\cos\theta \end{vmatrix} = 4r^3\sin\theta\cos\theta$$

Substituting, we get

$$\Gamma(k_1)\Gamma(k_2) = 4\int_0^{\pi/2}\partial\theta\int_0^\infty(\cos\theta)^{2k_1-1}(\sin\theta)^{2k_2-1}r^{2k_1+2k_2-1}e^{-r^2}\,\partial r$$

$$= 4\int_0^{\pi/2}(\cos\theta)^{2k_1-1}(\sin\theta)^{2k_2-1}\,\partial\theta\int_0^\infty r^{2k_1+2k_2-1}e^{-r^2}\,\partial r$$

To evaluate the rightmost integral, put $r = \sqrt{y}$, yielding $\partial r = \partial y/2\sqrt{y}$. Hence,

$$2\int_0^\infty r^{2k_1+2k_2-1}e^{-r^2}\,\partial r = \int_0^\infty y^{k_1+k_2-1}e^{-y}\,\partial y = \Gamma(k_1+k_2), \quad \text{by definition.}$$

In other words,

$$\Gamma(k_1)\Gamma(k_2) = \Gamma(k_1+k_2)2\int_0^{\pi/2}(\cos\theta)^{2k_1-1}(\sin\theta)^{2k_2-1}\,\partial\theta$$

Finally, put

$$\cos\theta = \sqrt{x}\ ; \quad \text{hence,}\ -\sin\theta\,\partial\theta = \frac{\partial x}{2\cos\theta}$$

$$\cos^2\theta = x; \quad \text{hence,}\ \sin^2\theta = 1 - x$$

Substituting,

$$\frac{\Gamma(k_1)\Gamma(k_2)}{\Gamma(k_1+k_2)} = \int_0^1 x^{k_1-1}(1-x)^{k_2-1}\,\partial x$$

which verifies that $\int_0^1 f(x)\,\partial x = 1$.

The function $f(x)$ is usually called the standardized Beta density function, and the factor $\Gamma(k_1)\Gamma(k_2)/\Gamma(k_1+k_2)$ is called the Beta function and written $\beta(k_1,k_2)$, for $0 \leqslant x \leqslant 1$;

$$f(x) = \frac{1}{\beta(k_1,k_2)} x^{k_1-1}(1-x)^{k_2-1}; \quad 0 \leqslant x \leqslant 1; \quad \beta(k_1,k_2) = \frac{\Gamma(k_1)\Gamma(k_2)}{\Gamma(k_1+k_2)} \tag{D.4}$$

§A.3 Moments of the Standardized Beta Density Function

The rth moment is given by

$$\mu_r' = \frac{\Gamma(k_1+k_2)}{\Gamma(k_1)\Gamma(k_2)} \int_0^1 x^{k_1+r-1}(1-x)^{k_2-1}\,\partial x$$

$$= \frac{\Gamma(k_1+k_2)}{\Gamma(k_1)\Gamma(k_2)}\,\frac{\Gamma(k_1+r)\Gamma(k_2)}{\Gamma(k_1+k_2+r)} = \frac{\Gamma(k_1+k_2)\Gamma(k_1+r)}{\Gamma(k_1)\Gamma(k_1+k_2+r)} \tag{D.5}$$

From which we deduce that the mean

$$\mu_1' = \mathcal{E}(X) = \frac{k_1}{k_1+k_2} \tag{D.6}$$

and the second moment (about the origin) is

$$\mu_2' = \frac{k_1(k_1+1)}{(k_1+k_2)(k_1+k_2+1)}$$

Hence, the variance is given by

$$\sigma_x^2 = \mu_2' - (\mu_1')^2 = \frac{k_1 k_2}{(k_1+k_2)^2(k_1+k_2+1)} \tag{D.7}$$

We also need a measure of skewness defined by $\alpha_3 = \mu_3/\sigma^3$; where μ_3 is the third central moment of the density function (i.e., the third moment about the mean) given by

$$\mu_3 = \mathcal{E}(x-\mu_1')^3 = \mu_3' - 3\mu_1'\mu_2' + 2\mu_1'^3$$

But,

$$\mu_3' = \frac{k_1(k_1+1)(k_1+2)}{(k_1+k_2)(k_1+k_2+1)(k_1+k_2+2)}$$

hence

$$\mu_3 = \frac{2k_1k_2(k_2-k_1)}{(k_1+k_2+2)(k_1+k_2+1)(k_1+k_2)^3}$$

and

$$\alpha_3 = \frac{\mu_3}{\sigma^3} = \frac{2(k_2-k_1)}{(k_1+k_2+2)}\sqrt{(k_1+k_2+1)/(k_1k_2)} \tag{D.8}$$

§A.4 The General Beta Density Function

In general, the Beta probability density function is usually written as

$$f(y) = \frac{1}{(b-a)^{k_1+k_2-1}\beta(k_1,k_2)}(y-a)^{k_1-1}(b-y)^{k_2-1}; \quad a \leqslant y \leqslant b \tag{D.9}$$

where $\beta(k_1, k_2)$ is as defined in (D.4). Through a simple change of variables this function can be standardized. For let $y = a + (b-a)x$; then $\partial y = (b-a)\partial x$ and

$$f(x)\partial x = \frac{1}{(b-a)^{k_1+k_2-1}\beta(k_1,k_2)}$$

$$\times (b-a)^{k_1-1}x^{k_1-1}(b-a)^{k_2-1}(1-x)^{k_2-1}(b-a)\partial x$$

$$= \frac{1}{\beta(k_1,k_2)}x^{k_1-1}(1-x)^{k_2-1}\partial x$$

which is the standard density function. Furthermore, from (D.6),

$$\mathcal{E}(Y) = a + (b-a)\mathcal{E}(X) = a + (b-a)\frac{k_1}{k_1+k_2} \tag{D.10}$$

and from (D.7),

$$\operatorname{var}(Y) = (b-a)^2 \operatorname{var}(X) = (b-a)^2 \frac{k_1 k_2}{(k_1+k_2)^2(k_1+k_2+1)} \tag{D.11}$$

Also, the mode m is easily obtained from $(\partial f(y)/\partial y) = 0$ to yield

$$m = \frac{a(k_2-1) + b(k_1-1)}{k_1+k_2-2} \tag{D.12}$$

Notice that the mean $\mathcal{E}(Y)$, the variance var(Y) as well as the mode m are all function of a, b, k_1, and k_2. The parameters a and b fix the location and range of the density function, while k_1 and k_2 fix the shape of the curve.

PROBLEMS

1. The following activities describe the tasks of ship-boiler repair [Feiler (1976)]. You notice that the activities are not numbered following the standard convention of a low number preceding a larger one. Draw the project network in A-on-N and A-on-A modes and then renumber the activities to conform to that convention.

Activity	Description	Predecessor	Duration
1	Remove refractory material	7, 15	4, 5, 10
2	Repair inner air casing	1	1, 5, 14
3	Repair outer air casing	2	1, 5, 10
4	Repair under boiler	none	14, 27, 35
5	Rebrick	2, 17	5, 6, 14
6	Chemically clean	11, 17	4, 6, 8
7	Remove air registers	none	0, 0.5, 3
8	Install plastic refractory	19, 10	0.5, 0.5, 1
9	Clean and repair air register assemblies	7	10, 10, 20
10	Install air registers	5, 9	0.5, 1, 3
11	Rag for chemical cleaning	15	5, 6, 17
12	Remove drum internals	15	0.5, 0.5, 3
13	Repair drum internals	12	5, 12, 18
14	Install drum internals	13, 18	1, 1.5, 3
15	Initial Hydrostatic Test	none	1, 2, 5
16	Exploratory block	1, 12	7, 8, 18
17	Retube and poll	16	4, 6, 12
18	Preliminary hydrostatic tests	6	0, 0, 14
19	Final hydrostatic tests	14	0, 0, 3

The three durations attached to each activity are the optimistic, most likely, and pessimistic estimates, obtained by asking the experts. However, in this application a *triangular* DF was fitted to the data and had the following interpretation.

a. Optimistic duration: probability of $\leqslant 0.05$ of a shorter duration.
b. Most likely duration: the mode of the triangular distribution.
c. Pessimistic duration: probability of $\leqslant 0.05$ of a longer duration.

Based on the above information, you are asked to do the following:

i. Determine the CP based on the *expected* duration of each activity. (This would be the analog of the standard PERT calculation, except for the different DF assumed here.)
ii. Determine, through a Monte Carlo sampling experiment (crude or refined) the following parameters:

(a) The DF of project-completion time.

(b) The expected value of project duration; compare with the result obtained in (i) above. Comment briefly.

(c) The criticality index of all activities in the project. Based on these indices, discuss the actions you would propose to shorten the duration of the project.

2. Read the paper by Madansky (1960) and based on it deduce the inequality

$$g_j \leqslant e_j \quad \forall\, j=2,3,\ldots,n$$

where g_j is the PERT estimate of the expected length of the CP to node j and e_j is the correct value of that expected length.

3. Is the PERT estimate of the var(T_n) unbiased? If all of the activities of the network have symmetric density functions of finite range, would the PERT estimate be unbiased?

   Rationalize your answers by either proofs or examples.

4. Chapter 2 was concerned exclusively with problems of optimal time–cost trade-offs in DANs, under different assumptions on the functional relationship between duration and cost.

   Formulate a meaningful problem of time–cost trade-off for PANs, and discuss a possible approach to its resolution. (*Hint:* You may continue to assume a linear functional relationship between expected activity duration and average cost.)

5. Prove that in the case of independent arc durations,

$$f_j \triangleq \mathcal{E}(Z_j) = \sum_{z=\alpha_j}^{\beta_j} z p(Z_j = z)$$

simplifies to

$$f_j = \sum_{\alpha_j}^{\beta_j} z \left\{ \prod_{k=1}^{m} P_k(z - f_k) - \prod_{k=1}^{m} P_k(z^- - f_k) \right\}$$

where $\alpha_j = \max_{1 \leqslant k \leqslant r_j}(f_{i_k} + a_{i_k})$; a_{i_k} being the smallest duration of activity $(i_k j)$; $a_{i_k} \geqslant 0$; and $\beta_j = \max_{1 \leqslant k \leqslant r_j}(f_{i_k} + b_{i_k})$; b_{i_k} being the largest duration of activity $(i_k j)$; $b_{i_k} \geqslant a_{i_k}$. In essence, α_j is the earliest estimated realization of node j and β_j is the latest estimated realization of node j.

6. It is desired to determine the average duration of the project shown, assuming that all the activities possess a uniform discrete distribution

on the interval $[a,b]$. (For example, activity 4 has $a=3, b=10$; hence, it can be of duration 3, or 4, or ..., or 10 with probability $\frac{1}{8}$.)

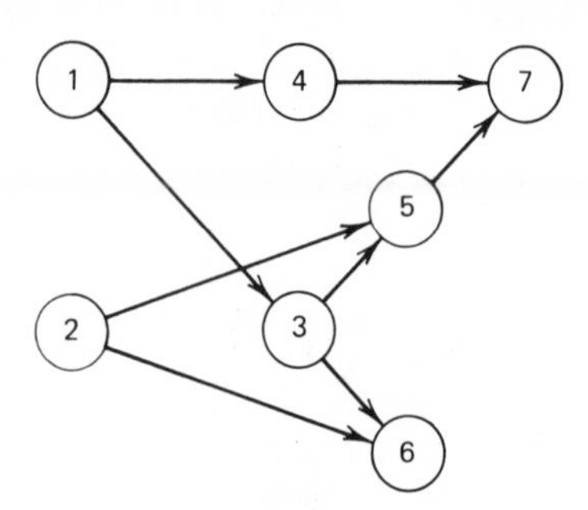

A-on-N Representation

Activity	a	b
1	2	6
2	4	11
3	5	8
4	3	10
5	8	17
6	1	5
7	7	12

Approximate the desired expected value by a series of four estimates: first, the PERT estimate, then the three methods of Sections 2.1–2.3.

7. In the network of Problem 6, suppose that we are willing to invest more computing effort in order to gain more information. In particular, suppose we are interested in approximating the DF of the project duration, but in order to lighten the computing burden we are willing to accept a two-point surrogate for the individual activity durations. (Once this "surrogation" is accomplished, one proceeds analytically to determine the DF of the duration.)

   You are asked to choose the surrogate estimates (justify your selection) and perform the analysis to obtain $F_7(\tau)$. Then, determine the expected value $\mathcal{E}(T_7)$. Compare the result obtained with the estimates obtained in Problem 6 above. Comment briefly.

8. Relative to the Monte Carlo approach to the study of PANs, the query may arise as to the *range* of the duration of the project and how it varies. This question may be of interest to a manager who does not want a "point estimate" of the project duration, but an estimate that

gives him an idea about the variation in total duration. Of course, managers are interested in "interval estimates," which effectively say that this project would take between a and b days with probability 0.95. What are your best estimates of a and b, and what sample size should be taken to guarantee the confidence level of 0.95?

9. For each of the following networks, identify the minimum set of activities on which one must "condition" (see Section 4.4) in order to reduce the network to a set of paths in parallel. Write the expression for the determination of the DF of the terminal node as a multiple integral over the conditional activities.

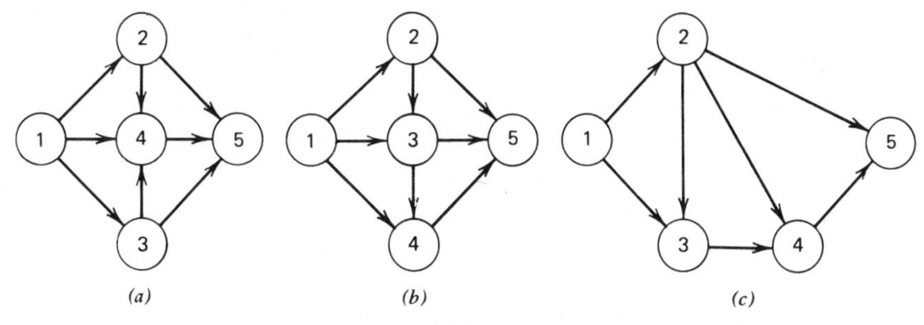

(a) *(b)* *(c)*

"Double wheatstone bridge"

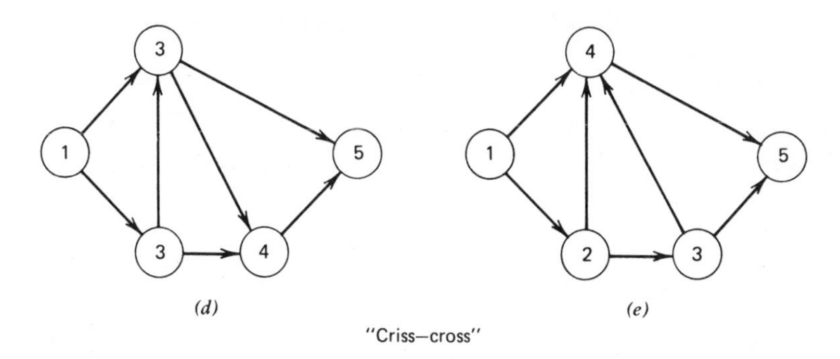

(d) *(e)*

"Criss–cross"

10. Consider a single activity whose duration has mean μ and range $b-a$. Suppose that it is subdivided into three activities in series of equal average duration ($=\mu/3$ each), such that the total range of the duration remains at $b-a$.

Determine the PERT estimate of the variance of the completion time of the activity before and after the subdivision and comment briefly.

11. Consider N independently distributed RVs $X_1,\ldots,X_N$ of known DF. Let $Z=\max(X_1,\ldots,X_n)$; whose DF can be deduced from those of the X values. Prove, from first principles, that $\mathcal{E}(Z) \geqslant \max(\mathcal{E}X_1, \mathcal{E}X_2,\ldots,\mathcal{E}X_N)$; in other words, prove that the expected value of the maximum of a set of RV's is no smaller than the maximum of the individual expected values.
12. Prove the assertion of the third approach (Section 2.3) to the approximation of the expected value of the project duration, namely, that: $s_n \leqslant w_n \leqslant e_n$.
13. Complete the analysis of the application of the antithetic variable approach to paths in parallel by constructing the antithetic variable to δ_τ in the range $\tau \in [0,0.5]$. Determine the mean and variance of the antithetic estimate of $F_4(\tau)$.
14. For each of the "generic" networks of Problem 9 you are asked to do the following (you may assume any DF of the activity durations):
    a. Apply the control variate technique, in which you first define the surrogate network, derive the analytical form of its DF, and then sample from the original network to determine the required DF.
    b. Apply the conditional sampling technique in which you first specify the "conditioned" arcs, derive the PDF of the completion time analytically, and then sample from the conditioned arcs.

    In either case, explain your rationale for choosing a particular sample size. Comment briefly on the results obtained.
15. Write the expression for the PDF of the terminal node of network (c) of Problem 9.
16. Detail the sequence of steps of iteration required to determine the duration of the following project. In case you cannot identify a subnetwork as belonging to one of the generic types discussed either in the text (Section 4.2) or in Problem 9, write the expression for determining the PDF of its equivalent arc.

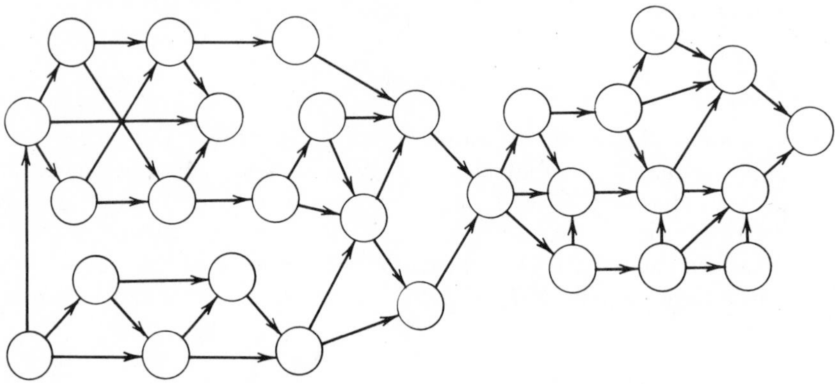

17. Apply the approximating procedure of Section 4.3 to the network of Fig. 4.10. Proceed in the following steps:
   a. Assume a particular continuous DF, or a family of continuous DFs, to the activities (say, all activities are normally distributed). Choose the parameters of the DF for each activity.
   b. Substitute a two-point surrogate for the DF.
   c. Perform the analysis to obtain the approximation to the PDF of T_{15}.

   As a test for the "goodness" of your approximation, compare the result with that obtained from a population-sampling approach. Comment briefly.

18. The treatment of Sections 1 and 2 suggest that the error in the estimation of the expected duration of a project increases with increased "parallelism" in the networks and that it is further compounded by the presence of dependence relationships among the paths to the terminal node n.

   You are asked to illustrate the degradation in the estimate of the expected duration and other probability statements by application to the following network.

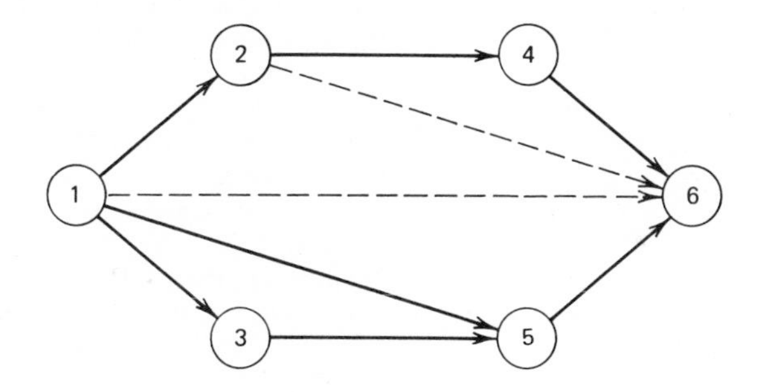

   All activities are exponentially distributed, $\mu e^{-\mu y}$, with parameter $\mu = 2$.
   a. Determine the "true" expected value $\mathcal{E}(T_6)$, of the duration of the project first *without* the dashed activities and then *with* the dashed activities. In both cases, determine the PERT estimate of $\mathcal{E}(T_6)$, and comment briefly. (*Hint:* Recall that RVs in series convolve, and RVs in parallel multiply.)
   b. For the complete network (i.e., including the dashed activities), determine the probability that the project will terminate no later than time 14 according to the PERT approach and then according to the "true" DF of T_6 determined in (a) above.

19. Determine, using any approach you choose, the criticality index of each activity of the project of Problem 6, assuming the durations to be

uniformly distributed between the limits shown (you may assume discrete durations). Tabulate your results in the format of Tables 4.7 and 4.8 and comment briefly.

REFERENCES

Abernathy and Demski, 1973.
Blanning and Rao, 1965.
Burt, Jr., Gaver, and Perlas, 1970.
Burt, Jr. and Garman, 1971.
Carruthers, 1968.
Clark, 1961.
———, 1962.
Clingen, 1964.
Coon, 1965.
Davies, 1973.
Donaldson, 1965.
Elmaghraby, 1967.
Feiler, 1976.
Frank, 1969.
Fulkerson, 1962.
Garman, 1972.
Hald, 1952.
Hammersley and Handscomb, 1967.
Hammersley and Mauldon, 1956.
Hartley and Wortham, 1966.
Healy, 1961.
Hoel, 1954.
Jewell, 1965.
King, 1971.
King, Wittevrongel, and Hezel, 1967.
King and Wilson, 1967.
King and Lukas, 1973.
Klingel, Jr., 1966.
Kotiah and Wallace, 1973.
Lindsey, 1972.
Lukaszewicz, 1965.
McBride and McClelland, 1967.
McClellan, Jr., 1969.
MacCrimmon and Ryavec, 1964.
Madansky, 1960.
Martin, 1965.
Moder and Rogers, 1968.
Moder, 1975.
Morris and Denison, 1967.
Murray, 1963.
Noettl and Brumbaugh, 1967.
Page, 1965.
Parzen, 1960.
Pocock, 1962.
Pospelov and Barishpolets, 1966.
Pritsker and Kiviat, 1969.
Rao, 1971.
Ringer, 1966.
———, 1969.
———, 1971.
Robillard and Trahan, 1977.
Shrieder, 1966.
Sielken, Hartley, and Arseven, 1975.
Sobczak, 1963.
Swanson and Pazer, 1971.
Tippett, 1925.
Van Slyke, 1963.
Ventura, 1965.
Welsh, 1965.
———, 1966.
Wilks, 1962.

5

GENERALIZED ACTIVITY NETWORKS (GANs)

Up to this point we have expounded on the structure and analysis of ANs that are models of one of two types of project: (a) deterministic, modeled by deterministic activity networks (DANs) and (b) probabilistic, modeled by probabilistic activity network (PANs). In retrospect, and with the benefit of hindsight, it must be clear to the reader that the conceptual models of both categories are severely limited from a *logical* point of view, and in many instances fall short of adequately representing a wide variety of projects, especially of the research and development kind. For example, in all the network models discussed thus far, all activities incident on a node must be completed before the node is realized—in particular, all activities and events of the network must be realized before the final event is realized. All activities emanating from a node must be undertaken; all the nodes must represent deterministically realizable events (in the A-on-A representation), and so on. In brief, there is an implicit determinateness in the existence of all events and activities, which may not be a correct image of the real-life situation. Uncertainty in PANs (exemplified by the PERT model) is not related to the eventual undertaking of these activities, but only to their duration.

The study of large scale systems, including large scale projects, abounds with different logical classes of events and activities, different initial and terminal conditions, and the like, which cannot be handled by the models discussed thus far.

For instance, consider the following situation: at a particular point of time, say event *1*, price quotations are to be submitted on three subsystems: *A*, *B*, and *C*. Furthermore, suppose that there is a reasonable chance of winning a contract on either one or two subsystems, *but not on all three* of them. Each eventuality, and there are seven of them, requires a different path of progress thereafter, including the case of winning no contract at all, which may necessitate a drastic reduction in labor force! How can such a "road map" of future progress be represented? Obviously, the CPM (or PERT) models are inadequate, unless we are willing to construct an activity network for each possible turn of events—a procedure that is at best clumsy and at worst computationally infeasible as well as self-defeating, since it fragmentizes the representation and thus prevents the possibility of seeing the "total picture."

As another example, consider a project which, because of urgency and character (perhaps as related to national security), has a great deal of redundancy built into it in the form of several parallel activities. However, the success of *anyone* of these activities (which are usually of research and development type, such as the creation of new technologies, with a great deal of uncertainty in their progress) is sufficient to terminate the other parallel activities and continue with the successful one alone.† In this case, the realization of the terminal node of the parallel activities is dependent on the activity that is completed the *earliest*, which is in direct violation of the basic axioms of CPM!

As a third and final example, consider a project in which, on reaching a certain stage of experimentation, progress is either continued along previously laid-out plans or a fundamental flaw in the product is discovered and

†The history of the Manhattan project, which was concerned with the development of the atom bomb during World War II, abounds with these considerations.

engineers are "back to their drawing boards".[†] Here is an instance of feedback, previously banned from the CPM–PERT models!

In an effort to cope with the need for such expanded representations, generalized activity networks (GANs) were introduced by Elmaghraby in 1964 motivated by the pioneering work of Eisner (1962). Since then, several significant contributions to the concept have been made, first by Pritsker and Happ (1966) and Pritsker and Whitehouse (1966), who laid the theoretical foundations for the GERT model (graphical evaluation and review technique), and subsequently by the same authors and others who expanded it in several directions, especially in the construction of special purpose simulation models (GERTS, where the "S" stands for simulation).

Despite the fact that GAN, GERT, and GERTS had their genesis in, and were motivated by, project planning studies, it is well to remember that these approaches have much wider applicability than this single field of application. In fact, these approaches have been used to model and analyze contract-bidding situations, population dynamic behavior, maintenance and reliability studies, vehicle traffic networks, accident causation and prevention, computer algorithms, and many others too numerous to mention.

In the present chapter we limit ourselves, by necessity, to an exposition of the basic concepts underlying these approaches and to their illustration. For more detailed exposition of the wide spectrum of their application, the reader should consult the references cited. This is even more forcefully recommended for the GERTS approach, which is briefly described in Section 3. Since the start of the writing of this manuscript, that simulation language has undergone several major changes (it now stands at GERTS III-Z) and has proliferated into three separate languages (viz., GERTS-R to handle resources, GERTS-Q to handle waiting lines, and SAINT to incorporate a variety of logical enrichments). And more is to come, to be sure, as the languages are expanded and refined to accommodate the insatiable desires of their users.

The present chapter contains four sections. Sections 1 and 2 give, respectively, the basic theory of the general GAN model and its specialization to the GERT model. In Section 3 the basic ideas of the simulation language, GERTS, are explained and illustrated. Finally, Section 4 expands the time–cost trade-off studies of Chapter 2 to GERT networks.

In a more fundamental sense, the present chapter is devoted to the expansion of the scope of the original AN models (of CPM and PERT) beyond their confining assumptions. It is an open invitation to the reader

[†]The design and construction of modern transportation systems utilizing advanced technology give rise to these considerations.

to free himself of the shackles of "traditional" approaches and to construct the models and their analytical tools that best describe his real-life problems.

Consequently, we would be remiss to close this introduction without mention of another approach that attempts to do just that, namely, the *decision-critical-path method.* Decision-CPM, as it is commonly abbreviated, is the name given by Crowston and Thompson (1967) to their analysis of DANs characterized by discrete multiple choices at some of their nodes. These may represent either a choice among activities to be undertaken next or a choice among sets of resources to be utilized by the activity itself. In the former instance, one or more of the prospective activities must be undertaken. Those activities not selected must "disappear" from the network—in the sense that all their precedence relations must be eliminated. In the latter case, it is evident that the planning of the allocation of the resources be intimately related to the scheduling of the activities, since the very duration of these activities is dependent on the resources to which the activities are allocated![†]

The reader will immediately recognize that such characterization bears curious resemblance to our treatment in Section 4.2 of Chapter 3 of project planning under "resource-related" activities. Little wonder that we concluded, as did Crowston and Thompson, that an ILP is the appropriate model for such decision situations.

As a matter of fact, our treatment and its expanded form is of a more general nature, since one may include constraints on the availability of some or all the resources—a limitation not considered by Crowston and Thompson. However, because of the specific objective and the exclusive concentration on the impact of the multiple choices on the scheduling aspects of the project, Crowston (1970) could propose a "reduction method" by which subnetworks containing no "decision nodes"—these are the multiple-choice nodes—could be "condensed" into equivalent paths. The resultant is a more efficient algorithm for the determination of the minimum project cost (or duration); see Crowston and Wagner (1970). The specialized nature of this algorithm, and the fact that the salient features of their concern were treated in Section 4 of Chapter 3, preclude any further discussion of this approach here. The interested reader may consult the abovementioned references for more detailed exposition, or the work by Crowston (1971) for an overall view of the approach.

[†]The authors also permit logical dependencies among activities, of the form: activity k may be undertaken only if j is undertaken, or, if j then k and conversely, and so on, where j and k need not belong to the same decision node.

§1. THE GAN MODEL

To free our models from the "curse of determinateness," the systems with which we are concerned may be characterized by states and transitions from one state to another. The probabilities of transition are assumed fixed and known a priori. From any state the system may move *with certainty* to one or more states *as well as probabilistically* to two or more other states. (Clearly, if *all* transitions are certain, we are back to the CPM–PERT models, which may now be considered as special cases of this generalized model.)

In performing the transition from one state to another the system may be viewed as undertaking a transformation, or experiencing a change in some of its entities. The entities of usual concern are time, cost, resources, location, and size. These are given in vector form. Any parameter may be changing deterministically or probabilistically; for instance, the time of transition may be a fixed constant (as was assumed in CPM) or a RV (as was assumed in PERT), and the cost of the transition (i.e., the cost of the activity) may be a constant or a RV.

The above two concepts lead to the definition of the generic element of a GAN; it is a branch, as shown in Fig. 5.1, consisting of a directed arc (i.e., an arrow) and its end points, the nodes, which represent states.

On the arc is defined a vector of parameters, u, of dimensionality $\geqslant 2$, of which we have shown only three in Fig. 5.1:

- p_u: the probability that arc u [i.e., the activity $(1,2)$] will be realized[†] given that node 1 is realized.
- Y_u: a RV representing the duration of arc u, assuming u is realized; Y_u is assumed having the probability density (or mass) function $h_u(\tau)$
- C_u: a cost function that may or may not depend on the duration Y_u (and hence may be a RV)

The first two parameters, probability and duration, are *always* specified.

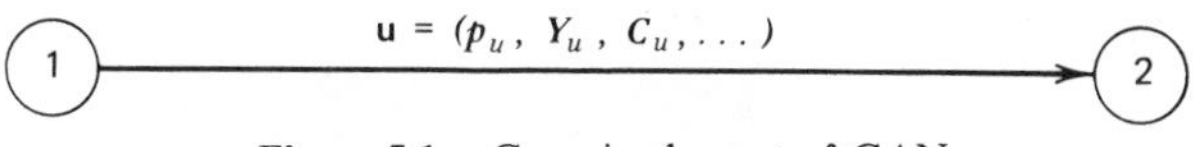

Figure 5.1. Generic element of GAN.

[†]This probability is not synonymous with the probability of realization of node *2* since there may be other paths to *2*.

Other parameters, such as cost and consumption of resources, may be specified in addition. If the activity occurs with certainty, the probability p_u may be dropped from explicit mention, although it is always implicitly recognized. If *all* arcs are realized with certainty, the GAN model reduces to the PERT model. If, furthermore, all durations $\{Y_u\}$ are fixed constants, the model further reduces to the more elementary CPM model.

The graph of a network is composed of nodes connected by arcs. Each node must have at least one arc incident on it and one arc emanating from it, except the origins and the terminals; an origin has arcs only starting from it and a terminal has arcs only incident on it. Although a network may possess more than one origin and more than one terminal, it is always more convenient to modify the network, through the use of dummy nodes and dummy activities, to have only one source and one terminal.

Consider the nodes of the network; each node may be considered as a "receiver" when it is at the head of the arrow and an "emitter" when it is at the tail of the arrow. For the time being, we define three kinds of receiving nodes (see Fig. 5.2) (later on, when we discuss the simulation models for GERT, these definitions are expanded):

1. *The logical "And" receiver*: the node (i.e., event) will be realized when all arcs leading into it are realized. Thus, the node will be realized at the

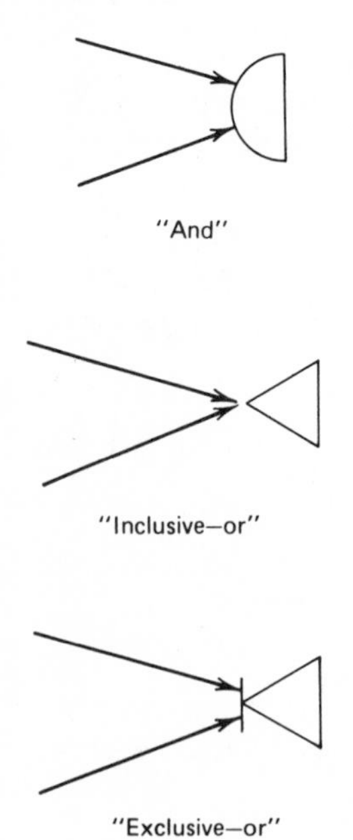

Figure 5.2. The three types of receiving nodes.

latest completion time of the activities leading into it. The models of CPM and PERT have all their receiving nodes of this type.
2. The *logical "inclusive-or" receiver:* the node will be realized if any one arc or combination of arcs leading into it is realized. Thus the node will be realized at the earliest completion time of the activities leading into it.
3. The *logical "exclusive-or" receiver*: the node will be realized if one, and only one, of the arrows leading into it is realized.

Note that in cases 2 and 3 the node need not be realized at all.

Turning next to the "emitter" side of a node, we define two types depending on whether all the activities *must* be undertaken on the realization of the node, or the occurrence of two or more of these activities is *probabilistic* in nature (see Fig. 5.3) (these are the "decision boxes" in Eisner's terminology; see Example 5.1 below):

1. *The "must-follow" emitter*: a node from which all emanating activities must be undertaken, sooner or later. The models of CPM and PERT have all their emitter nodes of this type.
2. *The "may-follow" emitter*: a node from which an emanating activity may be realized with a known probability less than 1. Exactly one uncertain activity emanating from the node will be realized if the node itself is realized, since the sum of fractional probabilities adds up to unity.

Of course, any receiver-type node of Fig. 5.2 can be combined with either of the two emitter types to specify completely the character of the node. Hence there are, in all, six possible nodes.

§1.1 The Basic Algebra

Graphical representation is usually undertaken with two objectives in mind. The first, and more elementary, is to assist in "visualizing" the total

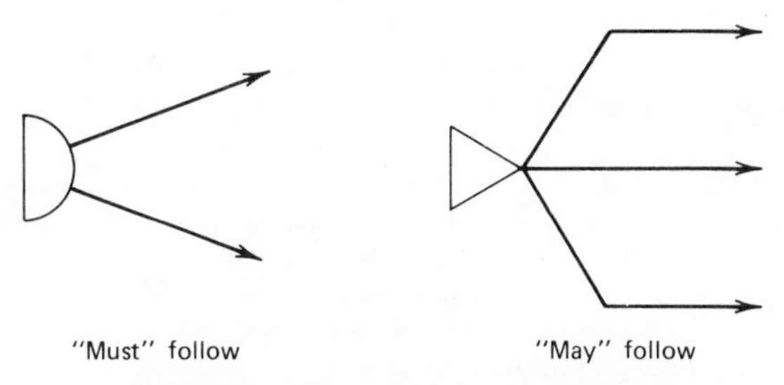

Figure 5.3. The two kinds of emitter nodes.

system. The reader of this text should be quite familiar by now with this objective since it has been a recurring theme throughout Chapter 1; see in particular Section 1 and the discussion of "lag restrictions." The second and more analytical objective is to assist in studying the modes of interaction among the various components of the system, and in deducing functional relationships based on such understanding. Such practice is by no means uncommon. For instance, electrical engineers have relied on "block diagrams" for almost half a century to analyze the dynamic behavior of control systems; civil engineers have relied on "stress diagrams" to analyze the forces in a truss; and mechanical engineers have relied on analog computer diagrams to analyze the dynamic behavior of shock absorbing systems.

This objective is typically accomplished through the *reduction* and *simplification* of the original graphs to yield an "equivalent element" that connects the "origin" to the "terminal(s)". The transfer characteristics of this equivalent element is then the desired relationship.

In essence, there are five basic graphs that can be viewed as "building blocks" from which a GAN is synthesized. Table 5.1 summarizes the basic algebra for reducing each type to the fundamental (equivalent) element shown in Fig. 5.1. For the sake of clarity of exposition in Table 5.1, we assumed that the vector u is of dimensionality two. In other words, $u=(p_u, Y_u)$; in which Y_u is generic of all *additive* parameters (i.e., the value for two arrows in series is equal to the sum of the two values).[†] As we see below, *multiplicative* parameters may also be handled. Generalization to higher-dimensional vectors is straightforward.

The following remarks should be helpful in understanding this set of rules.

1. By definition, each node (except terminal nodes) must have at least one arc (i.e., activity) emanating from it. In the case of more than one arc, it is possible that some arcs occur with certainty (hence their p_u values $= 1$) while others occur probabilistically (hence their p_u values are strictly >0 and <1). We *adopt the convention that the sum of the probabilities of the activities whose p_u values are less than 1 must add up to 1*. Consequently, the sum of "probabilities" of the set of arcs emanating from any node must add up to an integer $\geqslant 1$. Such requirement is neither

[†]When all the nodes of the network are of the "and" type on their receiving side, while probabilistic branching is permitted but with no subsequent "interaction" among the descendants of a probabilistic branch (i.e., the descendants have no node in common), a very special kind of GAN results. In particular, such a network can be represented as a *tree*. This type of AN has been studied by Pospelov and Barishpolets (1966) and by Robillard and Trahan (1977). The latter determined bounds on the time of realization of terminal nodes (always assuming random activity durations).

illogical nor infeasible. On the contrary, the converse is true, since it is meaningless to say that after the realization of a node the system may realize one activity only with probability 0.25, say. The natural question is what happens in the remaining 75% of the time. The above convention provides the answer to this question.

2. It is impossible to have a feedback loop on an "and"-type node because that would imply that an activity must be realized before it is realized!
3. The realization of an "and" node is dependent on the simultaneous realization of all the arcs leading into it, and the time of the equivalent arc is the *maximum* of a finite set of random variables. This is to be contrasted with the "inclusive-or" node whose realization is dependent on the realization of *at least one* of the arcs leading into it, and the time of the equivalent arc is the *minimum* of a finite number of random variables.
4. In GAN it is possible that some nodes are *never realized.* If arcs a, b, $c, \ldots,$ lead into a node, then:
   a. An "and" node may not be realized with probability $1-P(a \cap b \cap c \cap \ldots)$
   b. An "inclusive-or" may not be realized with probability $1-P(a \cup b \cup c \ldots)$
   c. An "exclusive-or" may not be realized with probability $1-P(a \otimes b \otimes c \otimes \cdots)$

   This should come as no surprise to the reader since it was precisely to accommodate such eventualities that GAN was developed in the first place.
5. Of all the newly defined nodes, by far the easiest to handle mathematically is the "exclusive-or" type. In fact, the mathematical analysis of GANs is greatly facilitated when all the nodes of the network are of this type. This is the special case of GERT, which was extensively treated by Pritsker and his colleagues, and which is described in more detail in Section 2. Then it will become evident that the reason for such facility is that we would be dealing with semi-Markov processes (SMP) whose theory is well established.

What is important to note at this time is that it is always possible to transform other nodes into "exclusive-or" nodes, at the expense of enlarging the network. For example, an "and" node can be replaced by two "exclusive-or" nodes, one representing the eventuality that the "and" node is realized, the other when it is not. Similarly, an "inclusive-or" node may be replaced by "exclusive-or" nodes provided that the incoming activities are transformed into the various possible realizations of subsets of the original activities.

Therefore, it seems that the price paid for the use of an already

Table 5.1. Basic Algebra for GAN

Original Network	Name	Equivalent Vector $\mathbf{V}_e^a$
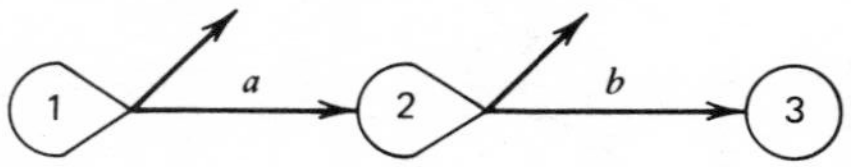 (a) 1 →a→ 2 →b→ 3	Series	$p_e = p_a p_b$ $Y_e = Y_a + Y_b$
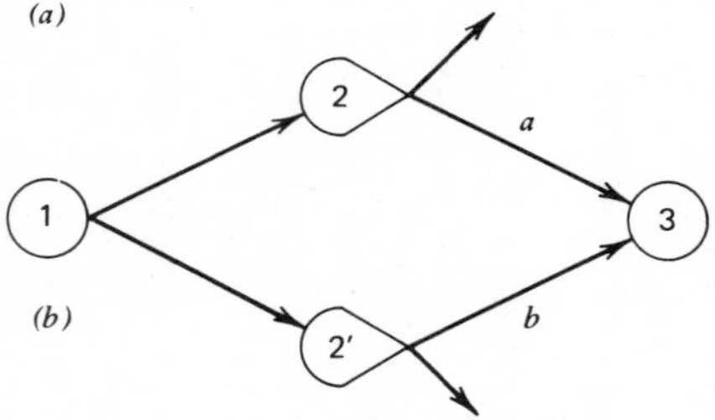 (b) 1 → 2 →a→ 3; 1 → 2′ →b→ 3	Parallel "and"	$p_e = P(a \cap b)$ $Y_e = \max(Y_a, Y_b)$
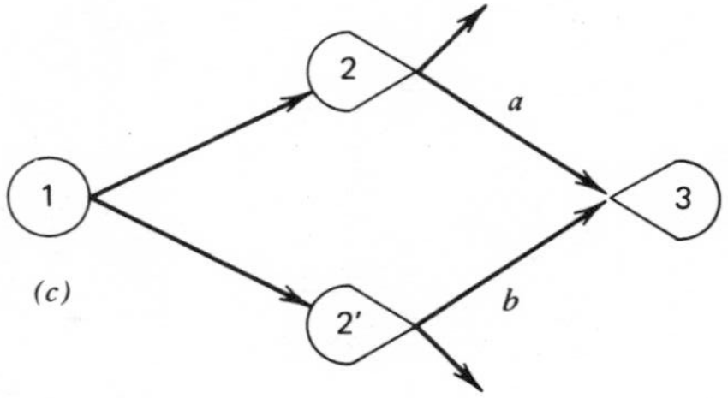 (c) 1 → 2 →a→ 3; 1 → 2′ →b→ 3	Parallel "inclusive-or"	$p_e = P(a \cup b)$ $Y_e = \min(Y_a, Y_b)$

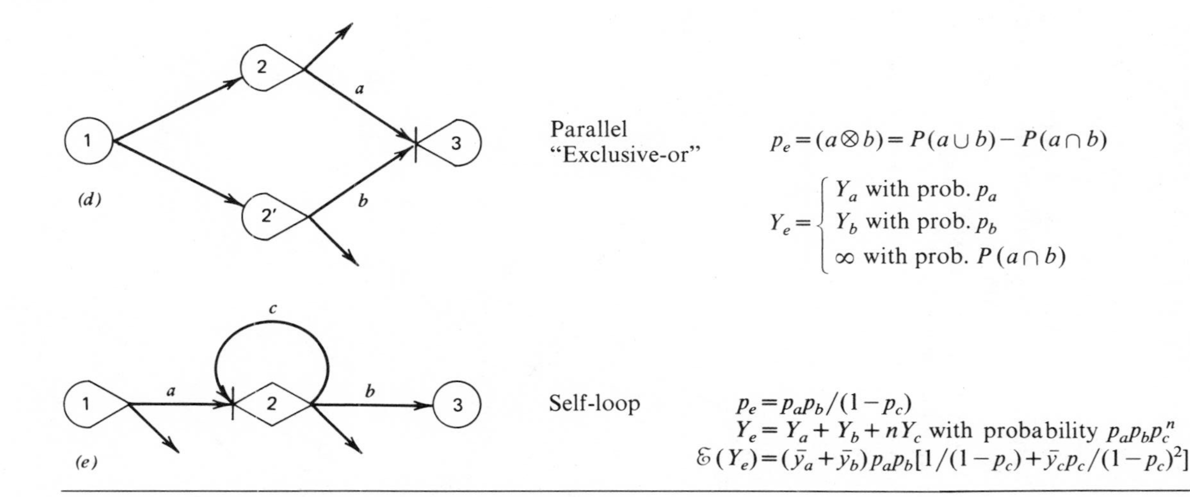

Parallel "Exclusive-or"

$$p_e = (a \otimes b) = P(a \cup b) - P(a \cap b)$$

$$Y_e = \begin{cases} Y_a \text{ with prob. } p_a \\ Y_b \text{ with prob. } p_b \\ \infty \text{ with prob. } P(a \cap b) \end{cases}$$

Self-loop

$$p_e = p_a p_b/(1 - p_c)$$
$$Y_e = Y_a + Y_b + nY_c \text{ with probability } p_a p_b p_c^n$$
$$\mathcal{E}(Y_e) = (\bar{y}_a + \bar{y}_b) p_a p_b [1/(1 - p_c) + \bar{y}_c p_c/(1 - p_c)^2]$$

---

[a]Whenever the durations themselves are stated, rather than their DF, it is understood that we refer to the random variables, as in (b)–(d). The symbol ⊗ is the "ring sum," defined as the occurrence of one, and only one, event.

established theory (that of SMP) is the enlargement of the original logic of the network. Whether such trade-off is advantageous can be answered only on the basis of empirical evidence.

To illustrate, consider the network of Fig. 5.4*a*, in which the two arcs in parallel, activities *a* and *b*, terminate in the "and" node *5*. It is obvious that event *6* will be realized if node *2* is realized (which occurs with probability p_1), *and* if *both* activities *a* and *b* are realized (which occurs with probability $Pr(a \cap b) = p_a p_b$). Hence, *6* will be realized with probability $p_1 p_a p_b$. The time to such realization is a RV given by $Y_1 + \max(Y_a, Y_b) + t_1$. On the other hand, if *2* is realized but activities *a* and *b* do not *both* occur, an event which occurs with probability $p_1(1 - p_a p_b)$, node *6* will not be realized and the process will reach node *8*. Finally, node *8* can be realized independently via the path *1, 7, 8*.

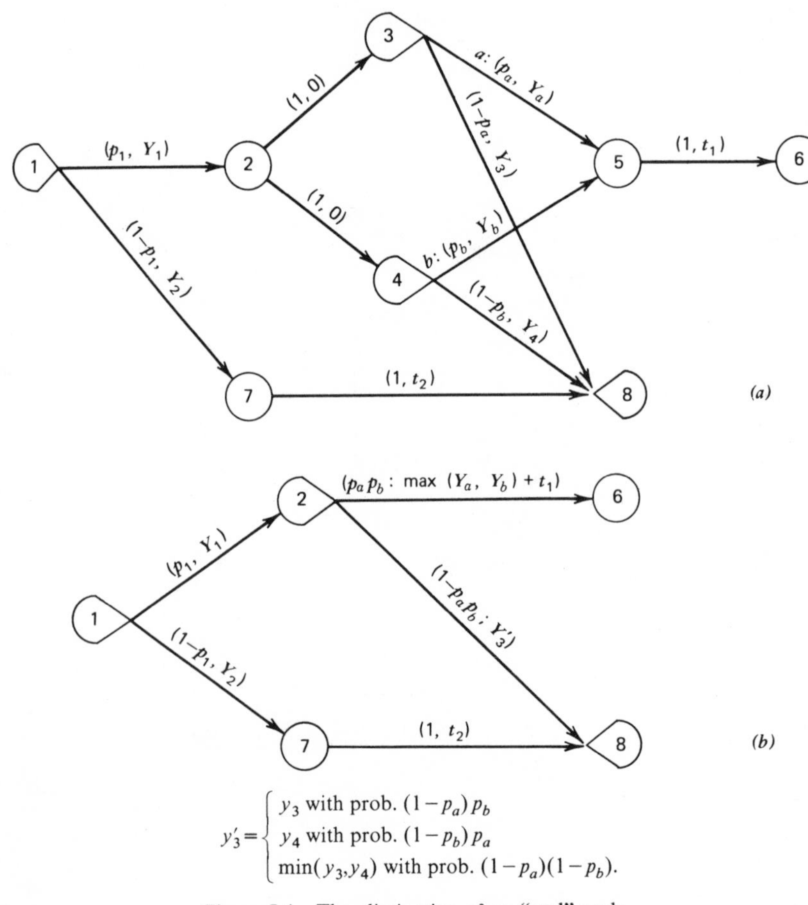

$$y_3' = \begin{cases} y_3 \text{ with prob. } (1-p_a)\,p_b \\ y_4 \text{ with prob. } (1-p_b)\,p_a \\ \min(y_3, y_4) \text{ with prob. } (1-p_a)(1-p_b). \end{cases}$$

Figure 5.4. The elimination of an "and" node.

Figure 5.4*b* represents the same logic as Fig. 5.4*a* after eliminating the "and" node *5*.

We leave it as an exercise to the reader to develop the equivalent representation in case node *5* is an "inclusive-or" node.

Example 5.1. Network For Research and Development. As our first example we treat a problem presented by Eisner (1962) in the context of research and development programs. To clarify the example, we remind the reader that Eisner used the term *decision box*, DB, to refer to nodes that lead to alternatives (our term *probabilistic branching*). His network is given in Fig. 5.5. Note that the realization of terminal events *A* and *B* depends not only on the outcome of DB*3* but also on the outcome of DB*2*. This exemplifies what Eisner called *conjunctive path dependency*, which arises when the planned work on one path depends on the answer to a particular DB on a different path. It led him to repeat node *2* and establish the dummy node 2_1, with the understanding that the two nodes are really one and the same. Thus, event *A* will be realized if the outcome at DB*3* is NO *and* the outcome at DB*2* is also NO, while event *B* will be realized if the outcome at DB*3* is NO *and* the outcome at DB*2* is YES.

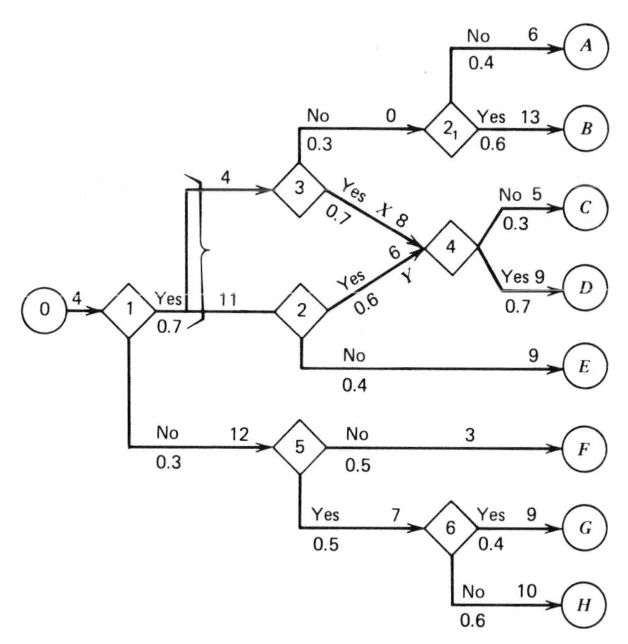

Figure 5.5. Eisner's network with a priori probabilities (number on arc denotes its duration and number below arc, probability).

The logical relationships governing the outcomes were given by Eisner as follows:

$$\text{outcomes}=\{[(A\cup B)\cup X]\cap(Y\cup E)\}\cup\{(G\cup H)\cup F\};$$

$$\text{where } (X\cap Y)=(C\cup D);$$

where "$\cup$" represents a disjunctive operation and "$\cap$" a conjunctive operation. Note the special logical relationship between branches X and Y and DB*4*; node *4* will be realized if *both* activities are realized. This is equivalent to the "and" relationship in our terminology.

Given the above structure, it is easy to deduce that the possible final outcomes are[†]

$$\begin{array}{lll} A \text{ and } E & (0.0840) & \\ B \text{ and } Y\rightarrow B & (0.1260); & \text{since } 4 \text{ cannot be realized} \\ C & (0.0882) & \\ D & (0.2058) & \\ E \text{ and } X\rightarrow E & (0.1960); & \text{since } 4 \text{ cannot be realized} \\ F & (0.1500) & \\ G & (0.0600) & \\ H & (0.0900) & \end{array} \tag{5.1}$$

The compound outcomes may be explained as follows. The outcome "A and E" occurs if DB*3* yields NO and DB*2* yields NO also; the outcome "B and $Y\rightarrow B$" occurs if DB*3* yields NO and DB*2* yields YES, in which case activity X will not be realized and, consequently, node *4* cannot be realized; finally, the outcome "E and $X\rightarrow E$" occurs if DB*3* yields YES and DB*2* yields NO, in which case activity Y will not be realized and consequently node *4* cannot be realized.

Our representation of the same logic is different, and is shown in Fig. 5.6. It differs from Eisner's representation in two major respects. First, we have succeeded in avoiding the duplication of DBs through the artifice of adding dummy nodes and arcs, which can always be done to represent any desired relationships among the outcomes of DBs. Obviously this would not have been possible without the definition of the various logical relationships among arcs incident on or emanating from a node, shown in Table 5.1. Second, the implied logical relationships are brought into sharper focus. For instance, event A cannot occur alone; it must occur in conjunction with event E, hence the terminal node $A\cap E$. Furthermore, it is now more evident that event B depends on the realization of activity Y, and so on.

[†]The numbers between parentheses are the probabilities of event realizations. Note that they add up to 1.0.

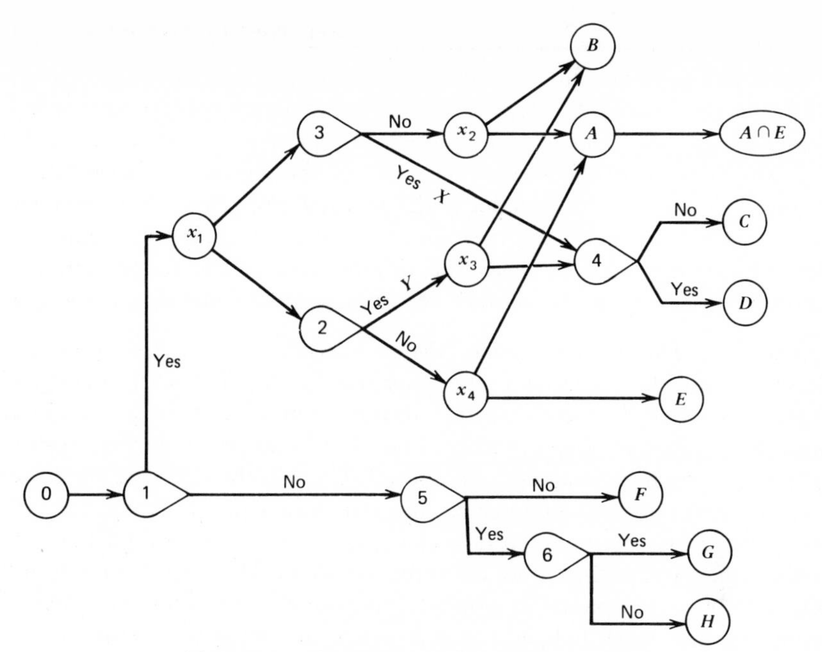

Figure 5.6. The GAN network for Eisner's project.

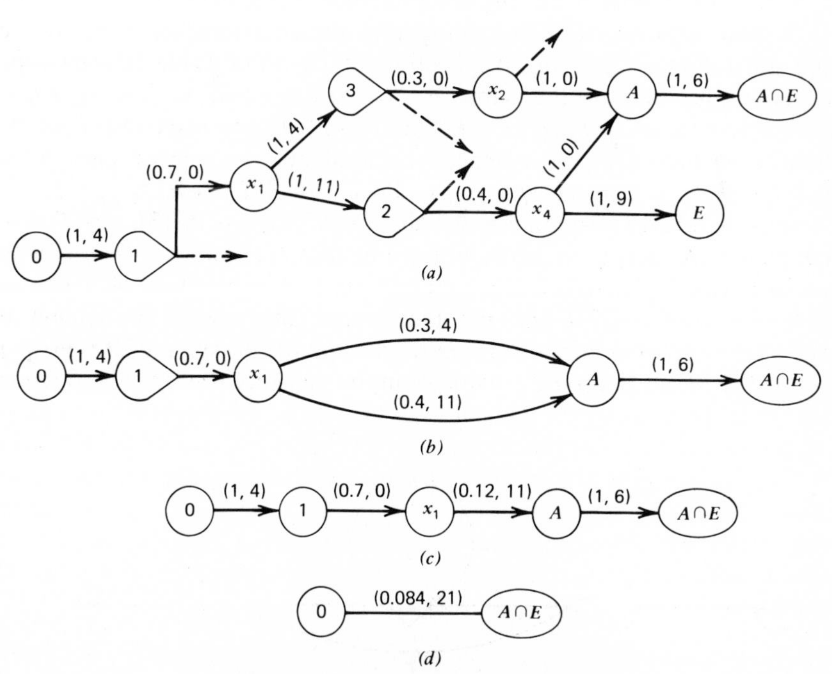

Figure 5.7. The reduction to arc $(0, A \cap E)$.

There is one origin, node 0, and eight sinks given in the list (5.1). We illustrate the graphical steps of reduction by deriving the equivalent element $(0, A \cap E)$. This is accomplished in four steps as shown in Fig. 5.7 *a–d*. The duration of the various activities are those given by Eisner. Step *b* in Fig. 5.7 reduces the parallel paths to their respective equivalent elements; step *c* further combines the parallel paths into one equivalent element. Finally, step *d* yields the desired result. [For a still different representation of Eisner's problem, see Pritsker and Whitehouse (1966).]

Example 5.2 Flow in Production Shop. Consider a shop that produces electromechanical equipment characterized by the following sequence. There is a series of operations that terminate at an inspection station I. Inspected units are dispatched to one of two areas: a further testing operation T or an adjustment operation J. Units in the test area are either accepted and sent to J or rejected and sent to a separate repair area, R. From the repair area material flows back to be tested. After adjustment in J, the units are packed and delivered to store. The verbal description detailed above is depicted in graphical form in Fig. 5.8. The first series of operations have been lumped together in one activity a terminating in I.

Study of Fig. 5.8 indicates that there are two forward paths from A to $F: \pi_1 = afg$ and $\pi_2 = abcg$. Clearly, $t_{\pi_1} = t_a + t_f + t_g$, occurs with probability $p_{\pi_1} = p_f$. On the other hand, the determination of the expected duration of π_2 is best achieved by determining first the equivalent element to the subpath ITJ (see Fig. 5.9*a*). Application of rule (*e*) of Table 5.1 yields the expected duration of the subpath ITJ to be equal to $\Sigma_{n=0}^{\infty}(t_b + t_c + nt_l) p_b p_c p_l^n = (t_b + t_c) + t_l p_d/(1 - p_d)$, since $p_c = 1 - p_d$, and occurs with probability p_b, where $t_l = t_d + t_e$. Hence, the total duration of the path π_2 is $t_{\pi_2} = t_a + (t_b + t_c) + t_l p_d/(1 - p_d) + t_g$, and occurs with probability $p_{\pi_2} = p_b$.

Thus, assuming $a = (25)$, $b = (0.7; 6)$, $c = (0.7; 4)$, $d = (0.3; 3)$, $e = (4)$, $f = (.3; 6)$, $g = (2)$, then $t_{\pi_1} = 33$ hrs, with probability 0.3; and $t_{\pi_2} = 40$ hrs, with probability 0.7. Therefore, the expected time of completion of a unit is $33 \times 0.3 + 40 \times 0.7 = 37.9$ hrs. In other words, the rate of production is approximately 53 units per year per production line (assuming 2000 hrs per year). This result is also obtainable from the sequence of steps of reduction

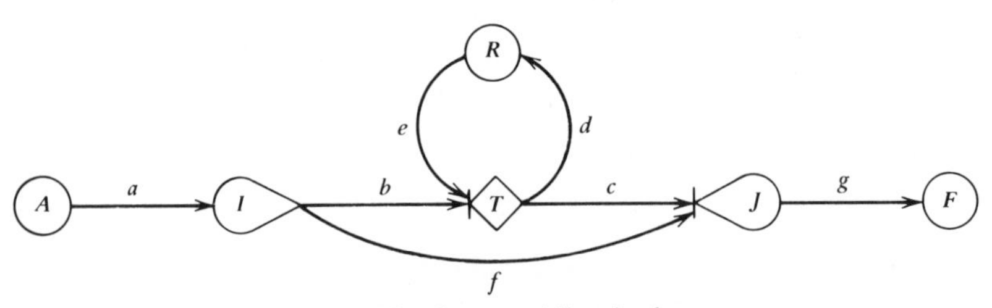

Figure 5.8. Pattern of flow in shop.

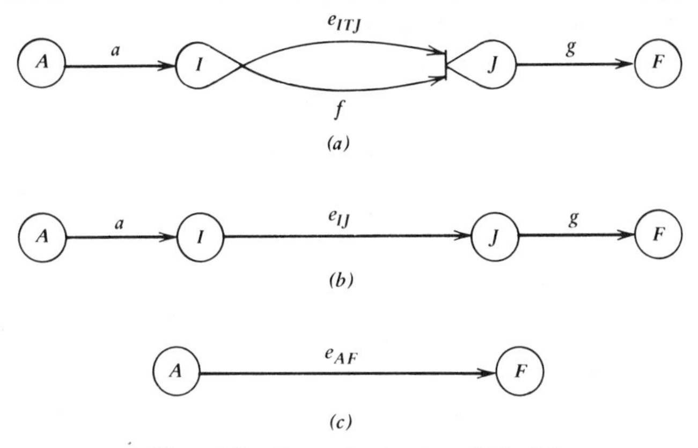

Figure 5.9. Steps of reduction of Fig. 5.8.

shown in Fig. 5.9*a–c*, in which we would first determine the equivalent element e_{IJ}, then the equivalent element e_{AF}. The duration of the equivalent element *ITJ* is (see above) $t_b + t_c + t_l p_d/(1-p_d) = 13$. Hence, the expected duration of the equivalent element e_{IJ} of Fig. 5.9*b* is $0.3 \times 6 + 0.7 \times 13 = 10.9$. This, added to $t_a + t_g = 27$, yields the same result of 37.9 as obtained before.

§2. THE GERT MODEL

When all the nodes of an AN are of the "exclusive-or" type on their receiving side, we have a GERT model. This acronym, bestowed on the model by Pritsker, its developer and main researcher, stands for *graphical evaluation and review technique*. Since we have previously indicated that other nodes can be reduced to "exclusive-or" nodes, at least in theory, it is easily seen that the model is quite general.

Its appeal, however, stems from two other important considerations. First, as was briefly mentioned before, such a network is then a representation of a semi-Markov process (SMP). Second, the network itself, after a simple transformation of variables, is really a signal flowgraph (SFG). Both subjects (SMPs and SFGs) are rich in mathematical structure, thus enhancing the analytical power of the model. Most importantly, a link has been established between two fields of study: (a) the operational aspects of large scale projects and (b) stochastic processes. Such linking between seemingly diverse fields of human study is always welcome, since it usually inaugurates new and fertile horizons for more fruitful research in both fields.

§2.1. GERT and Semi-Markov Processes

A SMP is a stochastic process that makes its transition from any state i to any other state j (which may be i) according to the transition-probability matrix of a Markov process, but whose time between transitions is a RV that may depend on both i and j. The duration will be denoted by Y_{ij}, which is assumed to have a probability density (or mass) function denoted by $h_{ij}(y)$:

$$h_{ij}(y) \geqslant 0, \qquad \int_0^{\infty} h_{ij}(y)\,dy = 1$$

In discrete-time Markov processes, $h_{ij}(y)$ is degenerate since the time of transition is a constant (which is usually defined as the unit of time).

It may be of some assistance to the reader to visualize a SMP in the following fashion. When the system reaches state i it immediately decides on its next state j. This decision is made according to the probabilities (p_{ij}) of the transition matrix. Now the system chooses to wait (i.e., "hold") in state i for some time y, where y is chosen randomly according to the DF of Y_{ij}.

Perhaps the most widely studied SMPs (or, for that matter, stochastic processes in general) are those that possess the so-called ergodic property, that is, those achieving statistical equilibrium after being operated for a "long" period of time. In such systems, the influence of the initial state disappears and the system reaches a situation for which the absolute state distribution is independent of time (i.e., stationary). To be sure, a great deal of the analysis of SMPs has concentrated on precisely the *stationary* properties of such systems.

In contrast, ANs possess both "start" and "terminal" nodes, with all intermediate states being transient. In the analysis through GERT, interest is focused on the behavior of the system, *given* that it is initiated in a particular "start" state—there is no question of ignoring the initial state, because its effect is never "washed out."

This observation has led some authors to conclude that the basic theorems of SMPs are "not of interest" in the study of GERT. While such a contention may partially apply in the case of limit theorems on the stationary properties of SMPs,[†] it certainly does not in the *transient* analysis of such processes. In fact, with the above definitions and interpretations in mind, it is not difficult to see immediately that a GERT network is a SMP with one or more absorbing states.[‡] The state is the event; the

[†]Actually, SMP theory yields the probability of the realization of the terminals directly—a valuable piece of information in GERT!

[‡]In the vernacular of Markov processes, an absorbing state (or a "trapping" state) is a state in which the system remains permanently once it is entered.

transition probability p_{ij} represents the probability that activity (ij) will be realized; and the duration of the activity, Y_{ij}, plays the role of the "holding time" in SMPs.

The study of SMPs in relation to GERT is relevant from another point of view. While the main motivation to the development of the concepts of GANs was the graphical representation of ANs, the reader should realize that GERT use is not confined to this subject. As a modeling technique it is equally applicable to the modeling of problems in queuing, inventory, computer programs, reliability, quality control, and several other fields. In such contexts, stationary behavior is often of interest, and the identity between GERT and SMPs and their graphic representation as SFGs (discussed below) becomes even more forceful, as is amply demonstrated below.

§2.2 Semi-Markov Processes and Signal Flowgraphs

Signal flowgraphs (SFGs), sometimes referred to simply as *flowgraphs*, are graphic representations used for the modeling and analysis of linear systems. They originated in the study of electrical networks in the early 1950s, but have since been proposed for the modeling of numerous systems in many fields of science and engineering.† They derive their utility from the fact that many systems, irrespective of content, can be modeled as a set of linear equations to which the methodology of SFG theory is directly applicable.

In Appendix E we give a brief resumé of SFG theory and the analytical tools it utilizes. If the reader is not familiar with these basic concepts, he is urged to read that appendix before proceeding any further.

An observer watching the system in state i can only infer the unconditional probability density function of the waiting time at state i, denoted by $w_i(y)$, and given by

$$w_i(y) = \sum_{j=1}^{n} p_{ij} h_{ij}(y), \quad y \geqslant 0$$

Let $\bar{h}_{ij}$ denote the conditional mean holding time between i and j, defined by

$$\bar{h}_{ij} = \int_0^\infty y h_{ij}(y)\, dy$$

and let $\bar{h}_i = \sum_{j=1}^{n} p_{ij} \bar{h}_{ij} = \int_0^\infty y w_i(y)\, dy = \bar{w}_i$ denote the unconditional mean holding time at i. Let $W_i(t)$ denote the PDF of the unconditional waiting

†For examples, see the book by Howard (1971).

time at i,

$$W_i(t) = \int_0^t w_i(y)\,dy$$

and let

$$\tilde{W}_i(t) = 1 - W_i(t)$$

denote the complementary DF. Finally, let $\phi_{ij}(t)$ denote the (conditional) probability that the system is in state j at time t given that it was in state i at time zero. Note the difference between ϕ_{ij} and p_{ij}: the latter is the (conditional) probability of a single transition whose duration is the RV Y_{ij}, while the former is the (conditional) probability of *being* in state j at time t starting from state i at time zero, which may involve several transitions. We refer to ϕ as the *interval transition probability* (ITP).

The ITP may be stated in a recursion equation as follows. Let δ_{ij} denote Kronecker delta: $\delta_{ij} = 1$ if $i = j$ and $\delta_{ij} = 0$ otherwise. Then either the system started in state j and has remained there all the time, and the probability of that eventuality is clearly $\delta_{ij}\tilde{W}_j(t)$, or it started at some state $i \neq j$, made a transition to some state k after "holding" at i for a period $\tau < t$, and then transferred from k to j in the remaining $t - \tau$ (in one or more transitions); and the probability of this event is $[p_{ik}\int_0^t h_{ik}(\tau)\phi_{kj}(t-\tau)\,d\tau)]$. Since k is any state, which may be either i or j, we must sum this latter probability over all possible intermediate states to obtain

$$\phi_{ij}(t) = \delta_{ij}\tilde{W}_j(t) + \sum_{k=1}^{n} p_{ik}\int_0^t h_{ik}(\tau)\phi_{kj}(t-\tau)\,d\tau \tag{5.2}$$

The convolution under the sign of integration suggests the utilization of the Laplace transform theory.† Let

$$g(s) = \int_0^\infty g(t)e^{-st}\,dt \tag{5.3}$$

be the (one-sided) Laplace transform of $g(t)$. Transforming both sides of

†For a comprehensive study of transform theory, see the classic book of Van Der Pol and Bremmer (1955); for a concise and lucid treatment of the relevant theory, see the work by Howard (1971), Section 11.4.

Eq. (5.2) we obtain

$$\phi_{ij}(s) = \delta_{ij}\tilde{W}_j(s) + \sum_{k=1}^{n} p_{ik}h_{ik}(s)\phi_{kj}(s)$$

$$= \delta_{ij}\tilde{W}_j(s) + \sum_{k=1}^{n} u_{ik}(s)\phi_{kj}(s) \tag{5.4}$$

where

$$u_{ik}(s) \triangleq p_{ik}h_{ik}(s) \tag{5.5}$$

Thus, the device of Laplace transformation has reduced the set of integral equations of (5.2) to the set of *linear* equations of (5.4). But these latter can be represented by SFGs! The generic element of these equations is shown in Fig. 5.10. In particular, we must define n^2 variables $\{\phi_{ij}(s);\ 1 \leqslant i, j \leqslant n\}$, if the probabilities of the so-called virtual transitions† p_{ii} are >0, or define $n(n-1)$ variables $\{\phi_{ij}(s),\ 1 \leqslant i,j \leqslant n;\ i \neq j\}$ if $p_{ii}=0$ for all i.

The SFG representation of Eq. (5.4) reveals to the naked eye two important facts. First, that the $\phi_{ij}(s)$ values are dependent on the products $p_{ij}h_{ij}(s)$, not on the individual quantities. On some reflection, this is seen to stand to reason. Second, that this system of linear equations has a rather peculiar structure, which simplifies to a considerable degree the direct evaluation of the ITPs.

Before delving into the more detailed analysis of this recursive equation, which we do presently, we remark that the dependence of $\phi_{ij}(s)$ on the product $p_{ij}h_{ij}(s) \triangleq u_{ij}(s)$ is of a more fundamental significance than is apparent at first glance. For $h_{ij}(s)$ may be viewed as the "moment generating function" (MGF) of the RV Y_{ij} representing the "holding time" of the SMP or, in the language of ANs, the duration of activity (ij). In this light, we can deduce some of our previous results and many others in a more direct and elementary way.

Indeed, we immediately recognize from Eq. (5.4) that $u_{ij}(s)$ is the transmittance along arc (ij). Recalling the rules of SFGs, it becomes

Figure 5.10. Generic element of SFG.

†These are the transitions from any state to itself, as distinguished from "real" transitions, which indicate movement to a different state.

evident that for GERT networks in which all nodes are of the "exclusive-or" type, the most direct manner for analyzing such networks must be through the use of these transmittances. The building blocks of such analysis are the three elementary networks of: activities in series, activities in parallel, and feedback loops. In Problem 1 we ask the reader to derive the expressions for the equivalent element in each case.

To illustrate the ease and utility of this approach, consider once more the production-flow problem of Example 5.2. Since all durations are constants, we obtain the z-transform instead of the Laplace transform of the DFs. The transmittances along the various arcs are as shown in Fig. 5.11. The equivalent transmittance $u_{AF}(z)$ may be determined directly from Mason's rule (see Appendix E) and is given by (we drop the explicit dependence on z for ease of writing):

$$u_{AF} = u_a u_g \left[u_b u_c / (1 - u_d u_e) + u_f \right]$$

However, for the sake of deeper understanding of the mechanisms of transmittances and SFGs, we derive this expression in a step-by-step approach.

Let x_A be the value of the (fictitious) variable at node A. Since this is the "forcing function," we eventually put $x_A = 1$. Define x_I, x_T, x_R, x_J, and x_F as the variables at their respective nodes. Then following the elementary rules of SFG theory, we have that

$$x_I = x_A u_a$$

$$x_T = x_I u_b + x_T u_d u_e = x_A u_a u_b + x_T u_d u_e$$

which yields

$$x_T = \frac{x_A u_a u_b}{1 - u_d u_e}$$

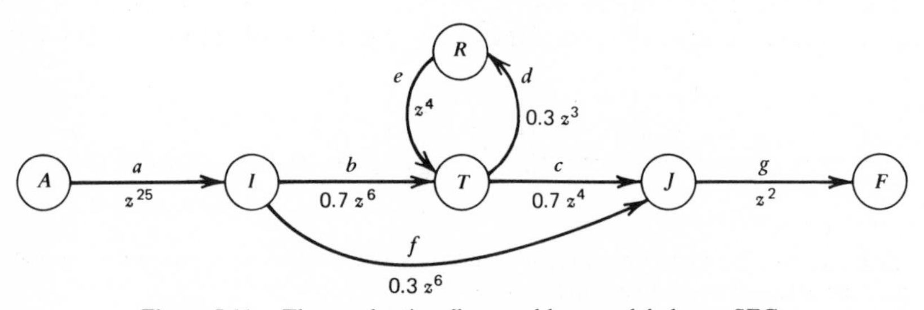

Figure 5.11. The production flow problem modeled as a SFG.

Continuing,

$$x_J = x_T u_c + x_I u_f$$

$$= \frac{x_A u_a u_b u_c}{1 - u_d u_e} + x_A u_a u_f$$

$$= x_A u_a \left[\frac{u_b u_c}{1 - u_d u_e} + u_f \right]$$

Finally,

$$x_F = x_J u_g = x_A u_a u_g \left[\frac{u_b u_c}{1 - u_d u_e} + u_f \right]$$

Hence, the equivalent transmittance between A and F is exactly as the expression stated before.

Now substituting the numerical values of the various probability and time parameters we obtain

$$u_{AF}(z) = z^{27} \left[\frac{0.49 z^{10}}{1 - 0.3 z^7} + 0.3 z^6 \right] = p_{AF} y_{AF}(z)$$

The rightmost equality follows from the definition of the transmittance u_{AF}. Thus, the probability

$$p_{AF} = u_{AF}\big|_{z=1} = 1$$

as it should be, since the system must terminate in state F sooner or later. Consequently, $u_{AF}(z)$ is in fact the MGF $y_{AF}(z)$. Therefore, to obtain the expected duration from A to F, we evaluate

$$\left. \frac{\partial}{\partial z} y_{AF}(z) \right|_{z=1} = \left\{ 27 z^{26} \left[\frac{0.49 z^{10}}{1 - 0.3 z^7} + 0.3 z^6 \right] + z^{27} \right.$$

$$\left. \left. \times \left[\frac{(1 - 0.3 z^7)(4.9 z^9) - 0.49 z^{10}(-2.1 z^6)}{(1 - 0.3 z^7)^2} + 1.8 z^5 \right] \right\} \right|_{z=1}$$

$$= 37.9$$

which is precisely the result obtained previously.

We now pick up the thread of discussion of Eq. (5.4). First we treat the stationary performance of SMPs. To obtain the matrix representation of Eq. (5.4), let

$\mathbf{\Phi}(s)$: the matrix $[\phi_{ij}(s)]$
$\mathbf{U}(s)$: the matrix $[u_{ij}(s)]$; where $u_{ij}(s)=p_{ij}h_{ij}(s)$
$\tilde{\mathbf{W}}^d(s)$: the diagonal matrix whose ith entry is $\tilde{W}_i(s)$
$\mathbf{M}^d$: the diagonal matrix whose ith entry is the unconditional holding time $\bar{w}_i$

From Eq. (5.4) we have

$$\mathbf{\Phi}(s)=\tilde{\mathbf{W}}^d(s)+\mathbf{U}(s)\mathbf{\Phi}(s)$$

or

$$\mathbf{\Phi}(s)=\left[\mathbf{I}-\mathbf{U}(s)\right]^{-1}\tilde{\mathbf{W}}^d(s) \tag{5.6}$$

The flowgraph representation of the latter equation is given in Fig. 5.12a.

To derive the stationary performance, the behavior of the system as $t\to\infty$ is given by

$$\mathbf{\Phi}_\infty \triangleq \lim_{t\to\infty}\mathbf{\Phi}(t)=\lim_{s\to 0}s\mathbf{\Phi}(s)$$

where the second equality follows from the final value theorem of Laplace transforms. Using Eq. (5.6) we can write this limit in the form

$$\mathbf{\Phi}_\infty=\lim_{s\to 0}s\left[\mathbf{I}-\mathbf{U}(s)\right]^{-1}\lim_{s\to 0}\tilde{\mathbf{W}}^d(s) \tag{5.7}$$

But from the definition of $\tilde{W}_i(t)=1-W_i(t)=1-\int_0^t w_i(r)\,dr$ we have that

$$\tilde{\mathbf{W}}^d(s)=\frac{\mathbf{I}-\mathbf{W}^d(s)}{s} \tag{5.8}$$

where $\mathbf{W}^d(s)$ is the diagonal matrix whose ith entry is $w_i(s)$ and $\mathbf{I}$ is the identity matrix. Since $\lim_{s\to 0}\mathbf{W}^d(s)=\mathbf{I}$, both numerator and denominator of Eq. (5.8) are equal to zero. The evaluation of this limit is by L'Hôpital's

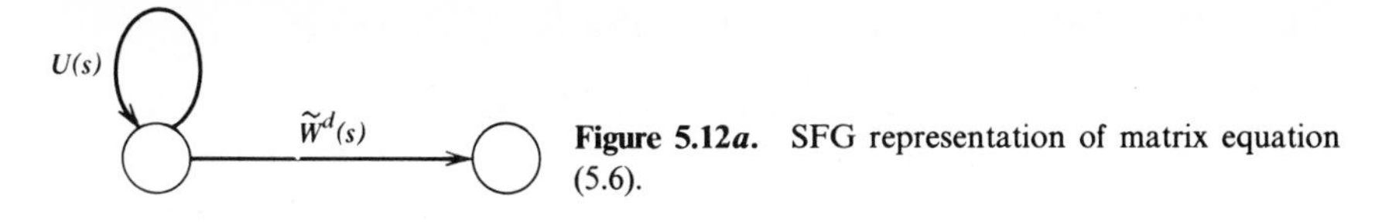

Figure 5.12*a*. SFG representation of matrix equation (5.6).

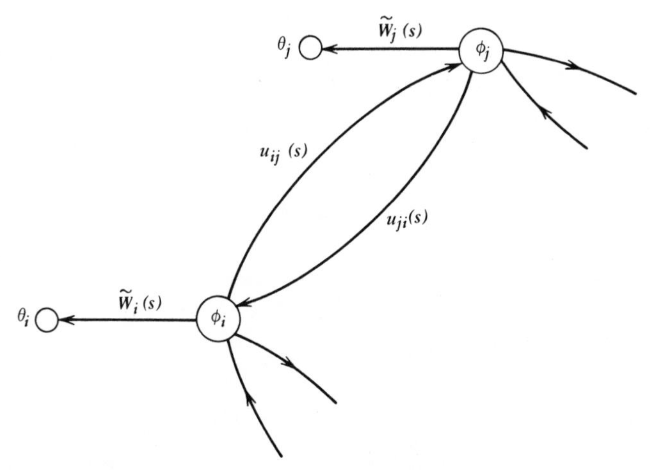

Figure 5.12*b*. Portion of the SFG representation of steady-state equations (5.11).

rule

$$\lim_{s\to 0} \tilde{\mathbf{W}}^d(s) = \left. \frac{\frac{d}{ds}\left[\mathbf{I}-\mathbf{W}^d(s)\right]}{\frac{d}{ds}\left[s\right]} \right|_{s=0}$$

$$= -\left.\frac{d}{ds}\mathbf{W}^d(s)\right|_{s=0}$$

$$= -\left.\frac{d}{ds}\int_0^\infty \mathbf{W}^d(t)e^{-st}\,dt\right|_{s=0}$$

$$= \int_0^\infty t\mathbf{W}^d(t)\,dt = \mathbf{M}^d$$

where $\mathbf{M}^d$ is defined above. Substituting in Eq. (5.7), we get

$$\boldsymbol{\Phi}_\infty = \lim_{s \to 0} s\left[\mathbf{I} - \mathbf{U}(s)\right]^{-1}\mathbf{M}^d \tag{5.9}$$

Therefore, except for the constant diagonal matrix $\mathbf{M}^d$, the limiting conditional transition probability is just the limit of the product $s[\mathbf{I}-\mathbf{U}(s)]^{-1}$.

It is intuitively clear, as well as can be rigorously proven, that the limiting matrix $\boldsymbol{\Phi}_\infty$ is of equal rows. Hence, we can talk about ϕ_j, the limiting *unconditional* transition probability that after a long period of time the system would be in state j. As can be seen from Eq. (5.9), this must be given by the product of a matrix $\mathbf{E} \triangleq \lim_{s \to 0} s[\mathbf{I}-\mathbf{U}(s)]^{-1}$, whose rows are identical, and $\mathbf{M}^d$. Howard (1964) proved that the jth entry of any row of $\mathbf{E}$ is simply

$$e_j = \frac{\pi_j}{\sum_j \pi_j \bar{h}_j},$$

where the vector $\boldsymbol{\Pi} = (\pi_1, \pi_2, \ldots, \pi_n)$ is the vector of the steady-state probabilities of the "imbedded Markov chain" with transition matrix $\mathbf{P} = [p_{ij}]$. Consequently, we have

$$\phi_j = \frac{\pi_j \bar{h}_j}{\sum_{i=1}^{n} \pi_i \bar{h}_i}; \qquad j = 1, 2, \ldots, n \tag{5.10}$$

If attention is focused on aspects of the stationary behavior of the system, then we may, indeed, capitalize on the above results to simplify the SFG. In particular, we now define only n variables $\phi_1, \phi_2, \ldots, \phi_n$, which are interrelated by the equations [cf. Eqs. (5.4)]:

$$\phi_j(s) = \delta_{ij} \tilde{W}_j(s) + \sum_{k=1}^{n} u_{kj}(s)\phi_k(s) \tag{5.11}$$

A facsimile of the corresponding SFG is shown in Fig. 5.12*b*.

Example 5.3. To illustrate the above theory consider the "machine-minding" example treated by Whitehouse using GERT (1973b, p. 343 ff.) The statement of the problem runs as follows: there are $m=2$ identical machines, supported by $n=1$ similar "spare" machine. There is one repairman who is capable of repairing one machine at a time. A machine fails in a completely random fashion at an average rate $\lambda = 1$. The time of service of a machine is assumed exponentially distributed with mean $= \frac{1}{2}$. Hence, the rate of service is $\mu = 2$.

Let the state of the system be the number of machines failed. Then there are four states: *0*, *1*, *2*, and *3*; where state *0* is "all machines operative," state *1* is "one machine failed," and so on. The rates of "arrivals" to and "departures" from service are obviously state dependent. Let r denote the number of machines failed, and let λ_r denote the rate of arrivals when r machines are "failed" and μ_r, the corresponding rate of service. Clearly, the following conditions apply:

1. If $r=0$, then $\mu_0=0$ (since no machines are being repaired), and $\lambda_0=2$.
2. If $r=1$, then $\mu_1=2$, while $\lambda_1=2$ still, because the spare machine is brought into action.
3. If $r=2$, then $\mu_2=2$ still, since there is only one repairman, but $\lambda_2=1$ because there is one machine still operative.
4. If $r=3$, then $\mu_2=2$ still, but $\lambda_3=0$ because there are no machines operative.

Summarizing, we have the following table of the values of λ_r and μ_r:

r	0	1	2	3
λ_r	2	2	1	0
μ_r	0	2	2	2

(5.12)

Next, we must determine the p_{ij} and $h_{ij}(y)$ for all i and j. Complete randomness in arrivals and departures result in the holding time h_r being exponentially distributed with mean $1/(\lambda_r+\mu_r)$ for all r; in other words, $h_r(y)=(\lambda_r+\mu_r)\exp-(\lambda_r+\mu_r)y$. We immediately deduce that

$$w_r(y)=\sum_k p_{rk}h_{rk}(y)=h_r(y)$$

Since the Laplace transform of the exponential density function $\alpha e^{-\alpha y}$ is $\alpha/(s+\alpha)$, and recalling the definitions of $W_r(t)$ and $\tilde{W}_r(t)$ and the table of (5.12), one can easily construct the following table:

r	$h_r(y)=w_r(y)$	$h_r(s)$	$W_r(t)$	$\tilde{W}_r(t)$	$\tilde{W}_r(s)$
0	$2e^{-2y}$	$\frac{2}{s+2}$	$1-e^{-2t}$	e^{-2t}	$1/(s+2)$
1	$4e^{-4y}$	$\frac{4}{s+4}$	$1-e^{-4t}$	e^{-4t}	$1/(s+4)$
2	$3e^{-3y}$	$\frac{3}{s+3}$	$1-e^{-3t}$	e^{-3t}	$1/(s+3)$
3	$2e^{-2y}$	$\frac{2}{s+2}$	$1-e^{-2t}$	e^{-2t}	$1/(s+2)$

(5.13)

The final step in these preliminary calculations is to determine the transition probabilities p_{ij}, which are known from elementary queuing considerations. Now, from state *0* the system moves with certainty to state *1*; while from state *3* the system moves with certainty to state *2*. Hence, $p_{0,1}=1=p_{3,2}$, and $p_{0j}=0$ for $j\neq 1$ and $p_{3j}=0$ for $j\neq 2$. In either state $r=1,2$, the system moves to state $r+1$ with probability $p_{r,r+1}=\lambda_r/(\lambda_r+\mu_r)$ and to state $r-1$ with probability $p_{r,r-1}=\mu_r/(\lambda_r+\mu_r)$. All other transition probabilities are equal to zero. These statements are conveniently summarized in the matrix of transition probabilities [recalling the values in the table of (5.12)] as follows:

Matrix of Transition Probabilities

State	0	1	2	2
0		1		
1	$\frac{1}{2}$	0	$\frac{1}{2}$	
2		$\frac{2}{3}$	0	$\frac{1}{3}$
3			1	
Π	0.2	0.4	0.3	0.1

(5.14)

The matrix of (5.14) is the matrix of transition probabilities of the "imbedded Markov chain"—in other words, the transition probabilities if our attention is solely devoted to the points of transition from one state to another. The last row of (5.14) is the stationary probability vector $\mathbf{\Pi}$.

We are now in a position to draw the SFG corresponding to the SMP. It is shown in Fig. 5.13. The graph is a representation of the (steady-state) system of linear equations of (5.11). For example, $\phi_1(s)$ is given by

$$\phi_1(s)=\delta_{i1}\tilde{W}_1(s)+\sum_j u_{j1}(s)\phi_j(s);$$

$$=\delta_{i1}\tilde{W}_1(s)+u_{01}\phi_0(s)+u_{21}\phi_2(s)$$

$$=\frac{\delta_{i1}}{s+4}+1\times\frac{2}{s+2}\phi_0(s)+\frac{2}{3}\times\frac{3}{s+3}\phi_2(s)$$

To determine the unconditional state probabilities, $\{\phi_j\}$, we need the diagonal matrix $\mathbf{M}^d$, which is easily determined from the matrix of (5.13)—recalling that $\bar{h}_r=\bar{w}_r=\int_0^\infty yw_r(y)dy$. We have $\bar{h}_0=\frac{1}{2}$, $\bar{h}_1=\frac{1}{4}$, $\bar{h}_2=\frac{1}{3}$, and $\bar{h}_3=\frac{1}{2}$. Hence, $\Sigma_{i=0}^3\pi_i\bar{h}_i=0.35$, and from (5.10),

$$\phi_0=\frac{0.2\times 0.5}{0.35}=\frac{2}{7};\phi_1=\frac{2}{7};\phi_2=\frac{2}{7};\phi_3=\frac{1}{7}$$

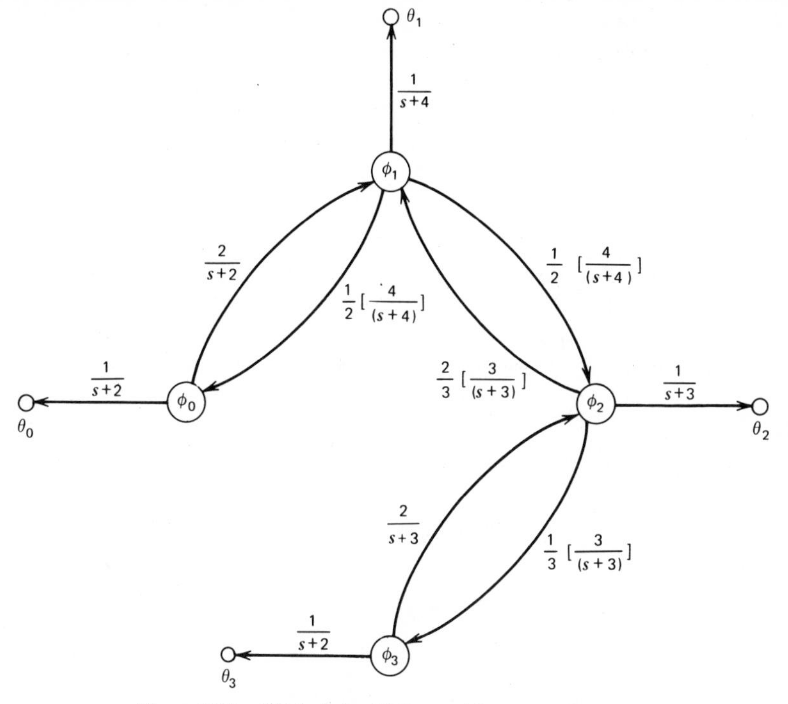

Figure 5.13. SFG of the SMP; machine-minding example.

These are precisely the values obtained by Whitehouse† (1973b, p. 348) for the steady-state probabilities of occupancy of the various states.

It is instructive to note the difference between the stationary probabilities $\mathbf{\Pi}$ of the imbedded Markov chain and the probabilities ϕ of the SMP: the former implicitly assume a unit holding time at all states, while the latter take into account the varying holding times.

Apart from the determination of the stationary probabilities of ergodic SMPs, interest is usually devoted to problems of conditional probabilities, counting, occupancy, and so on. Unfortunately, a direct attack on these problems through the development of joint and conditional probability functions of the same genre as the function ϕ above proves to be not rewarding at all—the elegance of the theoretical development notwithstanding. The reason for this unhappy situation is that the treatment requires the inversion of matrices that are themselves Laplace and z-

†The reader is alerted to the fact that the transmittances in Whitehouse's book are derived as $\mathcal{E}(e^{st})$ as opposed to our $\mathcal{E}(e^{-st})$.

transforms of time functions—a no minor feat. Additionally, there is still required, thereafter, the inverse transformation to the time domain, which is often extremely difficult to accomplish except in the most simple cases.

Fortunately, this bleak picture is brightened by the following crucial observation: most of these pertinent questions are more conveniently handled (after taking the necessary precautions) through the analysis of SMPs with one or more absorbing states—that is, by reinterpreting the SMP as a nonergodic system.

For these reasons, and in order not to depart from our main concern in this treatise (to-wit, ANs), we forfeit the detailed analysis of ergodic SMPs in response to the questions raised above. The interested reader may consult the work by Howard (1971, Chapter 11). Rather, we turn our attention next to *transient SMPs,* that is, processes with one or more absorbing states.

Suppose we define the joint probability $\phi_{ij}(k,t)$ as the probability that the process will occupy state j at time t and that it will have made k transitions, given that at time zero it was in state i;

$$\phi_{ij}(k,t) = Pr\left\{ \begin{array}{l} \text{number of transions } = k \text{ and state at time } t \text{ is} \\ j\text{; initial state is } i \text{ at time } 0 \end{array} \right\}$$

Next, assume that the process has only one absorbing state, denoted by r. Let $p_{ir}(k,t)$ be the probability that a process started in state i at time 0 and will require k transitions and duration t to reach the trapping state r. The difference between $\phi_{ir}(k,t)$ and $p_{ir}(k,t)$ is that $\phi_{ir}(k,t)$ is the probability that state r is entered *by* time t (i.e., on or before time t), while $p_{ir}(k,t)$ is the probability that state r is entered *at* time t; that is, $\phi_{ir}(k,t)$ is just the integral of $p_{ir}(k,t)$ with respect to time,

$$\phi_{ir}(k,t) = \int_{\tau=0}^{t} p_{ir}(k,\tau)\,d\tau \tag{5.15}$$

The two functions, ϕ and p, are functions of a continuous variable, the time t, and a discrete variable, the number of transitions k. Consequently, we define the Laplace–z transform (i.e., exponential–geometric transform) to be the result of transforming the function geometrically on k and exponentially on t; thus

$$\phi_{ij}(z,s) \triangleq \sum_{k=0}^{\infty} z^k \int_0^{\infty} dt\, e^{-st} \phi_{ij}(k,t); \tag{5.16}$$

and similarly for $p_{ij}(z,s)$.

We may now transform (5.15) to obtain

$$\phi_{ir}(z,s) = \frac{1}{s} p_{ir}(z,s)$$

This expression contains the clue to the procedure to be followed in treating transient SMPs. Consider the flowgraph of the SMP in the vicinity of the absorbing state r; it should appear as shown in Fig. 5.14. Note that because $p_{rr} = 1$ (r absorbing), no branches connect r with other states of the process. The transmittance of the loop around r is just

$$u_{rr}(s) = p_{rr} h_{rr}(s) = h_{rr}(s) = w_r(s)$$

Therefore, the effect of this loop on transmittances through node r is to multiply them by $1/[1 - w_r(s)]$. As far as transmittances to the output node, θ_r, associated with node r are concerned, we can remove this loop if we can change the transmittance on the output branch from $\tilde{W}_r(s)$ to

$$\frac{\tilde{W}_r(s)}{1 - w_r(s)} = \frac{1}{s} \tag{5.17}$$

in accordance with Eq. (5.8). The equivalent SFG at node r is also shown in Fig. 5.14.

Utilizing the result of Eq. (5.17), it is apparent that all we have to do to find $p_{ir}(z,s)$ is to remove the output branch associated with node r in Fig. 5.14, a branch that represents integration, and then find the transmittance of the original flowgraph from node i to node r—a result that is intuitively appealing in view of r being a trapping state.

To illustrate, suppose we are interested in the following "counting" problem of determining, in the above machine-minding example (of Fig. 5.13), the probability mass function of the number of times all the machines are operative (i.e., system in state *0*) before complete system

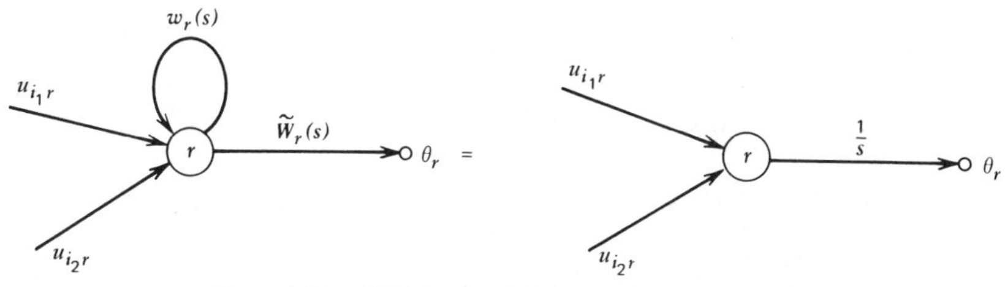

Figure 5.14. SFG in the vicinity of absorbing node r.

failure occurs (i.e., state *3* is visited), given that the system starts with all machines operative (i.e., in state *0*). In such a case, it is convenient to consider state *3* as an absorbing state, and since we are interested in the system returning to state *0*, we "tag" the arc leading into state *0* with z.† Of course, all output branches from *3* are eliminated, and the result is the SFG of Fig. 5.15. We immediately have, by Mason's rule,

$$p_{0,3}(z,s)=\frac{(2/(s+2))(\frac{1}{2})(4/(s+4))(\frac{1}{3})(3/(s+3))}{1-(2/(s+2))(z/2)(4/(s+4))-(\frac{2}{3})(3/(s+3))(\frac{1}{2})(4/(s+4))}$$

$$=\frac{4/(s+2)(s+3)(s+4)}{1-4z/(s+2)(s+4)-4/(s+3)(s+4)}$$

From the definitional equation (5.16), it is clear that $p_{ij}(z,0)$ is the z transform of the marginal probability distribution; the so-called "moment-generating function" of the count in question.‡ In the above example, we have

$$p_{0,3}(z,0)=\frac{1/6}{1-z/2-1/3}=\frac{1/6}{2/3-z/2}=\frac{1}{4}\left[1+\frac{3}{4}z+\left(\frac{3}{4}\right)^2z^2\right.$$

$$\left.+\cdots+(3/4)^k z^k+\cdots\right] \tag{5.18}$$

From (5.18) we conclude that with probability $\frac{1}{4}$ the system will have a complete breakdown without ever having returned to state *0* and that the probability of a complete breakdown after having been fully operative k times is $\frac{1}{4}(\frac{3}{4})^k$. As to be expected, the probability of the system eventually breaking down is

$$p_{0,3}(1,0)=\tfrac{1}{4}\left[1+\tfrac{3}{4}+\left(\tfrac{3}{4}\right)^2+\cdots\right]=1.$$

We continue with the remaining two classes of problems cited above and use the machine-minding example as our vehicle for illustration.

By "conditional probabilities" is meant statements of the form of the

†The identity between counting transitions and "tagging" the relevant branches has been established by several authors, for example, see the work by Howard (1971, Section 6.1).

‡Formally, one must "normalize" $\phi_{0,3}$ by the probability of the transition $\phi_{0,3}$, which is equal to unity in this case.

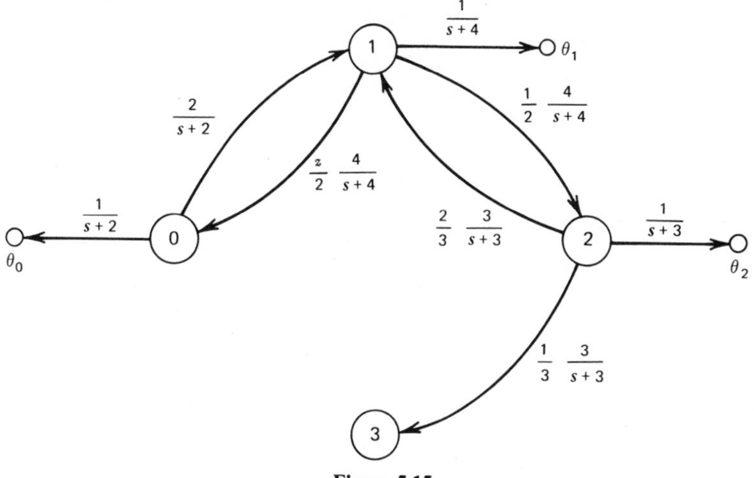

Figure 5.15.

probability that a particular event occurs given that the system is in state i at time 0. For instance, in the example we may be interested in the probability of complete system breakdown (state *3*) before being fully operative (visiting state *0*), given that the system starts with one machine failed (in state *1*). In such a case, one may consider states *0* and *3* as absorbing states and then determine the probabilities of transition from *1* to *3*. (That this representation fulfills the objective is evident from the fact that if the system ever visits state *0* it would be trapped there forever and never reach state *3*).

The corresponding SFG is shown in Fig. 5.16. Note that all branches emanating from states *0* and *3* have been eliminated. The required probability is easily obtained,

$$\phi_{1,3}(s) = \frac{\left(\frac{1}{2}\right)(4/(s+4))\left(\frac{1}{3}\right)(3/(s+3))}{1-\left(\frac{1}{2}\right)(4/(s+4))\left(\frac{2}{3}\right)(3/(s+3))} = \frac{2/(s+3)(s+4)}{1-4/(s+3)(s+4)}$$

It is easy to see that the probability that the system will be inoperative before returning to state *0* is $\phi_{1,3}(0)=\frac{1}{4}$. The probability $1-\phi_{1,3}(0)=\frac{3}{4}$ is the probability that the system returns to the "fully operative" condition (state *0*) before it experiences complete breakdown (state *3*).

Finally, problems of "duration" or "occupancy" are related to questions involving determination of the PDF of the time to the occurrence of a specific event, given that the system starts at time zero in some known

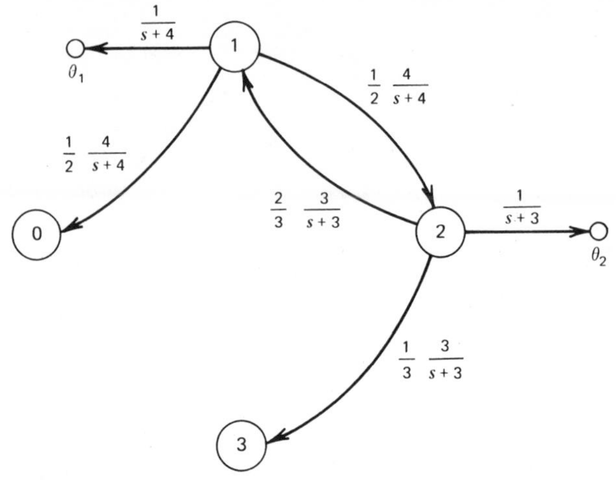

Figure 5.16. To evaluate the conditional probability that *3* will be visited before *0* when starting in *1*.

state and a particular condition is satisfied. For instance, using once again the above example, it is pertinent to ask what the PDF is of the time it takes the system to reach the "regeneration point," state 0, given that the system was never inoperative (i.e., that state *3* was not visited). In such a case it is convenient to make state *3* an absorbing state and create a dummy state, *0′*, which is an absorbing state representing the return to state *0* for the first time, and determine the PDF of the time it takes the system to traverse from *0* to *0′*.

The SFG representation of the system is as shown in Fig. 5.17, from which we have that

$$\phi_{0,0'}(s)\Big|_{s=0} = \frac{(2/(s+2))\left(\frac{1}{2}\right)(4/(s+4))}{1-\left(\frac{1}{2}\right)(4/(s+4))\left(\frac{2}{3}\right)(3/(s+3))}\Bigg|_{s=0} = \frac{3}{4}$$

hence

$$p_{0,0}(s) = \phi_{0,0'}(s)/\left[\phi_{0,0'}(s)\right]\big|_{s=0} = \frac{4}{3}\,\frac{\dfrac{4}{(s+2)(s+4)}}{1-\left[4/(s+3)(s+4)\right]} \qquad (5.19)$$

The inversion of (5.19) is rather complicated, but one can determine

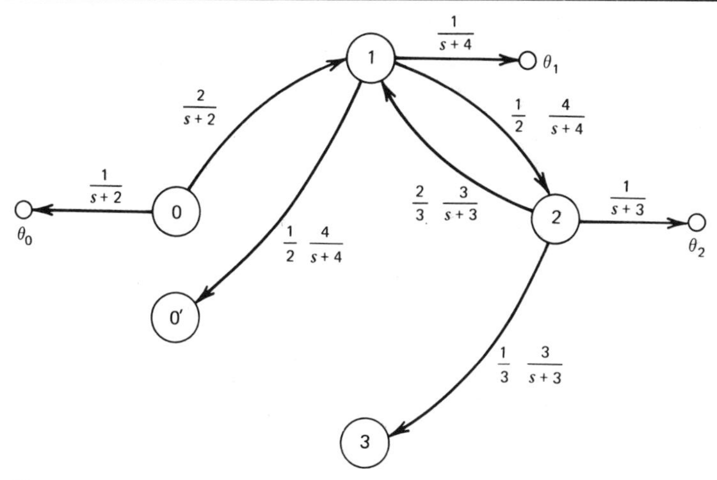

Figure 5.17. To evaluate the PDF of duration of first return to *0* before ever visiting *3*.

moments fairly easily. For instance, the average time to return to full operational posture is given by

$$-\frac{d}{ds}p_{0,0'}(s)\bigg|_{s=0}=\frac{25}{24}\text{ time units.}$$

We recapitulate the development thus far. We have demonstrated that the GERT model is really a model of a SMP. Then we have shown that by the straightforward use of standard transform methods, the graphical representation of the SMP yields a SFG. We have thus traveled the full circle: from the GERT graphical representation to the SMP mathematical structure that underlies such representation to the SFG graphical representation. We therefore conclude that GERT networks are in fact the SFG representation of SMPs!

This statement is not quite correct. We should have stated that the *original* GERT networks are the SFG representation of SMPs, because the acronym "GERT" has been expanded to cover several system simulation programs in which the nodes are *not* of the "exclusive-or" type, and in which several new classes of logical relationships have been introduced. A more detailed discussion of this simulation language and its uses are given in Section 3 below.

A pertinent comment is in order at this juncture. In the above analysis we opted to display in graphical form the SFG of the set of ϕ-equations of

(5.11) and the transient SMPs derived therefrom to conform to the various questions raised on occupancy, counting and conditional probabilities. However, this is not the only graphical representation possible of the same SMP—other representations are not only possible but may, in fact, be more desirable, depending on the type of questions asked. We remind the reader that SFGs are the graphical representation of any set of algebraic linear equations. Hence, if one defines other entities of the SMP and derives a set of algebraic linear equations involving these variables, such equations may be represented by a SFG. As an example of a different representation of a similar "machine minding" problem, see Elmaghraby (1970, p. 114 ff.). There, the variables of interest were the $p_j(t)$ values, where $p_j(t)$ is the probability of the system being in state j at time t. The set of linear equations were derived from the standard birth-and-death differential-difference equations.

§2.3. A Postscript: Discrete RVs, Multidimensional Vectors; and Multiplicative RVs

The above discussion has left three issues still unanswered: (a) how one should cope with *discrete* probability DFs; (b) what is to be done in case of several RVs defined on the activities so that the vector u is of dimensionality 3 or more; and (c) what model is constructed in the case of *multiplicative* RVs defined on the activities. We discuss these three issues next.

Discrete Random Variables.

In this case we utilize the z-transform instead of the standard Laplace transform. As is well known, the z-transform may be interpreted as a form of Laplace transform when integrals are taken as Lebesgue–Stieltjes; see Helm (1959) and Beightler, Mitten, and Nemhauser (1961). Let $\mathcal{L}[g(t)]$ denote the Laplace transform of $g(t)$ as defined in Eq. (5.3), and let $\mathcal{L}_z$ denote the z transform when $g(t)$ is defined only at discrete points in time $0,1,2,\ldots$. Then, indeed, the Laplace transform of Eq. (5.3) will be a function of e^{-s}. Now define $z=e^{s}$; then $\mathcal{L}_z$ can be written as a function of z^{-1}, the so-called "z-transform." Alternatively, we could have defined the z-transform ab initio as:

$$\mathcal{L}_z[g(t)] = \sum_{t=0}^{\infty} g(t)z^{-t} \triangleq g(z)$$

In any event, the z-transform has certain properties that are analogous to the standard Laplace transform properties:

1. If $y(t)=\Sigma_{\tau=0}^{t} g(\tau)x(t-\tau)$, then $y(z)=g(z)x(z)$. This is the discrete analog of the convolution integral, and is referred to as the *convolution sum*.
2. $\mathcal{L}_z[g_1(t)g_2(t)]=g_1(z)*g_2(z)$; that is, the $\mathcal{L}_z$ of the product of two time functions is the convolution of their respective z-transforms.
3. $\mathcal{L}_z[g(-t)] \triangleq \bar{g}(z)=g(z^{-1})$
4. $\mathcal{L}_z[g(t+\tau)]=z^{-\tau}\mathcal{L}_z[g(t)]-\Sigma_0^{\tau} g(t)z^{-\tau}$, and $\mathcal{L}_z[g(t-\tau)]=z^{\tau}\mathcal{L}_z[g(t)]+\Sigma_0^{\tau} g(t)z^{\tau-t}; \tau>0$
5. $\lim_{t\to 0} g(t)=\lim_{z\to\infty} g(z)$ (the initial value theorem)
   $\lim_{t\to\infty} g(t)=\lim_{z\to 1}(1-z^{-1})g(z)$ (the final value theorem)

The Case of Several Parameters.

Assuming that the parameters behave additively similar to Y_u,[†] we simply define the Laplace transform (or the z-transform) of the parameters in a manner similar to the definition of $h_{ij}(s)$. For example, if the cost of an activity is included in the vector u, assume that it is a RV with a (continuous) density function $g_u(c)$. Then we define the Laplace transform

$$g_u(r)=\int_0^{\infty} g_u(c)e^{-rc}\,dc$$

and the transmittance of the arc of Eq. (5.10) is now given by

$$u_{ij}(s,r)=p_{ij}h_{ij}(s)\,g_{ij}(r)$$

From this point onward the analysis proceeds as before.

The Case of Multiplicative Variables.

"Multiplicative variables" mean that the variables associated with arcs in series are multiplied, not added. Examples of such behavior occur in the context of "explosion" in materials planning (in which a unit of assembly requires several units of each subassembly, each unit of which, in turn, requires several units of each subsubassemblies, etc.), or in the context of epidemics in which a "carrier" of the disease may infect several individuals.

It has been suggested [Pritsker and Happ (1966)] that in such circumstances the appropriate transform to use is the Mellin transform [Sneddon (1973), pp. 262–288; Epstein (1948)]. Little is known about the application of this transform in the context of ANs. This presents a fertile area of investigation to researchers in this field.

[†]In the case of multiplicative behavior, see the following paragraph.

§3. THE GERT SIMULATOR (GERTS)

The previous section established the following facts: (a) that the (original) GERT model is a specialization of the GAN model to the case of all nodes being of the "exclusive-or" type, (b) that such specialization in fact represents a model of a SMP, and (c) that the GERT model is synonymous with the SFG representation of the SMP. The ultimate conclusion is that GERT is the graphical representation of a SMP.

The most salient characteristic of a SMP is that *it possesses no "memory"*: the probability of transition from a state i to another state j at any time t depends *only on being in state i at that particular time t, but is independent of the history of arrival at i.*

"Memoryless" behavior may seem an oddity among real-life systems; one would be hard put, indeed, to give examples in which past history plays no role in the current system's decisions. However, the reader should recognize the difference between "what is" in reality and the question of "how it can be modeled"; in many instances the system behaves *as if* it lacks memory and hence can be modeled as such. This explains the abundance of models of stochastic processes that *assume* memoryless behavior in telephony, population, epidemiology, and the like.

However, the question that persists is how systems *with memory* can be modeled and analyzed. The answer obviously lies in the abandonment of the analytical model of SMPs and the appeal to simulation models.[†] This is the rationale behind the development of GERTS and also explains why the concept of GERT has advanced well beyond its original formulation—in particular, from a graphical representation of a SMP to a general modeling instrument via simulation (see previous section). In undergoing such metamorphosis, GERT has become a general and valuable tool in systems modeling and analysis. Indeed, the price paid for such generality is the abandonment of analyticity in favor of experimentation through simulation. The price is well worth paying, since the capability to expand the horizon of systems study to encompass stochastically evolving systems with memory is of great value. And in the hands of a knowledgeable analyst, the method of simulation is a potent tool, indeed.

We devote this section to a brief exposition of the basic concepts of the GERT simulation, or GERTS for short. By so doing we hope to stimulate the reader to undertake a more in-depth study of this powerful tool.

[†]An alternative approach is to cling to the SMP model by expanding the state description to accommodate past history. This approach quickly becomes computationally infeasible as the number of states grows to astronomical figures, and the matrix of transition probabilities becomes extremely sparse to defy any attempt at the objective estimation of the probabilities.

Fortunately, a complete and rather lucid documentation of the program is available.[†] This section relies heavily on such documentation.

We recall the basic concepts of GANs which are also applicable to GERTS (see Section 1): (a) for nodes, a "receiving side" and an "emitting side", with both deterministic and probabilistic branching and (b) for branches, additive parameters such as time and cost, with the possibility of feedback.

The other concepts previously introduced are still retained in GERTS but are greatly expanded in their logical content; hence, we prefer to re-enumerate them within the framework of GERTS. Before we embark on the description of these concepts, we wish to define two terms used in GERTS that we had no reason to define before. Since GERTS is a simulation, we need to define what we mean by *a simulation run*. It is the realization of a terminal node or a set of sink nodes if several are so specified to constitute *a completion*. Notice that, because of probabilistic branching, it is possible that a run is completed with only a small fraction of the activities and nodes realized. Furthermore, GERTS offers the capability of restricting a run to be completed either when any of the terminal nodes are realized, or only after a specified number of sink nodes are realized. In this respect, GERTS makes a careful distinction between *sink* nodes and *terminal* nodes. In particular, a *sink node* is a node with no arcs emanating from it—it has incident arcs only on its receiving side. But a sink node need not be a terminal node, in the sense that it does not define a run either individually or in combination with other nodes. We discuss this point later (see p. 364). On the other hand, a *terminal node* is a sink node that defines the completion of a run, either individually or jointly with others.

We have previously referred to the occurrence of an event as the "realization" of a node. The simulation GERTS refers to the same event as the *release* of the node. We use the word *realization* when we wish to refer to the satisfaction of all the conditions specified for the realization of the event represented by the node—in other words, when we wish to focus attention on the receiving side of the node. And we use the word *release* when we are focusing attention on the activities emanating from the node —that is, when we wish to refer to the emitting side of the node. Thus, while the two words mean essentially the same, namely, the realization of the event, we use them to focus attention on either aspect of the node.

The main features of GERTS may be summarized under two main

[†]*The GERT IIIZ User's Manual*, A. A. B. Pritsker and C. E. Sigal, Aug. 1974, available from Pritsker and Associates, Inc., 1201 Wiley Drive, West Lafayette, Indiana, 47906, U.S.A.

classifications: (a) node characteristics and (b) arc characteristics. We now proceed with such summary.

§3.1 Node Characteristics.

There are five capabilities offered by GERTS.

1. *The capability to change conditions for the realization of a node*: a node may be realized the *first* time after $\nu_1 \geqslant 0$ incidencies, but its realization in *subsequent* times may require $\nu_2 \geqslant 1$ incidencies, where ν_2 may be different from ν_1. The capability offers an additional flexibility in the following sense:

   $\nu_1 = 0$: for start (or initiation) nodes that require no incidence to be realized;

   $\nu_2 = \infty$: for sink or terminal nodes realized only the first time in any run of the network, or for intermediate nodes that are prohibited from being realized except once.

2. *The capability to restrict realization*: in particular, the capability to specify three properties: (a) the conjunctive character of incidence (of two or more *different* activities) required to realize a node (evidently, this property can be invoked only in the case of multiple incidence), (b) the option to halt all ongoing but incomplete activities incident on the node at the time of the node's realization, and (c) both properties in (a) and (b).

Thus, if $\nu_1 > 1$—meaning that more than one input realization is required—one may wish to restrict the realization of the node to the incidence of ν_1 *different* activities, as opposed to ν_1 realizations of the same activity.

The symbol used to indicate the conjunction of activities is A (for the logical "and" condition). The symbol used to indicate the halting of on-going activities is H. Both properties are indicated by the symbol U (for "union").

To illustrate, consider the network of Figure 5.18. Node *2* is a source node and nodes *9* and *10* are terminal nodes. On release of *2* at time 0, activities $(2,3)$, $(2,4)$, and $(2,5)$ will start. Eventually, nodes *3*, *4*, and *5* will be realized. Now node *6* requires two *different* incidencies for *its* realization—this is the significance of the letter "A" on the node. This prohibits the realization of *6* due to two realizations of arc $(7,6)$, which is possible due to the feedback arc $(7,3)$. The H on node *8* specifies that at the time of realization of *8* due to either activity $(4,8)$ or $(5,8)$, the other ongoing activity is to be halted. Finally, the realization of node *9* requires

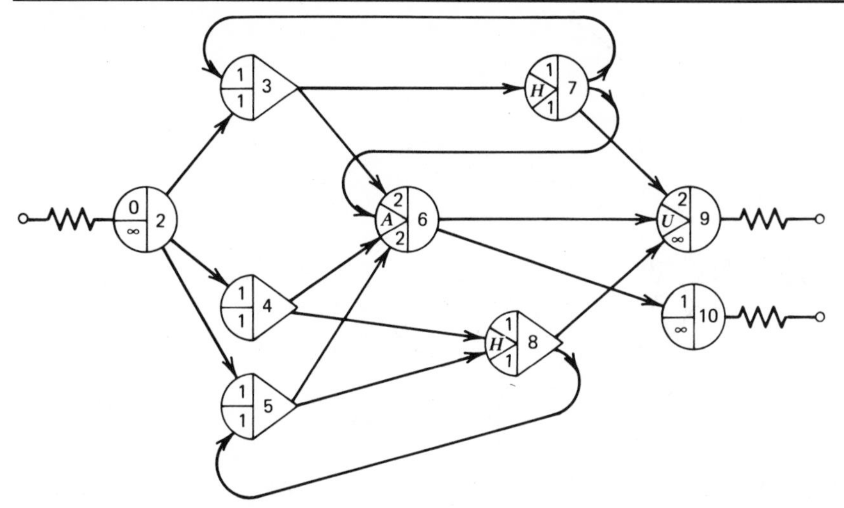

Figure 5.18. The node functions A, H, and U.

the completion of any two *different* activities incident on it. Furthermore, at the time of its realization, the third activity should be halted if it is still in progress. These two conditions are specified by the symbol U on the node. The termination of a run will be determined by the realization of either *9* or *10*, whichever occurs first.

3. *The capability of dynamic network modification*: this is always conditioned on the realization of specified activities, and operates in the following fashion. The receiving side of the node remains invariant; hence, the conditions for its realization remain the same. However, the emitting side is replaced by a different logic. This capability permits the insertion of literally any desired combination of logical contingencies.

The specification of this condition is modeled graphically as follows: (a) give the activity causing the modification a number, say a (usually inserted in a small square adjacent to the arc), (b) state the modified version of the emitting side of the node and give it a new node number, say j, and (c) draw a dotted arrow from the original node, say i, to the new modification j, and let it bear the same number a as the activity causing the modification.

To illustrate, consider the network of Fig. 5.19. Suppose that the "normal" operation of the system is to proceed along activities $(3,5)$ and $(4,6)$, which would release nodes *5* and *6*, respectively, to be followed by node *9* and finally *11* [when the cycle is repeated three times through the feedback

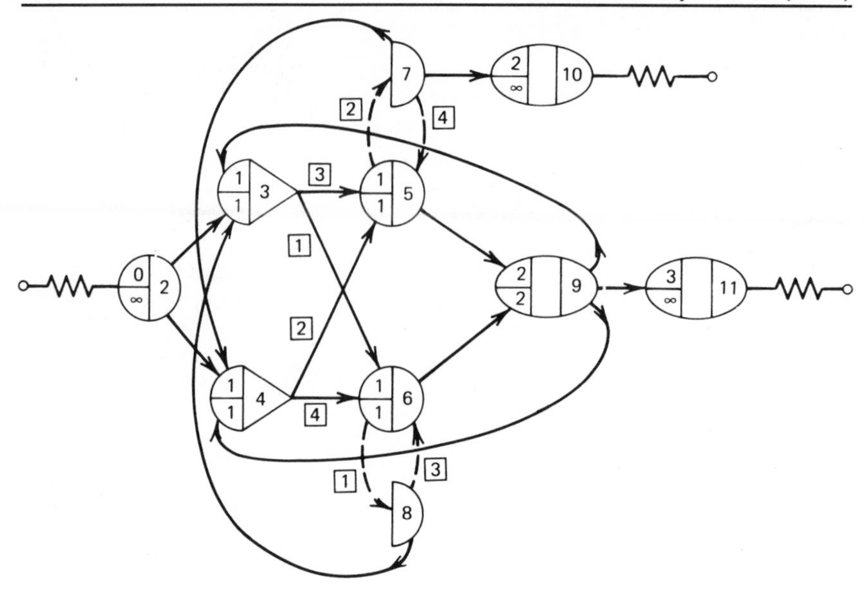

Figure 5.19. To illustrate network modification.

loops (*9,3*) and (*9,4*)]. However, due to the probabilistic outcomes from nodes *3* and *4*, the "abnormal" activities (*3,6*), bearing number [1], and (*4,5*), bearing number [2], may be realized. Consider first the case (*3,6*) is realized. Node *6* is modified to node *8*, whose output is (*8,3*), returning the system to *3*. On the other hand, if (*4,5*) is realized, *5* is modified to *7*, whose outputs are activities (*7,10*) and (*7,4*); the latter returns the system to *4*. Two realizations of *10* would terminate the run.

An interesting aspect of this network is the capability to return the system to the *status quo ante* when the "normal" activities, [3] and/or [4], are realized. Thus the realization of activity [3] nullifies the previous modifications (if any) and returns *8* to *6*, and similarly for activity [4]: *7* returns to *5*.

Network modification can be carried to any elaborate degree desired. Thus, one may choose to modify the modification, not by returning to the *status quo ante*, as illustrated in Fig. 5.19, but by changing to a new configuration. This is the so-called serial modification. We alert the reader that great care should be taken in its use because of the strict dependence of the modification on the order in which activities are completed. For example, consider Fig. 5.20. If activity [1] is realized followed by activity [2], then node i is first replaced by j then by k. However, if activity [2] is realized first followed by activity [1], node j is first replaced by k and then

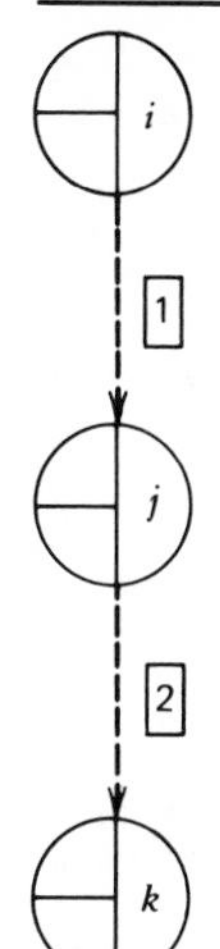

Figure 5.20. Serial modification.

i is replaced by *j*. Hence, the *sequence* of occurence of the activities determined the final outcome! This is because GERTS does not test for previous modifications of a node.

4. *The capability to accumulate different types of statistics*: an important function of nodes is to collect statistics on times and costs (or any other entities similarly defined), as well as on activity counts. Nodes having this function are called *statistics nodes*. Sink nodes are automatically considered statistics nodes—hence, there is no need for separate designation. The desired statistics are accumulated at the time of realization of the node. The types of statistics are as described below:

Code	Description
F	Time of *first* realization
A	Time of *all* realizations
B	Time *between* realizations
I	*Interval* of time between realization of a previous "mark node" in the network and the node collecting the I statistics (mark nodes are described below).
D	Time *delay* from the completion of the first activity incident on the node until the node itself is realized.[a]

[a]Such delay times necessarily occur at nodes requiring multiple activity realizations.

The I statistic imply the designation of some nodes as *mark nodes*. These simply establish points of reference for time (and cost) interval calculations. Source nodes are automatically mark nodes; and if no mark node is specified, the reference time is taken from the release time of the source node.

In summary, therefore, it is seen that GERTS admits three kinds of node: (a) statistics nodes (which include all sink nodes), (b) mark nodes (which include all start nodes), and (c) ordinary receiver-emitter nodes.

5. *Sink nodes may, or may not, be terminal nodes*: this property permits the creation of nodes that serve solely to specify an activity that causes network modification.

For instance, consider the network of Fig. 5.21. The path *2,3,4,5* leads to terminal *5*. However, if activity (*2,6*) is realized in place of (*2,3*), node *6* will be released and activity (*6,7*), designated as number [2], will be realized. This causes the modification of node *4* to node *8*, and termination will occur through node *9*. Note that *7* is a sink node that is not a terminal node.

§3.2 Activity Characteristics.

There are three new capabilities, which can be gleaned from the generic representation of an activity, shown in Fig. 5.22, specifying at most five sets of data: (a) the probability p of undertaking the activity, (b) the parameters FT, PS of the duration of the activity, (c) the parameters CS, CV of its cost, (d) the activity number, [a], and (e) the activity class type, /k\. The three capabilities are:

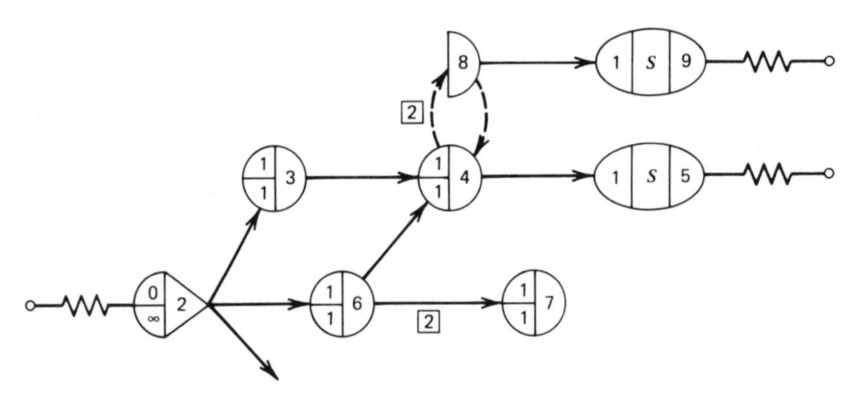

Figure 5.21. Sink nodes and terminal nodes.

1. *The choice of duration from among several distributions*: the specification of the duration is quite flexible, and is determined by the function type, FT, and the parameter set, PS. For the FT, one may choose from among eleven options. These include nine statistical distribution types and two methods for assigning constant times. The function types and their code number in GERTS are shown in the following table.

   Distribution types:

Code	Distribution
1	Constant[a]
2	Normal
3	Uniform
4	Erlang[b]
5	Lognormal
6	Poisson
7	Beta
8	Gamma
9	Beta distribution fitted to three parameters as in PERT
10	Triangular distribution fitted to three parameters

[a]Assignments of constant in specified parameter set (code 1); assignment of constant equal to PS divided by a scale factor (code 11).
[b]The exponential distribution can be obtained from the Erlang distribution by inserting a specific code in the specification of the Erlang distribution.

   The nine distribution types utilize parameter sets, PS, which form part of the input and are identified through the value of PS in the designation of the activity. For instance, we may have: FT = 2 and PS = 3. This means, in the code of GERTS, that the activity time is Normally distributed with parameters given in the parameter set 3. This latter will contain: (a) the mean value, (b) the standard deviation, (c) the maximum value, and (d) the minimum value. Alternatively, an activity may have FT = 1 and PS = 4, which means that the activity time is constant with the desired constant specified in the parameter set number 4. This latter contains only the constant desired.
2. *Capability of costing the activity*: the cost parameters are specified by the pair (CS, CV), in which CS stands for "setup cost" and CV for "variable cost". Setup cost is a fixed cost incurred each time the activity is undertaken. The variable cost varies linearly with the duration of the activity. The total cost of the activity is equal to $CS + CV \times T$, where T is its duration.

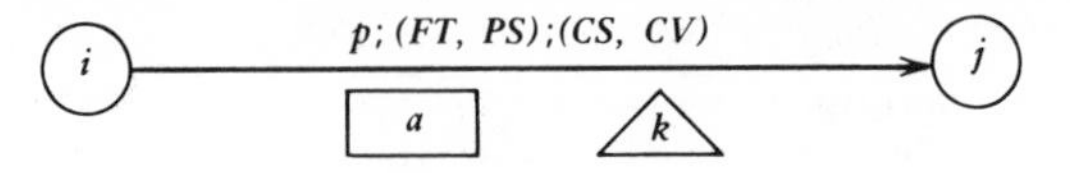

Figure 5.22. Generic designation of an activity in GERTS:
FT: Function type
PS: Parameter set
CS: Cost of setup
CV: Variable cost
a: Activity number
k: Counter type

3. *The capability of different counter types*: the "counter type" is used to give a common label to a set of activities, thus identifying them as belonging to the same set. The need for identifying such grouping stems from the desire to count the number of times certain activities occur

Table 5.2. Summary Characteristics of GERTS

1. Branches are characterized by:
   a. a start node;
   b. an end node;
   c. an activity number, [a];
   d. a time required to perform the activity represented by the branch—the time is specified by defining a function type (FT) and a parameter set specification (PS);
   e. a probability (p) of being selected for branching; the explicit statement of p may be foresaken if $p=1$;
   f. a setup cost (CS) for starting the activity, and a cost per unit time (CV) for performing the activity;
   g. a counter type to identify the branch as belonging to a particular group of branches, △k.
2. Nodes are characterized by:
   a. number of completions of incoming activities required to release the node for the first time, (ν_1);
   b. number of completions of incoming activities required to release the node after the first time, (ν_2);
   c. requirement that different activity completions are to be counted toward nodal releases (AND specification), (A);
   d. halting of activities incident to the node when the node is released, (H);
   e. combination of options (c) and (d) above, (U);
   f. branching method for initiating the activities enamating from the node (DETERMINISTIC or PROBABILISTIC);
   g. their function in the network (source, sink, statistics, or mark).
3. Network can be modified based on the completion of an activity.
4. A set of simulation runs can be traced and automatic analysis of the network model can be obtained.

Table 5.2 (*Cont'd*)

Node Characteristics

Input

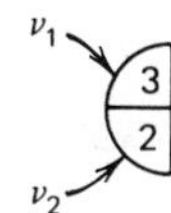

Number of incident activity completions required to release the node for the first time is 3.

Number of incident activity completions required to release the node after the first time is 2.

In determining if node is released, *all* activity completions must be *different*. This is called the "*A*" or AND specification and refers to all releases of the node.

An "*H*" capability specification halts all ongoing activities incident to the node when the node is released.

A "*U*" capability specification is a union of the "*A*" and "*H*" specifications.

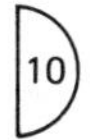

DETERMINISTIC Output for Node *10*. All emanating activities are started.

Output

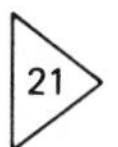

PROBABILISTIC Output for Node *21*. Only one of the emanating activities is started.

Table 5.2. (*Cont'd*)

Network Functions

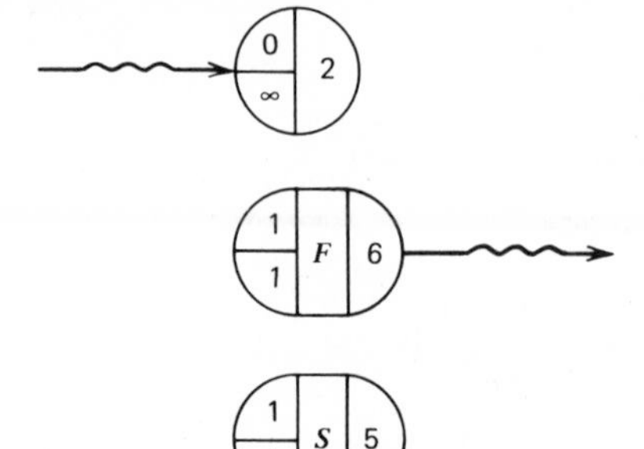

Source node *2* initiates output branches at time zero. Note that there is no node *1*.

Sink node *6* is used to determine network realization.

Node *5* is used for statistical purposes: $S = F, A, B, I, D$, or M.

All possible combinations of the above input and output sides are permitted.

Branch Characteristics

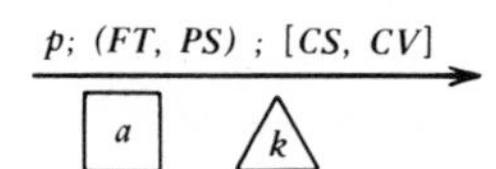

p = Probability of selecting the branch given the node from which it emanates is released. If $p = 1$, it normally is not shown on the network

(FT, PS) = Time descriptors: a function type FT and a parameter specification PS. Function type options include sampling from statistical distribu– tions and two constant methods all using the parameter specifications PS.

[a] = Activity number used to identify activities for network traces and modifications.

△k = Counter type identifying a group of activities as one type for statistical purposes.

$[CS, CV]$ = Cost descriptors: setup cost CS and variable cost CV.

Network Modifications

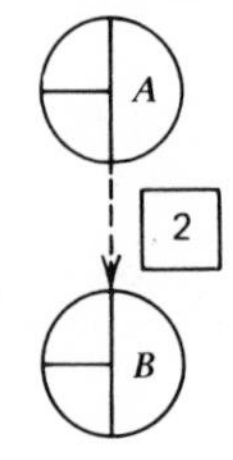

Branching occurs from node *B* when node *A* is released if activity [2] was completed before statistical nodes are realized. A counter type is increased by 1 every time any of its associated activities is completed. Each statistical node observes the number of activities of each type completed prior to its realization, and frequencies of occurrence are noted. Of course, it is possible to associate a unique counter type with each activity of interest, thus keeping statistics on individual activities.

Table 5.2 summarizes the GERTS symbolism and its features.

Example 5.4. Application of GERTS to Industrial Sales Negotiation Cost Planning for Corporate Level Sales. As an illustration of some of the modeling capabilities of GERTS we report on an interesting application in the domain of sales negotiations cost planning. The application is due to Bird, Clayton, and Moore (1973) and is summarized by Pritsker and Sigal (1974) (who also obtained a different set of simulation runs).

The study was motivated by the following three "observables" about corporate sales efforts in the industrial field: (a) industrial marketing accounts for approximately $1.3 trillion in sales, a not insignificant portion of the GNP, (b) most large industrial sales to private businesses or government agencies are negotiated (thus, the sellers must formulate sales strategy), and (c) most managers are not aware of how much selling costs them, despite the general feeling that sales costs in relation to other costs have been rising for a number of years, especially on sales of whole systems of industrial equipment.

The purpose of the study of Bird, Clayton, and Moore was to furnish the marketing manager with a planning technique for estimating the expected marketing costs as well as the probabilities of the firm's success or failure when entering into negotiations for a corporate level sale *for a given sales strategy*. Indeed, the technique can be extended to evaluate *alternate strategies*, and thus can be used as a tool of design as well as a tool of analysis.

From the discussion that follows the reader can also infer the suitability of GERTS in analyzing research and development activities, which are very much akin to negotiations activities.

The specific application reported here is relative to a study of the negotiation activities involved in bidding on the construction of a gasoline production plant for a major oil company. The price range of the bid was between $1.5 and $1.75 millions. The model is presented from the standpoint of one vendor (bidder)—incidentally, one who eventually won the contract. The structure of the network as well as the time and cost parameters were all derived from historical data recorded in company files or from employees' knowledge of the company's operation.

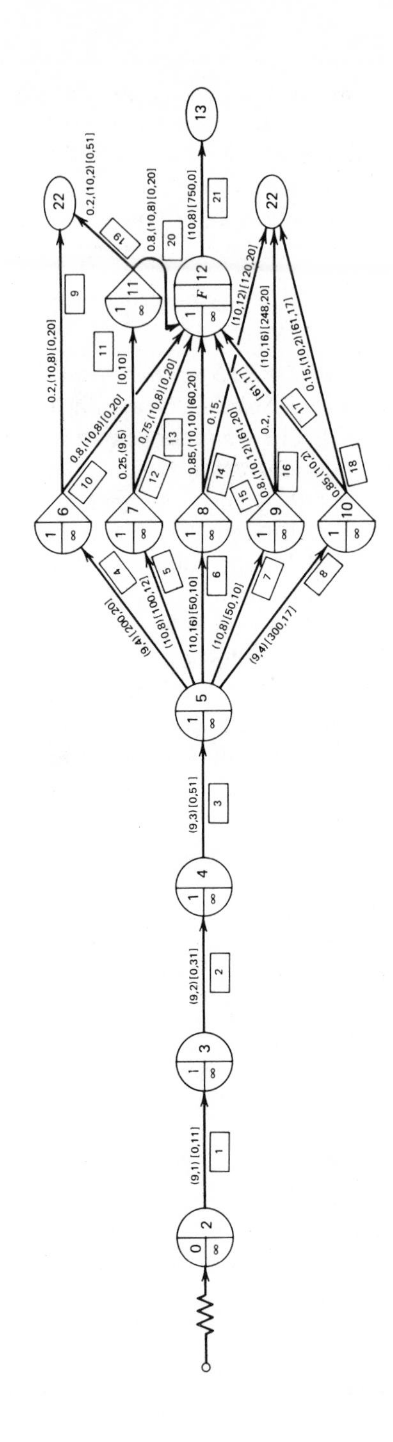

0
2
∞
1
3
4
5
6
7
8
9
10
11
F
12
13
22
22
(9,1) [0,11]
1
(9,2) [0,31]
2
(9,3) [0,51]
3
(9,4) [200,20]
4
(10,8) [100,12]
5
(10,16) [50,10]
6
(10,8) [50,10]
7
(9,4) [300,17]
8
0.2,(10,8) [0,20]
9
0.8,(10,8) [0,20]
10
[0,10]
11
0.25,(9,5)
12
0.75,(10,8) [0,20]
13
0.85,(10,10) [60,20]
14
0.15,
15
0.8,(10,12) [61,20]
0.2,
16
(10,16) [248,20]
0.85,(10,2)
[61,17]
17
0.15,(10,2) [61,17]
18
(10,12) [120,20]
0.2,(10,2) [0,51]
19
0.8,(10,8) [0,20]
20
(10,8) [750,0]
21

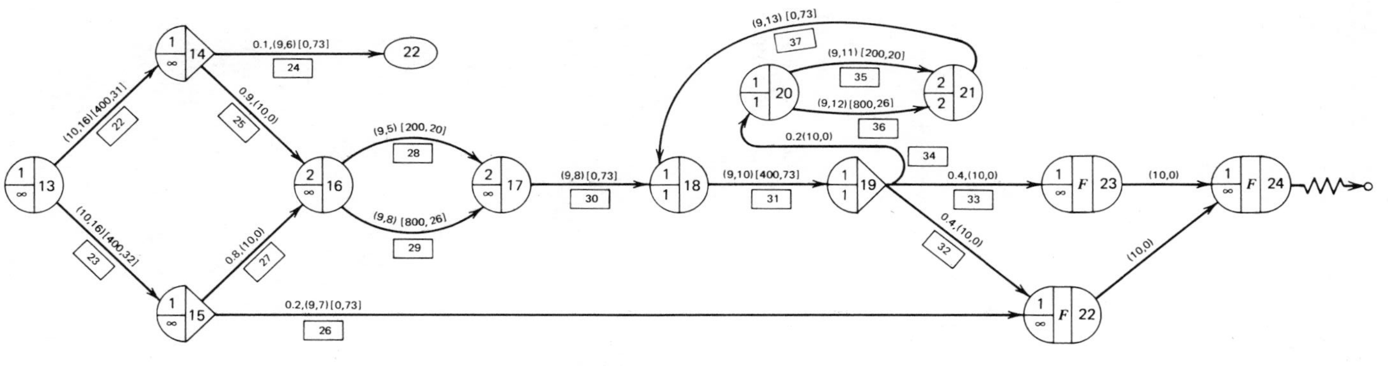

Figure 5.23. GERTS IIIZ model of sales negotiation activities.

Figure 5.23 shows the GERTS network that could have been used by the vendor, and Table 5.3 gives the details of the activities. It would be helpful if the reader were to follow a path to a terminal node—say the path *2,3,4,5,6,12,13,14,16,17,18,19,20,21,18,19,22,24*. The network contains one start node, node *2*, and one terminal node, node *24*. Node *22* represents failure to secure the contract. Node *23*, on the other hand, represents success in gaining the contract. Statistics on costs and times are collected on nodes *12,22,23,24*. (Node *12* is also an important landmark event since it signifies receipt of five "favorable" reports from: engineering, production, finance, marketing, and purchasing.)

Before discussing the results of the simulation, we make the following remarks on the model itself.

Nodes *6*, *7*, *8*, *9*, and *10* are decision nodes, and the estimates of probabilities may be obtained either on the basis of the history of "similar" projects or on the basis of subjective estimates relating only to *this* bid. For instance, consider node *6*; its output is probabilistic and indicates that the probability of a "negative" report from engineering is 0.2 and of a "positive" report, is 0.8. These figures may be based on a long history of engineering reports on similar bids or on the chief engineer's subjective estimate of the feasibility of this particular bid.

Node *19* has three possible outputs: (a) contract denied (activity 32), (b) contract awarded (activity 33), and (c) request modification (activity 34). One might question the validity of assigning a constant 0.2 probability estimate to requests for design modification (activity 34) regardless of the number of times the "final negotiation" (activity 31) is conducted. However, the actual negotiation experiences of the vendor used in this example indicate that the number of previous negotiation conferences has little, if any, effect on the outcome of the next negotiation session.†

Finally, all the statistics nodes are F nodes, denoting that "first" release statistics are collected. Alternatively, one may prefer to collect delay statistics at node *12*, which would describe the waiting time from the first successful report to the fifth successful report.

The simulation was repeated for 1000 runs, and the results are summarized in Table 5.4 and Figs. 5.24 through 5.29.

Statistics on node *12* reveal that all five preliminary reports are favorable 48.4% of the time—from which we deduce that 51.6% of the time the "no sale" condition (i.e., node *22*) will be realized due to unfavorable preliminary report(s). Since the probability of realization of node *22* is shown in Table 5.4 to be 83.5%, we conclude that failure to secure a contract is 31.9% of the time due to "other" unfavorable outcomes—in

†In Problem 11 we ask the reader to introduce network modification to permit such change in probabilities as a function of the number of "final negotiation" sessions.

particular, a negative outcome at nodes *14*, *15*, or *19*. Finally, sale is realized 16.5% of the time (=100%−83.5%), or once in six trials!

Time and cost statistics are equally revealing. Figure 5.28 shows the time statistics histogram for node *23* (a successful sale). Note that 90.9% of the time a successful sale is accomplished in less than 220 working hours, given that a sale is successful. There is a 51.5% chance that a successful sale will be accomplished in 160 hrs or less, given that a sale will be successful.

Since the path to node *23* must pass by node *12*, it is illuminating to study the time statistics of this latter node. Table 5.4 shows that the average time to the realization of node *12* is approximately 38 hrs, with a rather very small variation (coefficient of variation $\nu=0.1265$). Since the average time to the realization of node *23* is approximately 170.6 hr, with a rather large variation (coefficient of variation $\nu=0.231$), we immediately conclude that the overwhelming majority of the time of securing the contract is spent *after* the preliminary reports are all favorable. More significantly, this portion of the total activities (between nodes *12* and *23*) is the *most uncertain*; it may consume as little as 55.8 hrs (=110.5−54.7), but as much as 336.3 hrs (=364.6−28.3), or six times as long! The time to the realization of node *22* (the "no-sale" case) exhibits the same curious phenomenon, albeit not to the same pronounced degree. Here the average time is much smaller (ca. 65 hrs), the variation is much higher (coefficient of variation $\nu=0.859$), but the duration may vary between 110 hrs and 364.5 hrs, or three times as long!

An explanation for both occurrences may be found in the chance of repeated looping for redesign from the negotiation conference, represented by activity *34*. The probability of multiple design modification looping is quite small, but it does exist.

This explanation is also confirmed by a careful study of costs associated with a successful sale; see Fig. 5.29. The large majority of costs are clustered around the mean value of $11,332. However, there still exists the possibility of a high cost of $23,927. In fact, the same degree of uncertainty is also revealed; the costs of the path between nodes *12* and *23* may be as low as $5,665 (=9,203−3,538), and as high as $21,187 (=23,927−2,740)!

Figure 5.26 presents the probability distribution of the costs incurred given the condition that the sale is lost. Note that the most likely cost (the mode of the distribution) approximates the minimum cost, which is evidently due to the early cessation of activities because of one or more unfavorable preliminary reports. The high costs can be explained, as above, by the high costs associated with multiple looping prior to the failure of the negotiations. Recall from Table 5.4 that the lowest cost of a lost sales was $2360 in 1000 runs, while the maximum cost was $18,531.

Figure 5.27 (the distribution of total cost) confirms the high probability

Table 5.3. Activity Descriptions of Example 5.4

Activity			Time in hours			Cost in dollars	
No.	Activity Description	Distribution	(a)	(b)	(c)	(d)	(e)
1	Sales call by company salesman	Beta	2.00	1.25	4.0	0	11.0
2	Sales report to marketing vice president	Beta	0.50	0.25	1.0	0	31.0
3	Marketing vice president reports to president	Beta	0.25	0.25	0.50	0	51.0
4	Preliminary engineering report	Beta	24.0	16.0	40.0	200.	20.0
5	Preliminary production report	Constant	8.0	N/A	N/A	100.	12.0
6	Preliminary financial report	Constant	16.0	N/A	N/A	50.	10.0
7	Preliminary marketing report	Constant	8.0	N/A	N/A	50.	10.0
8	Preliminary purchasing report	Beta	24.0	16.0	40.0	300.	17.0
9	Negative engineering report examined	Constant	8.0	N/A	N/A	0	20.0
10	Favorable engineering report examined	Constant	8.0	N/A	N/A	0	20.0
11	Production subcontracting investigated	Beta	24.0	8.0	40.0	0	10.0
12	Favorable production report examined	Constant	8.0	N/A	N/A	0	20.0
13	Favorable financial report examined	Constant	10.0	N/A	N/A	60.	20.0
14	Negative financial report examined	Constant	12.0	N/A	N/A	120.	20.0
15	Favorable marketing report examined	Constant	12.0	N/A	N/A	61.	20.0
16	Negative marketing report examined	Constant	16.0	N/A	N/A	248.	20.0

17	Favorable purchasing report examined	Constant	2.0	N/A	N/A	61.	17.0
18	Negative purchasing report examined	Constant	2.0	N/A	N/A	61.	17.0
19	Unfavorable production subcontract examined	Constant	2.0	N/A	N/A	0	51.0
20	Favorable production subcontract examined	Constant	8.0	N/A	N/A	0	20.0
21	Corporate-level planning conference	Constant	8.0	N/A	N/A	750.	0
22	Sales call by marketing vice president and salesman	Constant	16.0	N/A	N/A	400.	31.0
23	Engineering call by engineering vice president and project engineer	Constant	16.0	N/A	N/A	400.	32.0
24	Unfavorable sales call conference	Beta	3.0	1.0	18.0	0	73.0
25	Favorable sales call no conference (dummy)	Constant	0.0	N/A	N/A	0	0
26	Unfavorable engineering call conference	Beta	3.0	1.0	16.0	0	73.0
27	Favorable engineering call no conference (dummy)	Constant	0.0	N/A	N/A	0	0
28	Marketing negotiation plan formulation	Beta	24.0	8.0	40.0	200.	20.0
29	Engineering design plan formulation	Beta	80.0	40.0	160.0	800.	26.0
30	Corporate-level strategy conference	Beta	2.0	1.0	8.0	0	73.0
31	Negotiation conference with buying firm	Beta	6.0	2.0	16.0	400.	73.0
32	No sale (dummy)	Constant	0.0	N/A	N/A	0	0
33	Contract awarded (dummy)	Constant	0.0	N/A	N/A	0	0
34	Modifications requested by buyer (dummy)	Constant	0.0	N/A	N/A	0	0
35	Modification of marketing negotiation plan	Beta	12.0	4.0	20.0	200.	20.0
36	Modification of engineering-design plan	Beta	40.0	20.0	80.0	800.	26.0
37	Corporate meeting to reconcile modifications	Beta	1.0	0.5	4.0	0	73.0

Table 5.4. Final Results for 1000 Simulations

			Count $\left(\begin{matrix}\text{hr.}\\ \$\end{matrix}\right)$				
Node	Probability	Mean	Standard Deviation	No. Observations	Minimum	Maximum	Node Type
12	0.4840	37.9336	4.7969	484	28.3379	54.7067	F
Cost (000)		3.0509	0.1457	484	2.7404	3.5381	
22	0.8350	64.8294	55.7592	835	20.5685	296.3474	F
Cost (000)		5.1210	3.2937	835	2.3600	18.5305	
23	0.1650	170.6643	39.3216	165	110.4653	364.6016	F
Cost (000)		11.3322	2.2574	165	9.2027	23.9275	
24	1.0000	82.2922	66.2885	1000	20.5685	364.6016	F
Cost (000)		6.1458	3.9005	1000	2.3600	23.9275	

of incurring low costs when the firm ends with no contract. The distribution is basically trimodal, with the costs of failure (which the reader will recall occurs 83.5% of the time) being represented by the first two modes and the cost of a successful negotiation being represented by the third mode. Table 5.4 reveals that any time the company seriously considers tendering a bid, about $6146 will be spent on the average, regardless of whether it makes the sale.

A major cost reduction for the industrial selling procedure used by this vendor is in the resequencing of certain activities. If time for tendering a

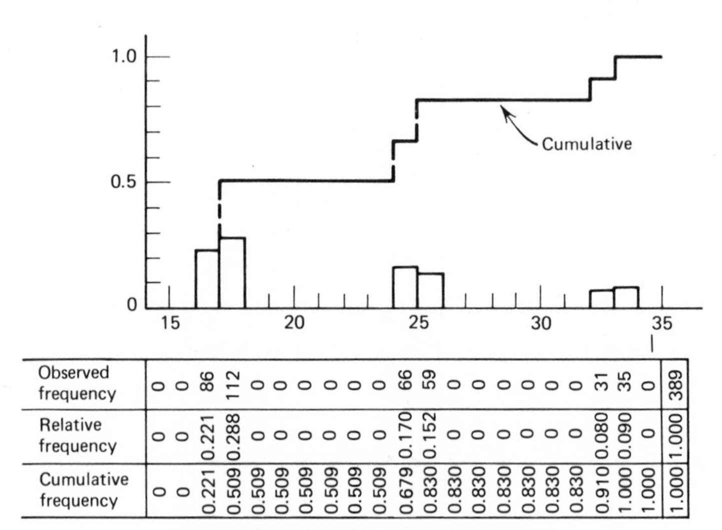

Observed frequency	0	0	86	112	0	0	0	0	0	0	66	59	0	0	0	0	0	0	31	35	0	389
Relative frequency	0	0	0.221	0.288	0	0	0	0	0	0	0.170	0.152	0	0	0	0	0	0	0.080	0.090	0	1.000
Cumulative frequency	0	0	0.221	0.509	0.509	0.509	0.509	0.509	0.509	0.509	0.679	0.830	0.830	0.830	0.830	0.830	0.830	0.830	0.910	1.000	1.000	1.000

Figure 5.24. Time histogram for node *12*.

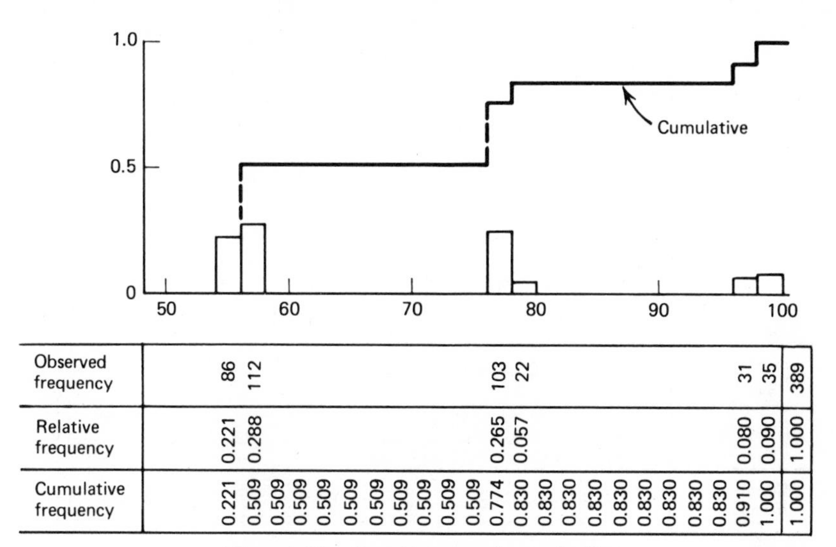

Observed frequency	Relative frequency	Cumulative frequency
86	0.221	0.221
112	0.288	0.509
		0.509
		0.509
		0.509
		0.509
		0.509
		0.509
		0.509
		0.509
		0.509
103	0.265	0.774
22	0.057	0.830
		0.830
		0.830
		0.830
		0.830
		0.830
		0.830
		0.830
		0.830
31	0.080	0.910
35	0.090	1.000
389	1.000	1.000

Figure 5.25. Cost histogram for node *12*.

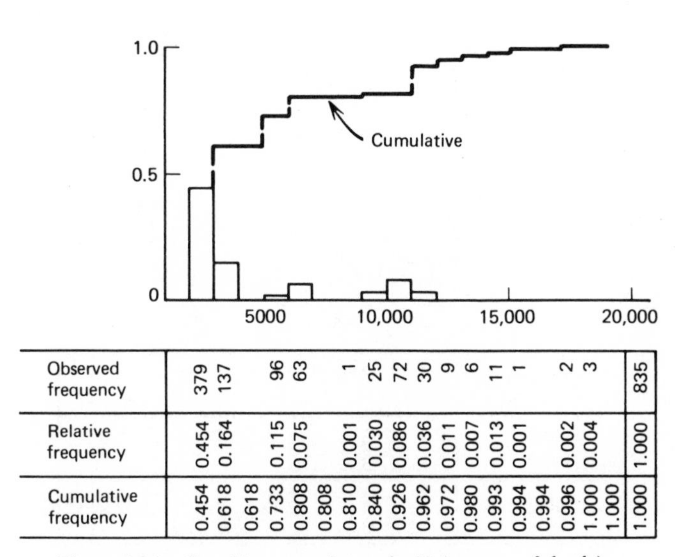

Observed frequency	Relative frequency	Cumulative frequency
379	0.454	0.454
137	0.164	0.618
		0.618
96	0.115	0.733
63	0.075	0.808
		0.808
1	0.001	0.810
25	0.030	0.840
72	0.086	0.926
30	0.036	0.962
9	0.011	0.972
6	0.007	0.980
11	0.013	0.993
1	0.001	0.994
		0.994
2	0.002	0.996
3	0.004	1.000
		1.000
835	1.000	1.000

Figure 5.26. Cost histogram for node *22* (unsuccessful sale).

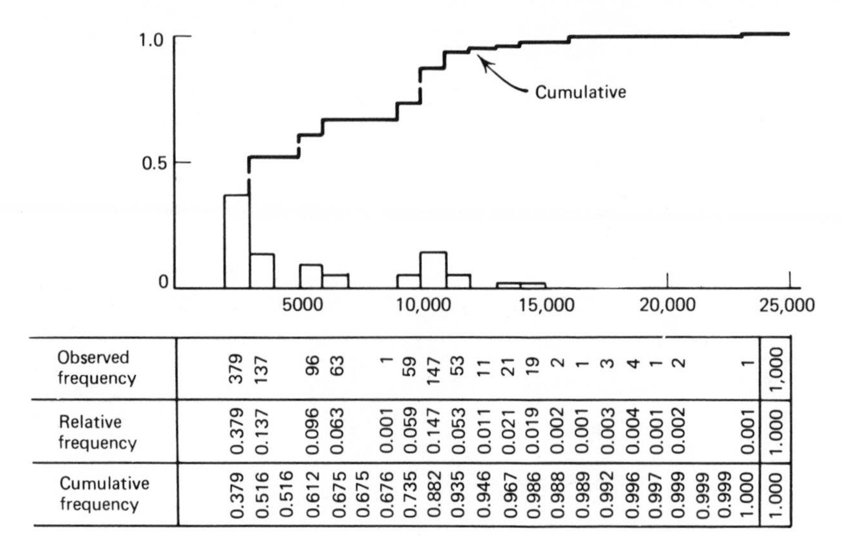

Observed frequency	379	137		96	63		1	59	147	53	11	21	19	2	1	3	4	1	2			1	1,000
Relative frequency	0.379	0.137		0.096	0.063		0.001	0.059	0.147	0.053	0.011	0.021	0.019	0.002	0.001	0.003	0.004	0.001	0.002			0.001	1.000
Cumulative frequency	0.379	0.516	0.516	0.612	0.675	0.675	0.676	0.735	0.882	0.935	0.946	0.967	0.986	0.988	0.989	0.992	0.996	0.997	0.999	0.999	0.999	1.000	1.000

Figure 5.27. Cost histogram for node *24* (all sales attempted).

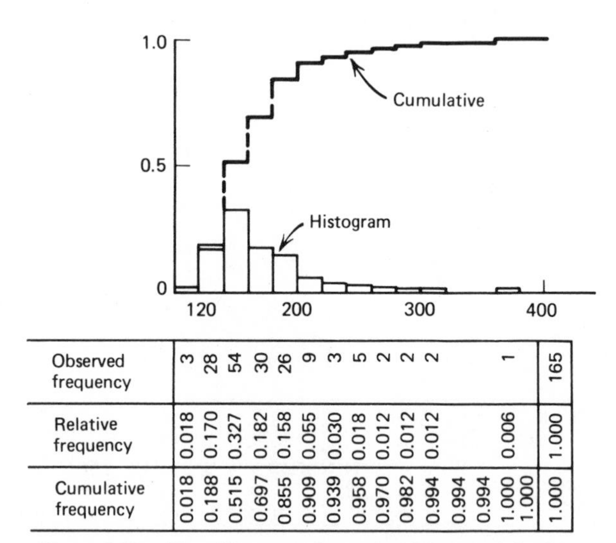

Observed frequency	3	28	54	30	26	9	3	5	2	2	2			1		165
Relative frequency	0.018	0.170	0.327	0.182	0.158	0.055	0.030	0.018	0.012	0.012	0.012			0.006		1.000
Cumulative frequency	0.018	0.188	0.515	0.697	0.855	0.909	0.939	0.958	0.970	0.982	0.994	0.994	0.994	1.000	1.000	1.000

Figure 5.28. Time histogram for node *23* (successful sale).

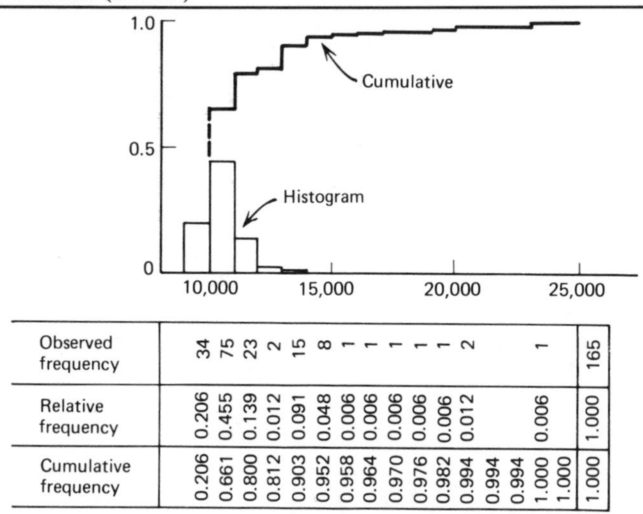

																	Total
Observed frequency	34	75	23	2	15	8	1	1	1	1	1	2			1		165
Relative frequency	0.206	0.455	0.139	0.012	0.091	0.048	0.006	0.006	0.006	0.006	0.006	0.012			0.006		1.000
Cumulative frequency	0.206	0.661	0.800	0.812	0.903	0.952	0.958	0.964	0.970	0.976	0.982	0.994	0.994	0.994	1.000	1.000	1.000

Figure 5.29. Cost histogram for node *23* (successful sale).

bid is flexible, which would be the case in the absence of a firm deadline, then a sequencing of the preliminary report activities would clearly reduce expected cost, although it would lengthen the company real time (= the actual elapsed working hours). It is not practical to sequence activities after the realization of node *12*. However, a sequencing of activities 4, 5, 6, 7, and 8 is feasible, in which case the first and only activity emanating from node *5* would be the activity with the highest probability of failure and the lowest cost, which for this vendor is the preliminary marketing report (activity 7). This report has the lowest costs and at least as high a probability of causing failure as any other preliminary activity. All the other preliminary activities could then either branch from a node indicating a completion of a favorable marketing report or continue in some sequential order. Merely sequencing the marketing report and allowing the other preliminary activities to emanate from its completion would increase real time by only 8 hr (since activity 7 takes a constant 8 hr). This would result in the company spending less than $1000 on the negotiation process each time a negative marketing report led to the cessation of negotiations. This situation occurs 20% of the time. The saving is significant in view of the fact that the firm now spends (almost always) a minimum of $2410 each time they start the sales negotiation procedure.

In summary, the GERTS approach described above provides the industrial vendor with a method of estimating: (a) the chances of success or failure when entering into negotiations for large-scale contracts, (b) cost and time statistics, which should be quite valuable as tools for analyzing the performance of the entire negotiating procedure, and (c) alternative strategies offering a chance of saving several thousands of dollars. These alternatives are readily apparent when using the GERTS approach, but may be difficult to recognize when using only conventional negotiation planning.

§4. OPTIMAL TIME–COST TRADE-OFFS IN GERT NETWORKS*

The problem of optimal time-cost trade-offs in DANS, the problem of "project compression," constituted the subject matter of Chapter 2, where we discussed methods of solution under various forms of individual activity time–cost relationships, such as linear, convex, concave, and discontinuous functions.

Simply stated, the problem is that of optimally investing a fixed sum of money to achieve the maximal reduction in the duration of the project. Alternatively, the problem may consist of calculating the minimum additional investment needed to reduce the duration of the project (or a subset thereof) by a given amount.

Under the assumption of deterministic activity durations, the analysis of Chapter 2 demonstrates the nontriviality of the question. Under the assumption of probablistic are durations (e.g., in PERT model), the problem increases in complexity.† The difficulty seems to stem from the fact that there is probabilistic variation in more than one variable simultaneously, and the variables are not necessarily independent of each other. The whole interactive system is further complicated by the precedence relations embodied in, as well as represented by, the AN.

In GANS of the GERT type, in which all nodes are of the "exclusive-or" type, the model is further complicated by the probabilistic (i.e., Markov) branching from any node. That is, not only is the duration of the activity a RV, but also its very existence is a chance event.

This section represents a first attempt at the modeling of the optimal allocation of funds to competing activities in GERT networks.

*Starred sections may be omitted on first reading.

†For an intuitive discussion of the difficulties involved and of its impact on the cost control and financial management of a project, see Section 4 of Chapter 1.

§4.1 Investment in Relation to Duration and Cost

In a manner reminiscent of the study of optimal time–cost trade-off in DANs, exemplified by the work of Fulkerson,[†] we shall assume that each activity (ij) in the network has an associated "normal" expected duration, u_{ij} and a "crash" expected duration, l_{ij}. The words "expected duration" are used advisedly to indicate the mathematical expectation of a RV. That is to say, under "normal" pace of operation the actual duration of the activity is a RV whose expected value is given by u_{ij}. On the other hand, under "crash" pace of operation, the actual duration of the activity is also a RV whose expected value is l_{ij}.

We further assume that the cost of performing the job varies with (i.e., is a function of) the actual duration of the activity. Specifically, the cost C_{ij} is a monotone nondecreasing function of the duration Y_{ij};

$$C_{ij} = \psi_{ij}(Y_{ij}) \tag{5.20}$$

Since Y_{ij} is assumed a RV, so will be C_{ij}. Denote its expected value by $\mathcal{E}(C_{ij}) = \gamma_{ij}$.

Assume the network to possess one "source," or "start" node; *1*, and several terminal nodes $n_1, n_2, \ldots, n_L$. Let

P_{n_i}: the probability of realizing node n_i
T_{n_i}: the duration to the realization of node n_i
C_{n_i}: the cost of reaching n_i, when it is realized

As always, p_{ij} is the probability that activity (ij) will be undertaken given that node i is realized, and Y_{ij} the duration of (ij) with expected value $\mathcal{E}(Y_{ij}) = \mu_{ij}$.

The following two results are well known in the theory of GERT and, in any case, can be easily proven: (a) the probability that an event j is realized is a function only of the transition probabilities $\{p_{ij}\}$ and (b) the expected time to the realization of any node as a function only of the transition probabilities $\{p_{ij}\}$ and the expected durations $\{\mu_{ij}\}$. Consequently, it is easy to obtain P_{n_i}, $\mathcal{E}(T_{n_i})$, and $\mathcal{E}(C_{n_i})$ as functions of the

[†]See Section 2 of Chapter 2.

given parameters, say:

$$P_{n_i} = g_1(\{p_{ij}\})$$

$$\mathcal{E}(T_{n_i}) = g_2(\{p_{ij}\}, \{\mu_{ij}\}); \quad \mu_{ij} = \mathcal{E}(Y_{ij}) \tag{5.21}$$

$$\mathcal{E}(C_{ni}) = g_3(\{p_{ij}\}, \{\gamma_{ij}\}); \quad \gamma_{ij} = \mathcal{E}(C_{ij})$$

Now suppose that an additional amount r_{ij} is invested in activity (ij) in order to shift its "level" from one (slow) level to a faster one. Then we expect the distribution of the activity duration to be shifted to the left by some amount depending on the sensitivity of the *average* duration to investment. (Notice that if such shift in average duration is not realized for any $r_{ij} > 0$, it would be futile to invest any sum in the first place.) Let the decrease in average duration be a function ϕ of the additional investment r_{ij}; hence, the new mean duration is given by

$$\hat{\mu}_{ij} = \mathcal{E}(\hat{Y}_{ij}) = \mu_{ij} - \phi(r_{ij}) \tag{5.22}$$

while the new expected cost, denoted by $\mathcal{E}(\hat{C}_{ij}(r_{ij}))$, is

$$\hat{\gamma}_{ij} = \mathcal{E}[\hat{C}_{ij}(r_{ij})] = \mathcal{E}[\psi_{ij}(\hat{Y}_{ij})] + r_{ij} \tag{5.23}$$

The actual cost of this activity at this level is still a function given by Eq. (5.20), whose expected value is given by Eq. (5.23). The relations among the various parameters are depicted in Fig. 5.30.

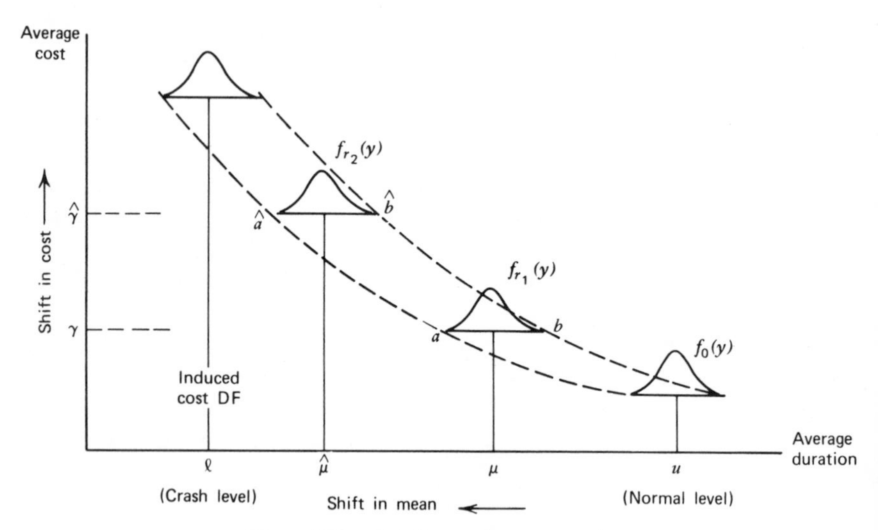

Figure 5.30. Relations between μ_{ij} and γ_{ij}.

Inserting the above expressions (5.22) and (5.23) into the standard GERT analysis, Eqs. (5.21), we obtain

$$P_{n_i} = g_1(\{p_{ij}\}) \quad \text{as before,}$$

$$\mathcal{E}(\hat{\mathrm{T}}_{n_i}) = g_2(\{p_{ij}\}, \{\hat{\mu}_{ij}\}) = \hat{g}_2(\{p_{ij}\}, \{\mu_{ij}\}, r_{ij}) \tag{5.24}$$

$$\mathcal{E}(\hat{C}_{n_i}) = g_3(\{p_{ij}\}, \{\hat{\gamma}_{ij}\}) = \hat{g}_3(\{p_{ij}\}, \{\gamma_{ij}\}, r_{ij})$$

Notice that the probability of realization of node n_i is not affected by the changes in cost and duration.

If the functions ϕ and ψ can be approximated in the region of interest by *linear* functions, then the means $\hat{\mu}$ and $\hat{\gamma}$ of Eqs. (5.22) and (5.23) become

$$\hat{\mu} = \mu + qr; \quad \frac{-\mu}{r} < q < 0 \tag{5.22a}$$

$$\hat{\gamma} = \mathcal{E}\left[m_0 + m\hat{Y}_{ij}\right] + r; \quad m > 0; \quad m_0 \geqslant 0;$$

$$= \gamma + br; \quad b = mq + 1 \tag{5.23a}$$

where $q < 0$ is the marginal decrease in the level of the activity per unit increase in investment in the activity and m is the linear "slope" of the function ψ.

§4.2 The Linear Case

For the linear case as given by Eqs. (5.22a) and (5.23a) it is possible to develop explicit expressions for $\mathcal{E}(\hat{\mathrm{T}}_{n_i})$ and $\mathcal{E}(\hat{C}_{n_i})$ in terms of the investments $\{r_{ij}\}$ in activities (ij).

Let π^k denote the kth path from starting node *1* to terminal node n_i. Let $P(\pi^k)$, or simply P^k, denote the probability that path π^k is realized, and let Y^k denote the duration of path π^k. Clearly, Y^k is the sum of the (independent) RVs $\{Y_{ij}\}$ of the durations of activities along π^k,

$$Y^k = \sum_{(ij) \in \pi^k} Y_{ij}$$

Hence,

$$\mathrm{T}_{n_i} = \sum_k \frac{P^k Y^k}{P_{n_i}} \tag{5.25}$$

In other words, it is a RV that is a convex combination of a finite number of RVs. Then

$$\mathcal{E}(\mathrm{T}_{n_i}) = \mathcal{E}\left(\sum_k \frac{p^k}{P_{n_i}} Y^k\right)$$

$$= \sum_k \frac{p^k}{P_{n_i}} \mathcal{E}(Y^k)$$

$$= \sum_k \frac{p^k}{P_{n_i}} \sum_{(ij)\in\pi^k} \mu_{ij}$$

Now replacing μ_{ij} with $\hat{\mu}_{ij} = \mu_{ij} + q_{ij} r_{ij}$, we obtain, after some simplification,

$$\hat{\mu}_{n_i} = \mathcal{E}(\hat{\mathrm{T}}_{n_i}) = \mathcal{E}(\mathrm{T}_{n_i}) + \sum_k \frac{p^k}{P_{n_i}} \cdot \sum_{(ij)\in\pi^k} q_{ij} r_{ij} \tag{5.26}$$

Similarly,

$$\mathcal{E}(\hat{C}_{n_i}) = \mathcal{E}(C_{n_i}) + \sum_k \frac{p^k}{P_{n_i}} \sum_{(ij)\in\pi^k} b_{ij} r_{ij} \tag{5.27}$$

A Remark on Additional Investment.

We wish to develop more precisely the concept of the additional investment r_{ij} in activity (ij).

Suppose that the duration Y [we drop the subscript (ij) for ease of notation] possesses a PDF $F_r(y)$, where the subscript r emphasizes the dependence of F on the level of additional investment $r[F_0(y)$ is the "raw" DF]. Hence, the expected cost is given by

$$\gamma = \int_a^b \psi(y)\, dF_0(y)$$

where a and b are the limits on the values of the duration of y.

If an investment r is made in that activity, the new average cost is given by

$$\hat{\gamma} = r + \int_{\hat{a}}^{\hat{b}} \psi(\hat{y})\, d\hat{F}_r(\hat{y})$$

which may be larger or smaller than γ.

Consider the case in which $\hat{\gamma} < \gamma$. This means that even after an activity (ij) has received the additional investment r, its expected cost is *less* than its cost before the additional investment. Then it must be true that the savings in cost due to the reduction in time more than offset the additional investment of r. Therefore, there must exist an initial investment r' corresponding to the minimal cost $\gamma°$,

$$\gamma° = \min_{r} \left[r + \int_{\hat{a}}^{\hat{b}} \psi(\hat{y})\, dF_r(\hat{y}) \right]; \quad 0 < r < \infty$$

We conclude that activity (ij) should be invested in by the additional amount r' immediately regardless of any desired reduction in the expected duration of the whole project.

As a result of the above analysis, it is henceforth assumed that the "zero level" of any additional investment incorporates r', such that the average cost of the activity is always increased. Furthermore, an additional investment in one or more activities is made for the explicit purpose of reducing the total duration to the realization of a specific terminal node.

The Basic Model.

Suppose that the status of the project (as revealed by the accomplishments in the activities), at some time $t > 0$ after its initiation, reveals that one or more activities have taken so long to accomplish that the probability of realizing the terminal node on or before a specified time is dangerously low. The question then arises as to *what is the optimal allocation of a fixed amount of funds K among the remaining activities such that the probability of realizing the terminal node on or before a specified time is maximized.*

Consider the case of terminal node n, and let the specified realization time be τ. Then we wish to

$$\begin{aligned} &\text{maximize} \quad z = Pr\left[\mathrm{T}_n(\{r_{ij}\}) \leqslant \tau\right] \\ &\text{s.t.} \quad \sum_{(ij)} r_{ij} \leqslant K \qquad (5.28) \\ &0 \leqslant r_{ij} \leqslant \bar{r}_{ij}, \quad (ij) \text{ in relevant subnetwork} \end{aligned}$$

where $\bar{r}_{ij}$ is an upper limit on the amount to be invested in activity (ij): $\bar{r}_{ij} \leqslant K$ and $\Sigma \bar{r}_{ij} > K$.

It is easy to translate the objective in the mathematical program of Eq. (5.28) from a statement on probability to a statement on duration, for it is

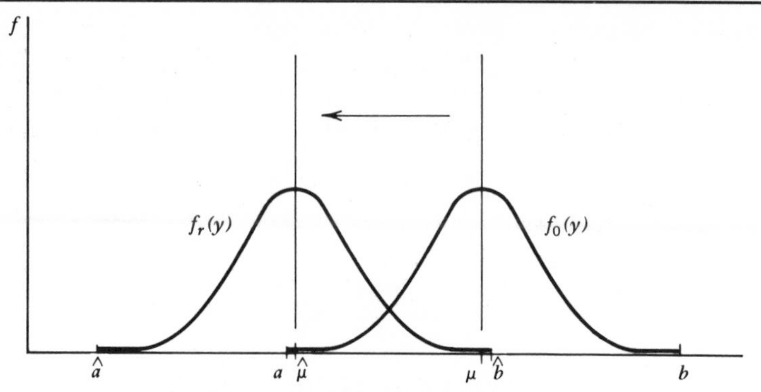

Figure 5.31. The shifting of $f(y)$.

clear that the duration to node n is a RV given by Eq. (5.25),

$$T_n = \sum_{(ij) \in \pi^k} \frac{P^k Y^k}{P_n}$$

By assumption, an additional investment $r > 0$ shifts the DF of Y "to the left" such that the new average duration $\hat{\mu}$ is strictly *smaller* than the original average duration μ (see Fig. 5.31). Consequently,

$$Pr[\hat{T}_n(r) \leqslant t] = F_r(t) > Pr[T_n \leqslant t] = F_0(t)$$

for all $t \geqslant t_1$ for some t_1 in the interval $0 \leqslant t_1 \leqslant \mu$. The proof of this inequality is easily obtained by contradiction, since assuming the contrapositive hypothesis would immediately result in $\hat{\mu} > \mu$, which we know is false.

Furthermore, by the continuity and linearity of $\hat{\mu}$ in r we deduce that $Pr[\hat{T}_n \leqslant t]$ increases with increasing r; that is, the probability, which is better written as $Pr[\hat{T}_n \leqslant t | \text{investment} = r]$, is itself a continuous and monotone nondecreasing function of r for any fixed t.

Hence, maximizing z in the criterion of (5.28) is equivalent to minimizing $\mathcal{E}(\hat{T}_n)$, and the new program can now be written as

$$\text{maximize} \quad \mathcal{E}(T_n) - \mathcal{E}(\hat{T}_n) \tag{5.29}$$

$$\text{s.t.} \quad \sum_{(ij)} r_{ij} \leqslant K; \quad 0 \leqslant r_{ij} \leqslant \bar{r}_{ij}: \quad (ij) \text{ in relevant subnetwork}$$

The Modified Model.

Placing the program in the equivalent format of (5.29) brings to light a subtle point that was not obvious hitherto. Namely, suppose that two investment values K_1 and K_2 with $K_1, K_2 \leqslant K$, yield the *same* maximal decrease in duration, but that K_1 is $< K_2$; are these two programs equivalent? The criterion functions of programs (5.28) and (5.29) would equate them, although it is intuitively clear that they are not equivalent. This logical deficiency can be easily corrected by demanding the maximum reduction in duration *per unit investment* (or, alternatively, the maximum probability of realization per unit investment). This leads to the program

$$\underset{r}{\text{maximize}} \frac{\mathcal{E}(\mathrm{T}_n) - \mathcal{E}\left[\hat{\mathrm{T}}_n(\{r_{ij}\})\right]}{\mathcal{E}\left[\hat{C}_n(\{r_{ij}\})\right] - \mathcal{E}(C_n)} \tag{5.30}$$

$$\text{s.t.} \quad \sum_{(ij)} r_{ij} \leqslant K; \quad 0 \leqslant r_{ij} \leqslant \bar{r}_{ij}: \quad (ij) \text{ in relevant subnetwork}$$

This is the program we propose to solve.

Let the moment-generating function (MGF) of the RV Y_v be denoted by $h_v(s) = \mathcal{E}[e^{sY_v}]$. Let the u function of activity v be given by $u_v(s) = p_v h_v(s)$.† Then, it is well known that the equivalent arc $(1, n)$, where 1 is the origin, has a u function which is easily obtained from the u functions of the network through standard GERT reduction methods. Furthermore,

$$P_n = u_n(0) \quad \text{and} \quad h_n(s) = \frac{u_n(s)}{u_n(0)}$$

in which we wrote the subscript n as a shorthand for the equivalent activity $(1, n)$. From the definition of $h_n(s)$ we deduce that

$$\mathcal{E}(\mathrm{T}_n) = \frac{d}{ds} h_n(s)\bigg|_{s=0}$$

which is a real number. After additional investments $\{r_{ij}\}$ have been made, and assuming linear relations as indicated in Eqs. (5.22a) and (5.23a), a similar analysis leads to

$$\mathcal{E}\left[\hat{\mathrm{T}}_n\right] = \frac{d}{ds} \hat{h}_n(s)\bigg|_{s=0}$$

†Recall the function $u(s)$ of Eq. (5.5).

We assert that $\mathcal{E}[\hat{\mathrm{T}}_n]$ is a *linear* function of $\{r_{ij}\}$. The proof of this assertion proceeds as follows.

$$\hat{h}_n(s) = \mathcal{E}(\exp s\mathrm{T}_{n_i}) = \int_0^\infty e^{s\hat{t}}\, d\hat{F}(\hat{t}), \quad \text{dropping the subscript } n$$

$$= \int_0^\infty d\hat{F}(\hat{t}) \sum_{m=0}^{\infty} \frac{(s\hat{t})^m}{m!}$$

$$= 1 + \int_0^\infty s\hat{t}\, d\hat{F}(\hat{t}) + \int_0^\infty d\hat{F}(\hat{t}) \sum_{m=2}^{\infty} \frac{(s\hat{t})^m}{m!}$$

Differentiating with respect to s and letting $s \to 0$ results in all terms vanishing except the term $\int_0^\infty \hat{t}\, d\hat{F}(\hat{t})$, as is indeed to be expected. But $\int_0^\infty \hat{t}\, d\hat{F}(\hat{t})$ is precisely what we denoted before by $\mathcal{E}(\hat{\mathrm{T}}_n)$, which we have already derived in Eq. (5.26) to be given by

$$\mathcal{E}(\hat{\mathrm{T}}_n) = \mathcal{E}(\mathrm{T}_n) - \sum_{(ij)} \alpha_{ij} r_{ij}; \quad (ij) \text{ in relevant subnetwork}$$

a linear function of $\{r_{ij}\}$ as asserted, where $\{\alpha_{ij}\}$ are constants.

A similar reasoning leads to the conclusion that $\mathcal{E}(\hat{C}_n)$ is also a linear function of $\{r_{ij}\}$. Thus, the mathematical program of (5.30) is retranslated into the *fractional linear program* (often called *hyperbolic program*)

$$\text{maximize} \quad \frac{\Sigma \alpha_v r_v}{\Sigma \beta_v r_v}; \tag{5.31}$$

$$\sum r_v \leqslant K; \quad 0 \leqslant r_v \leqslant \bar{r}_v$$

The solution of this fractional LP was treated in the work by Arisawa and Elmaghraby (1970) and is easily obtained as follows:

STEP 1. Find any nontrivial extreme point of the convex set of feasible solutions; for example; put $r_v^\circ = \bar{r}_v$ for $v = 1, 2, \ldots, m$, where

$$\sum_{v=1}^{m} \bar{r}_v \leqslant K \quad \text{but} \quad \sum_{v=1}^{m+1} \bar{r}_v > K \quad \text{and} \quad r_{m+1}^\circ = K - \sum_{v=1}^{m} \bar{r}_v$$

Substitute for these values in the objective function to yield the value z°.

STEP 2. Rank the variables in nonincreasing order of $\alpha_i - \beta_i z^\circ > 0$. Increase the investments sequentially according to their new rank until either the total available funds are exhausted or one runs out of ranked variables. The remaining variables are put $=0$. Go to Step 1. Stop when the basis repeats.

VARIATION OF $\mathcal{E}[\hat{\mathrm{T}}_n(K)]$ WITH $\mathcal{E}[\hat{C}_n(K)]$.

In order to emphasize the dependence of both values on the level of investment K, that fact has been highlighted in their designation.

The above reasoning leads to the conclusion that

$$\mathcal{E}\left[\hat{\mathrm{T}}_n(K)\right] = \mathcal{E}(\mathrm{T}_n) - \sum_{v=1}^{m} \alpha_v \bar{r}_v - \alpha_{m+1} r_{m+1}$$

for some m; $1 \leqslant m+1 \leqslant |A|$, and $r_{m+1} = K - \Sigma_1^m \bar{r}_v$; hence

$$= \mathcal{E}(\mathrm{T}_n) - \sum_{1}^{m} (\alpha_v - \alpha_{m+1}) \bar{r}_v - \alpha_{m+1} K$$

Similarly,

$$\mathcal{E}\left[\hat{C}_n(K)\right] = \mathcal{E}(C_n) + \sum_{v=1}^{m} (\beta_v - \beta_{m+1}) \bar{r}_v + \beta_{m+1} K$$

Consequently,

$$\mathcal{E}\left[\hat{C}_n(K)\right] = \mathcal{E}(C_n) + \sum_{1}^{m} (\beta_v - \beta_{m+1}) \bar{r}_v + \frac{\beta_{m+1}}{\alpha_{m+1}}$$

$$\times \left\{ \mathcal{E}(\mathrm{T}_n) - \mathcal{E}\left[\hat{\mathrm{T}}_n(K)\right] - \sum_{1}^{m} (\alpha_v - \alpha_{m+1}) \bar{r}_v \right\}$$

$$= \text{const} - \frac{\beta_{m+1}}{\alpha_{m+1}} \mathcal{E}\left[\hat{\mathrm{T}}_n(K)\right] \tag{5.32}$$

Therefore, we conclude that $\mathcal{E}[\hat{C}_n(K)]$ is a *linear* function of $\mathcal{E}[\hat{\mathrm{T}}_n(K)]$; the constant of proportionality is β_{m+1}/α_{m+1}, and the range in which Eq. (5.32) is valid in the interval

$$\sum_{1}^{m} \bar{r}_v \leqslant K \leqslant \sum_{1}^{m+1} \bar{r}_v$$

Example 5.5. In the network of Fig. 5.32 all activities are assumed (for simplicity) to be Normally distributed with specified (different) means but with the same variance. Suppose that for some reason (e.g., tardiness in delivering materials) the start of the project is delayed. Which activities should be reinforced in order to reduce the expected duration of the project in the most economical fashion, given that 10 units of capital are available?

The MGF of the Normal distribution with mean μ and variance σ^2 is given by

$$h(s) = \mathcal{E}\left[e^{st}\right] = \exp\left(\mu s + \frac{\sigma^2 s^2}{2}\right) \tag{5.33}$$

All of the pertinent data for the network are given in Table 5.5.

Evidently, we are interested in the event "success," which is terminal node *6* in the network. There are three paths from *1* to *6*:

$$\pi^{(1)} = 1, 3, 7$$

$$\pi^{(2)} = 2, 5, 7$$

$$\pi^{(3)} = 1, 4, 10, 5, 7$$

Let $u_v(s)$ denote the transmittance along activity v, to be determined presently. Then, applying Mason's rule yields directly the equivalent transmittance (dropping the explicit dependence on s)

$$u_e = \frac{\left[u_1 u_3 u_7 (1 - u_6 u_{10}) + u_2 u_5 u_7 + u_1 u_4 u_{10} u_5 u_7\right]}{\left[1 - u_3 u_9 - u_6 u_{10} - u_4 u_{10} u_5 u_9 + u_3 u_9 u_6 u_{10}\right]}$$

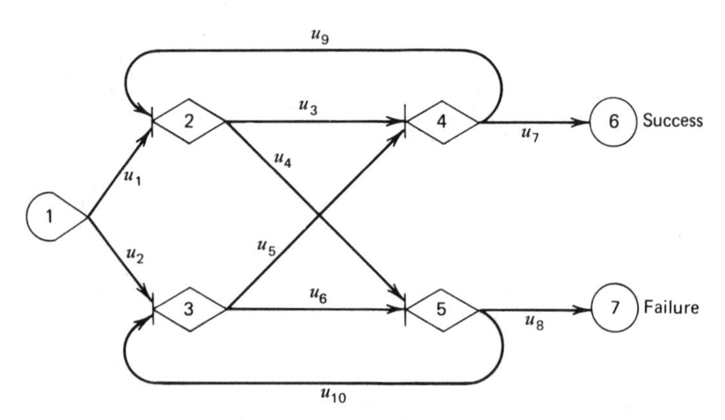

Figure 5.32. GERT network for Example 5.5.

Table 5.5. Parameters for Example 5.5

Activity	p_{ij}	μ_{ij}	$\underline{\mu}_{ij}$ [a]	$-q_{ij}$ [b]	m_{ij} [c]	γ_{ij} [d]	$\bar{r}_{ij}$ [e]
1=(1,2)	0.7	4	1.5	0.2	1.2	5.0	7.5
2=(1,3)	0.3	5	2.5	0.1	1.4	4.0	25.0
3=(2,4)	0.5	4	2.0	0.3	1.0	4.0	6.6
4=(2,5)	0.5	5	2.5	0.3	0.8	4.0	8.3
5=(3,4)	0.5	4	2.0	0.4	1.5	6.0	5.0
6=(3,5)	0.5	5	3.0	0.2	1.0	5.0	1.0
7=(4,6)	0.6	6	3.0	0.2	1.5	9.0	15.0
8=(5,7)	0.7	7	3.5	0.5	2.0	9.0	7.0
9=(4,2)	0.4	2	1.0	0.1	1.5	3.0	10.0
10=(5,3)	0.3	4	1.0	0.5	0.5	2.0	6.0

[a] $\underline{\mu}_{ij}$ is a lower bound on the expected duration of (ij).
[b] The parameter q is defined in Eqs. (5.22a) and (5.23a).
[c] The parameter m is defined in Eq. (5.23a).
[d] The parameter γ is defined in Eq. (5.21).
[e] $\bar{r}_{ij}$ is an upper bound on the investment in (ij).

This, in turn, immediately yields the probability of realization of node *6*, $P_6 = u_6(0)$, and the MGF of the time to the realization of *6* as $h_6(s) = u_6(s)/u_6(0)$.

To determine the u_v and the $\hat{u}_v$ values, consider as example activity *1* joining nodes *1* and *2*. From Table 5.5 we have that $\mu_1 = 4$ and $q_1 = -0.2$; hence, $\hat{\mu} = 4 - 0.2r_1$. Also, $b = mq + 1 = 0.76$; hence, $\hat{\gamma} = 5 + 0.76r_1$. Therefore, noting Eq. (5.33),

$$u_1(s_1) = 0.7\exp\left(4s_1 + \frac{\sigma^2 s_1^2}{2}\right)$$

and

$$\hat{u}_1(s_1) = 0.7\exp\left((4 - 0.2r_1)s_1 + \frac{\sigma^2 s_1^2}{2}\right)$$

The impact of the investment on cost can also be evaluated separately. However, it is more convenient to combine the two transmittance functions into one through the device of defining a different variable s_2. (That was the reason for writing s_1 instead of s in the above expressions of u_1 and $\hat{u}_1$.) Taking the cue from the evaluation of $\hat{\gamma}$ above, we conclude that the combined function is

$$u_1(s_1, s_2) = 0.7\exp\left[4s_1 + \frac{\sigma^2 s_1^2}{2} + 5s_2 + \frac{\sigma^2 s_2^2}{2}\right]$$

and

$$\hat{u}_1(s_1,s_2)=0.7\exp\left[(4-0.2r_1)s_1+\frac{\sigma^2 s_1^2}{2}+(5+0.76r_1)s_2+\frac{\sigma^2 s_2^2}{2}\right]$$

The other u and $\hat{u}$ values are determined in a similar manner. The $\hat{u}$ values are given in Table 5.6.

Substituting for $s=0$ to obtain P_6 we get

$$P_6=u_6(0)$$

$$=\frac{[0.7\times0.5\times0.6(1-0.5\times0.3)+0.3\times0.5\times0.6+0.7\times0.5\times0.5\times0.6\times0.3]}{[1-0.5\times0.4-0.5\times0.3-0.5\times0.5\times0.4\times0.3+0.5\times0.5\times0.4\times0.3]}$$

$$=\frac{0.3}{0.65}=0.4615$$

To determine the optimal allocation according to the model of Eq. (5.30) we need to determine the values $\mathcal{E}(\mathrm{T}_6)$, $\mathcal{E}(\hat{\mathrm{T}}_6)$, $\mathcal{E}(C_6)$, and $\mathcal{E}(\hat{C}_6)$. As always, expectations are determined through differentiation of the u_e function. For instance,

$$\mathcal{E}(\hat{\mathrm{T}}_6)=\frac{\partial}{\partial s_1}[u_e(s_1,s_2)]\bigg|_{s_1=0=s_2}/P_6$$

$$\mathcal{E}(\hat{\gamma}_6)=\frac{\partial}{\partial s_2}[u_e(s_1,s_2)]\bigg|_{s_2=0=s_1}/P_6\quad\text{(etc.)}$$

Table 5.6. $\hat{u}_v(s_1,s_2)$

Activity v	$\hat{u}_v(s_1,s_2)$
1	$0.7\exp[(4-0.2r_1)\,s_1+\sigma^2 s_1^2/2+(5+0.76r_1)s_2\,+\sigma^2 s_2^2/2]$
2	$0.3\exp[(5-0.1r_2)\,s_1+$ " $+(7+0.86r_2)s_2\,+$ "]
3	$0.5\exp[(4-0.3r_3)\,s_1+$ " $+(4+0.7r_3)s_2\,+$ "]
4	$0.5\exp[(5-0.3r_4)\,s_1+$ " $+(4+0.76r_4)s_2\,+$ "]
5	$0.5\exp[(4-0.4r_5)\,s_1+$ " $+(6+0.4r_5)s_2\,+$ "]
6	$0.5\exp[(5-0.2r_6)\,s_1+$ " $+(5+0.8r_6)s_2\,+$ "]
7	$0.6\exp[(6-0.2r_7)\,s_1+$ " $+(9+0.7r_7)s_2\,+$ "]
8	$0.7\exp[(7-0.5r_8)\,s_1+$ " $+(9+0.6r_8)s_2\,+$ "]
9	$0.4\exp[(2-0.1r_9)\,s_1+$ " $+(3+0.85r_9)s_2\,+$ "]
10	$0.3\exp[(4-0.5r_{10})s_1+$ " $+(2+0.75r_{10})s_2+$ "]

Performing these (rather laborious) calculations we end up with the program:

$$\text{maximize} \quad (0.155r_1+0.035r_2+0.384r_3+0.054r_4+0.101r_5 +0.0236r_6+0.261r_7+0.049r_9+0.107r_{10}) /(0.722r_1+0.482r_2+0.625r_3+0.266r_4+0.182r_5+0.092r_6 +0.913r_7+0.581r_9+0.163r_{10}) \tag{5.34}$$

s.t.

$$0\leqslant r_1\leqslant 7.5;\quad 0\leqslant r_2\leqslant 25.0;\quad 0\leqslant r_3\leqslant 6.6;\quad 0\leqslant r_4\leqslant 8.3;$$

$$0\leqslant r_5\leqslant 5.0;\quad 0\leqslant r_6\leqslant 1.0;\quad 0\leqslant r_7\leqslant 15.0;\quad 0\leqslant r_9\leqslant 10.0;\quad 0\leqslant r_{10}\leqslant 6.0$$

and

$$\sum_{\substack{j=1\\ j\neq 8}}^{10} r_j=10$$

Now we need to calculate a basic feasible solution. Put $r_9=10.0$ and the rest of $r_i=0$. Then

$$z^{(1)}=\frac{0.049\times 10}{0.581\times 10}=0.0843$$

Next, by Step 2, we need to rank the variables in nondecreasing magnitude of $(\alpha_i-\beta_i z^{(1)})=(\alpha_i-0.0843\beta_i)$. This ranking is shown in column 5 of Table 5.7.

Table 5.7. Ranking of the Activities

Variable	α_i	β_i	$\alpha_i-0.0843\beta_i$	Rank$^{(1)}$
r_1	0.155	0.722	0.094	3
r_2	0.035	0.482	−0.006	9
r_3	0.384	0.625	0.331	1
r_4	0.054	0.266	0.032	6
r_5	0.101	0.182	0.086	5
r_6	0.024	0.092	0.016	7
r_7	0.261	0.913	0.184	2
r_9	0.049	0.581	0.000	8
r_{10}	0.107	0.163	0.093	4

According to the ranking, the second basic feasible solution is $r_3=6.6$, $r_7=3.4$ and the rest of $r_i=0$. Therefore,

$$z^{(2)}=\frac{0.384\times 6.6+0.261\times 3.4}{0.625\times 6.6+0.913\times 3.4}=0.3244>z^{(1)}$$

Again, we need to rank the variables in the same manner as before since $z^{(1)}\neq z^{(2)}$. This ranking is shown in column 3 of Table 5.8.

According to the ranking, the third basic feasible solution is $r_3=6.6$, $r_{10}=3.4$, and the rest of $r_i=0$. Therefore,

$$z^{(3)}=\frac{0.384\times 6.6+0.107\times 3.4}{0.625\times 6.6+0.163\times 3.4}=0.6194>z^{(2)}$$

The ranking with respect to $z^{(3)}$ is in column 5 of Table 5.8, and according to the ranking, the fourth basic feasible solution is $r_{10}=6.0$, $r_3=4.0$, and the rest of $r_i=0$. Therefore,

$$z^{(4)}=\frac{0.384\times 4.0+0.102\times 6.0}{0.625\times 4.0+0.163\times 6.0}=0.6262>z^{(3)}$$

The ranking with respect to $z^{(4)}$ is in column 3 of Table 5.9. The basic feasible solution corresponding to $z^{(4)}$ is $r_{10}=6.0$ $r_3=4.0$. Therefore, z remains the same and, by Step 2, this value is the optimal. Therefore,

$$z^*=0.6262$$

and

$$r_3^*=4.0,\ r_{10}^*=6.0, \text{ and the rest of } r_i^*=0$$

Table 5.8.

Variable	$\alpha_i-0.3244\beta_i$	Rank$^{(2)}$	$\alpha_i-0.6194\beta_i$	Rank$^{(3)}$
r_1	−0.079	7	−0.292	8
r_2	−0.121	8	−0.264	7
r_3	0.181	1	−0.003	2
r_4	0.032	4	−0.111	5
r_5	0.042	3	−0.012	3
r_6	−0.006	5	−0.033	4
r_7	−0.077	6	−0.305	9
r_9	−0.139	9	−0.221	6
r_{10}	0.054	2	0.006	1

Table 5.9.

Variable	$\alpha_i - 0.6262\beta_i$	Rank[4]
r_1	−0.297	7
r_2	−0.267	6
r_3	−0.007	2
r_4	−0.112	5
r_5	−0.013	3
r_6	−0.034	4
r_7	−0.311	8
r_9	−0.315	9
r_{10}	0.005	1

Therefore, when 10 additional units of funds are available, one should invest in activities *3* and *10* in amounts of 4.0 and 6.0, respectively. The expected time of the project will be reduced by 2.178 time units, while the expected cost will be increased by 3.478 investment units.

APPENDIX E. Signal Flowgraphs

Signal flowgraphs (SFGs), sometimes referred to simply as *flowgraphs*, are an analytic tool often used in the modeling and analysis of linear systems. Although their use originated in the analysis of electrical networks, increased interest in SFGs derives from the importance of the analysis and synthesis of linear systems occurring in many fields of science and engineering. It is a known fact that, irrespective of their content, many systems can be modeled as a set of linear equations, to which the methodology of SFG theory is directly applicable.

The generic element is shown in Fig. E.1. It consists of a directed arc (i.e., an arrow) connecting two nodes, the *origin* x_1 and the *terminal* x_2. The arc is said to be of *transmittance* t_{12}, meaning that the value of the node x_1 is multiplied by t_{12} as it is transmitted through the arc such that the variable represented by the node x_2 is defined by $x_2 = x_1 t_{12}$.

In general, more than one arc may leave any node and more than one arc may terminate at any node. In the former case, the value of the node is multiplied by the transmittance of each arc emanating from it; while in the latter case, the value of the node is equal to the *sum* of all the inputs into it. These two properties define the 'equivalent transmittance' in the case of *parallel paths* (see Fig. E.2).

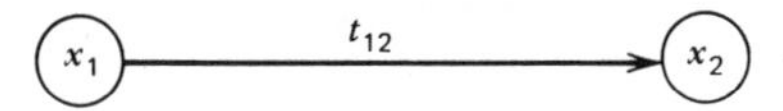

Figure E.1. Generic element of SFG.

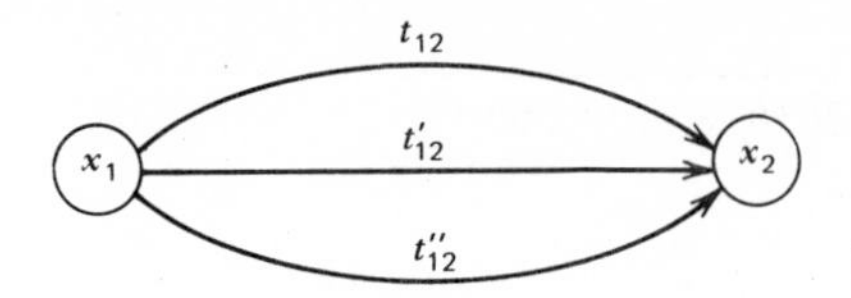

Figure E.2. Parallel paths: $x_2 = x_1(t_{12} + t'_{12} + t''_{12})$.

Now, in the analysis of SFGs one is usually interested in the transmittance between a selected subset of nodes. The graphic–algebraic procedure utilized to reduce a SFG to a residual graph showing only the nodes of interest is based on a few simple rules. These rules are derived from the fact that a SFG is basically a representation of a set of linear equations. And it is this very fact that bestows importance on SFG theory because the behavior of so many physical systems can be described by linear algebraic equations, as was amply demonstrated in Section 2 of the present chapter.

Instead of cluttering this appendix with a long list of such rules for graph reduction, and with proofs of their validity, we content ourselves with constructing the minimal required set so that the reader will grasp the salient features of the approach. We have already specified the rule for combining arcs in parallel. On the other hand, it must be obvious that for two arcs in series, (see Fig. E.3) the value of x_3 is given by $x_3 = x_2 t_{23} = x_1 t_{12} t_{23}$. From this and the previous rule, we conclude that *transmittances of arcs in series multiply and of arcs in parallel, add.*

We are often concerned with *feedback loops* (see Fig. E.4*a*) and, therefore, the value of the variable x_4. We develop the general rule as follows. From the graph we have

$$x_2 = x_1 t_{12} + x_4 t_{42}$$

$$x_3 = x_2 t_{23}$$

which, on substitution yields

$$= x_1 t_{12} t_{23} + x_4 t_{42} t_{23}$$

and finally

$$x_4 = x_3 t_{34}$$

$$= x_1 t_{12} t_{23} t_{34} + x_4 t_{42} t_{23} t_{34}$$

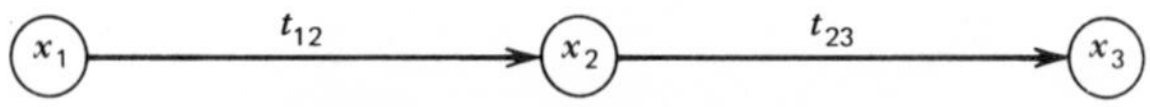

Figure E.3. Branches in series.

Notice that this last equation has the SFG representation of Fig. E.4*b*. It is then obvious that the transmittance from x_1 to x_4 with the feedback loop of Fig. E.4*a* is equivalent to the transmittance of the *path* π_{x_1,x_4} and the *self-loop* around x_4 of transmittance equal to the transmittance of the feedback loop of Fig. E.4*b*. Finally, the last equation reduces to

$$x_4 = x_1 \times \frac{t_{12}t_{23}t_{34}}{1 - t_{42}t_{23}t_{34}}$$

which implies that a *self-loop* of transmittance l is equivalent to $1/(1-l)$ multiplied by the transmittance of all the arrows incident *into* the node.

Although the above three rules (paths in parallel, paths in series, and feedback or self-loops) are the cornerstone rules by which one can reduce any SFG to its equivalent residual graph, the reduction of complicated graphs can be rather cumbersome and prone to error. For such graphs Mason (1953) proposed a rule for reduction which, despite its formidable appearance, is rather simple and straightforward.

In a network composed of several sources, sinks and intermediate nodes† the transmittance from a source r to a sink s is evaluated as follows. Let π_j denote the transmittance along the jth *forward* path between the designated source and terminal, $L_1, L_2, \ldots, L_n$ represent the transmittance of the loops (both feedback and self-loops) in the graph; and let

$$D = 1 - \sum_i L_i + \sum_{i,j} L_i L_j - \sum_{i,j,k} L_i L_j L_k + \cdots$$

where each multiple summation extends over *nonadjacent loops* (two loops are adjacent if they share at least one node). Let Δ_j be evaluated as D but with path π_j removed (which implies the removal of all nodes and branches along π_j). Then the desired transmittance $t_{r,s}$ is given by $t_{r,s} = (\Sigma_j \pi_j \Delta_j / D)$.

The reader is reminded, once again, that the common feature of all the analyses using SFG is that the systems in question can be represented by a set of *linear algebraic equations*. In some cases, these algebraic equations are arrived at by applying transform methods (Fourier, Laplace, probability generating, and z-transforms) to the original equations representing the dynamic behavior of the system. But once such transformation is accomplished, analysis proceeds on the basis of the SFG.

†By convention, a *sink* is a node with no arcs leaving it, and a *source* is a node with no arcs entering it. A SFG may have any number of sources and sinks but must have at least one of each.

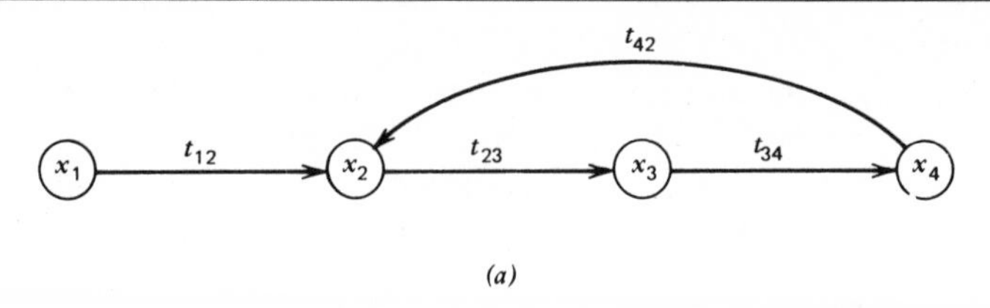

(a)

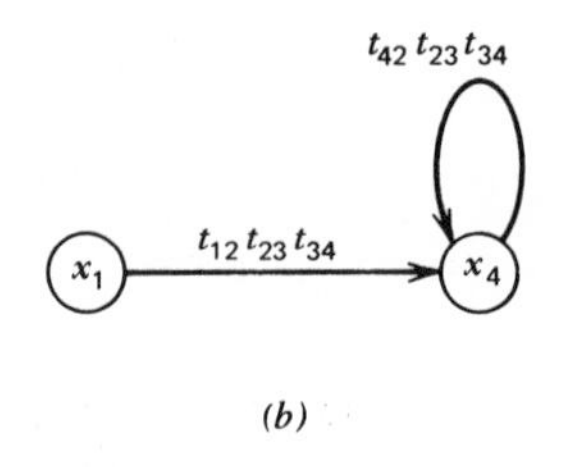

(b)

Figure E.4. Feedback and self-loops.

To illustrate, consider the following system of three equations in three variables: x_1, x_2, and x_3:

$$x_1 + 4x_2 - 2x_3 = 10$$

$$-3x_1 + x_2 + 4x_3 = 16$$

$$6x_1 - 3x_2 + 5x_3 = 36$$

which may be rewritten as:

$$x_1 = 10y - 4x_2 + x_3$$

$$x_2 = 16y + 3x_1 - 4x_3$$

$$x_3 = \tfrac{36}{5}y + \tfrac{6}{5}x_1 + \tfrac{3}{5}x_2$$

where y is a pseudovariable of known value 1. The SFG is as shown in Fig. E.5.

Since the external "forcing function" is the pseudovariable y, it is evident that the value of any variable is given by the value of the equivalent transmittance between the node y and the variable in question. Suppose, for example, we wish to determine the value of x_1. The loops of

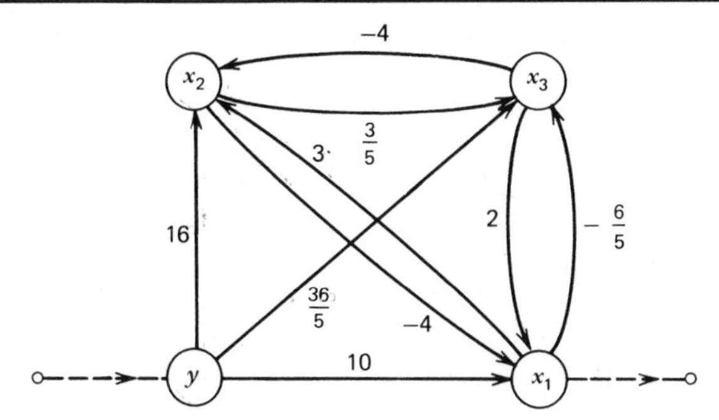

Figure E.5. Signal flow representation of the three simultaneous equations.

the SFG and their transmittances are as follows:

$$L_1 = x_1, x_2, x_1 \Rightarrow -12$$

$$L_2 = x_2, x_3, x_2 \Rightarrow -\frac{12}{5}$$

$$L_3 = x_3, x_1, x_3 \Rightarrow -\frac{12}{5}$$

$$L_4 = x_1, x_2, x_3, x_1 \Rightarrow \frac{18}{5}$$

$$L_5 = x_1, x_3, x_2, x_1 \Rightarrow -\frac{96}{5}$$

Therefore,

$$D = 1 - \sum_i L_i = \frac{167}{5}$$

since no two loops are independent of each other.

There are five forward paths from y to x_1, whose values are given by

$$\pi_1 = y, x_1 \Rightarrow 10; \qquad \text{and } \Delta_1 = 1 - L_2 = \frac{17}{5}$$

$$\pi_2 = y, x_2, x_1 \Rightarrow -64 \qquad ; \quad \Delta_2 = 1$$

$$\pi_3 = y, x_2, x_3, x_1 \Rightarrow \frac{96}{5} \qquad ; \quad \Delta_3 = 1$$

$$\pi_4 = y, x_3, x_1 \Rightarrow \frac{72}{5} \qquad ; \quad \Delta_4 = 1$$

$$\pi_5 = y, x_3, x_2, x_1 \Rightarrow \frac{567}{5} \qquad ; \quad \Delta_5 = 1$$

Consequently, the equivalent transmittance between nodes y and x_1 is given by

$$t_{y,x_1} = \sum_j \pi_j \frac{\Delta_j}{D} = \frac{594}{5} \div \frac{167}{5} = \frac{594}{167}.$$

The reader may wish to determine the values of the equivalent transmittances t_{y,x_2} and t_{y,x_3}, which give the values of the variables x_2 and x_3, respectively. To assist in the verification of such calculations, we offer the inverse of the matrix of coefficients and the complete algebraic solution of this problem:

$$A = \begin{bmatrix} 1 & 4 & -2 \\ -3 & 1 & 4 \\ 6 & -3 & 5 \end{bmatrix}; \; A^{-1} = \frac{1}{167}\begin{bmatrix} 17 & -14 & 18 \\ 39 & 17 & 2 \\ 3 & 27 & 13 \end{bmatrix};$$

$$A^{-1}b = \frac{1}{167}\begin{bmatrix} 594 \\ 734 \\ 930 \end{bmatrix} = \begin{bmatrix} x_1 \\ x_2 \\ x_3 \end{bmatrix}$$

Notice that D of the SFG is proportional to the determinant of the matrix of coefficients.

PROBLEMS

The analysis of Section 2 of the present chapter concentrated on demonstrating both the identity of the GERT model and the SFG representation of SMPs and dealt only tangentially with the algebraic expressions useful in solving such models (by whichever names they are called). The following set of problems complement the abovementioned theory and give the reader an opportunity to flex his muscle in this approach.

1. Construct a table of transmittances of "equivalent" arcs similar to Table 5.1 in which all the nodes are assumed of the "exclusive-or" type (GERT networks). In particular, prove that:
   a. for two arcs in series: $u_e(s) = u_a(s)u_b(s)$;
   b. for two arcs in parallel: $u_e(s) = u_a(s) + u_b(s)$;
   c. for a loop: $u_e(s) = u_a(s)u_b(s)/[1 - u_c(s)]$.

   In each case segregate the probability, p_e, from the holding time function, $h_e(s)$.

2. Demonstrate that the function $h_a(s)$ may be viewed as a moment-generating function of the "holding time." In particular, if the jth moment is denoted by μ^j and the jth cumulant by K^j, then

$$\mu^j = \frac{\partial^j}{\partial s^j}\left[(-1)^j h(s)\right]\Big|_{s=0} \quad \text{and} \quad K^j = \frac{\partial^j}{\partial s^j}\left[(-1)^{-j}\ln h(s)\right]\Big|_{s=0}$$

3. It is well known that if a branch is added to close a network, its transmittance is equal to the inverse of the equivalent branch of the network itself. Thus, if $u_e(s)$ is the equivalent transmittance of a network and the feedback arc $u_f(s)$ is introduced about the network, then $u_f(s) = 1/u_e(s)$.

$0 \rightarrow$ $u_e(s)$ $\rightarrow 0$

$u_f(s)$

Deduce this result from the "topology equation" for a closed network:

$$H(s) = 1 - \sum_i L_i + \sum_{i,j} L_i L_j - \sum_{i,j,k} L_i L_j L_k + \cdots = 0$$

where the L values and summations are as defined in Mason's rule (p. 397).

Apply the concept to deduce the equivalent transmittance of the following network:

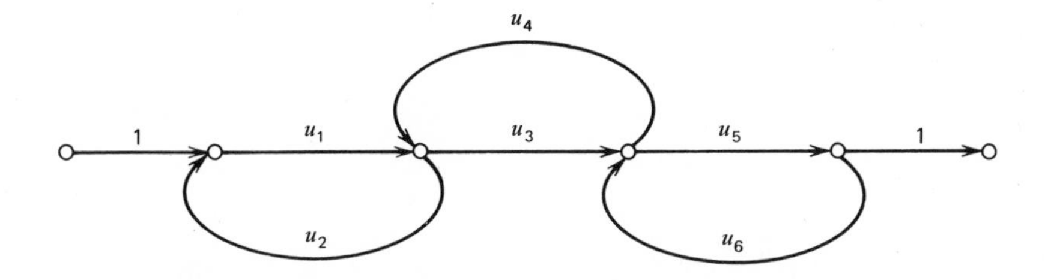

Verify that the result obtained by "closing the loop" and using the topology equation is identical to that obtained from applying Mason's rule.

4. Apply the topology equation of Problem 3 (or Mason's rule) to the following adaptation of the famous "Thief of Baghdad" puzzle

[Pritsker and Whitehouse (1966)], where the thief has been placed in a dungeon with three doors. One door leads to freedom, one door to a long tunnel, and a third door to a short tunnel. The two tunnels return the thief to the dungeon after varying amounts of time. If he returns to the dungeon, he attempts to gain his freedom again, but his past experiences do not help him in selecting the door that leads to freedom (i.e., there is no learning and hence the probabilities stay constant).

Construct the GERT model and deduce its equivalent transmittance, $u_e(s)$. Assuming that he chooses any of the doors with equal probabilities, and that; (a) the time to freedom is one day, (b) to traverse the short tunnel is 10 days, and (c) to traverse the long tunnel is 25 days, what is his expected time to freedom? What is the probability that it will take him more than 100 days to gain freedom?

5. Consider the exponential–geometric transform exemplified by Eq. (5.16). It can be viewed as the z transform of the Laplace transform. This concept can be applied to the u function of Eq. (5.5) to give an alternative approach to the modeling of SMPs, which is equally useful especially in counting problems.

Let

$$u(s,z)=\sum_{k=0}^{\infty} u(s,k)z^k$$

which is commonly referred to as the *u-generating function* and may be derived in a natural fashion if the SFG with transmittances $\{u(s)\}$ has some of its branches "tagged" with z. Then each time one of these branches is traversed, the transmittance will be multiplied by z. Hence, $u(s,k)$ may be viewed as the *conditional* u-function associated with the network, given that the branches tagged with z are traversed k times. Clearly, $u(s,k)$ can be obtained from $u(s,z)$ as

$$u(s,k)=\frac{1}{k!}\left.\frac{\partial^k u(s,z)}{\partial z^k}\right|_{z=0}$$

Normally, it is the conditional MGF that is of interest, and it is given by $h(s,k)=u(s,k)/p(k)$, where $p(k)=u(0,k)$. You are asked to verify and/or exemplify the following statements:

a. The u-generating function of a SFG with all tagged arcs removed is given by $u(s,0)$, and the MGF of the number of times the tagged arcs are traversed is $u(0,z)$.

b. To count the number of times a node is realized, the input and output sides of the node are split and a branch tagged with a z inserted between the two.

c. To determine the MGF of the time to the nth return to a node, the first step is to tag the node as in (b). What then?

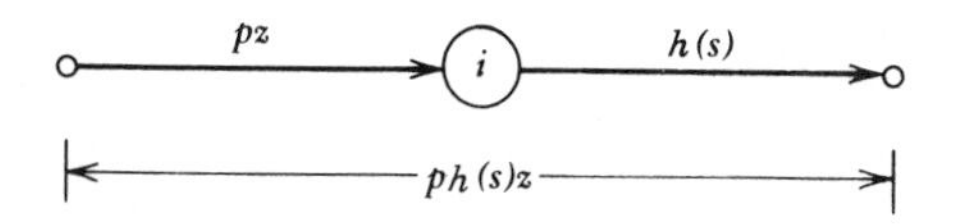

6. Using the results of Problems 3 and 5;
   a. Draw the SFG (or the GERT network) for determining the number of throws of a pair of dice required to obtain n consecutive "snake's eyes" (i.e., two ones). Deduce its equivalent transmittance, $u_e(s)$, assuming the dice to be fair.
   b. Determine the MGF of the event "success," which is defined as two consecutive "snake's eyes."
   c. Determine the MGF of the time to the kth "success," as defined in (b)? (see Whitehouse (1973); p. 259).
7. All durations of the following SFG network are negative–exponentially distributed, $\mu e^{-\mu t}$, with parameters μ as follows:

(Arc)	μ
(1,2)	1
(2,3)	2
(2,4)	4
(3,2)	2
(3,3)	1

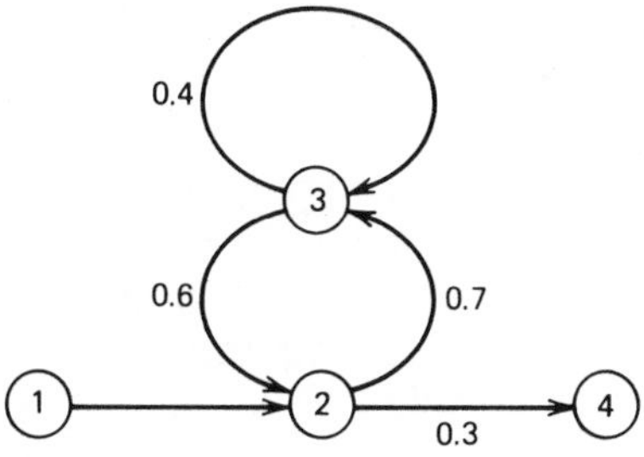

   a. Deduce the u-generating function for the equivalent element of the network, assuming no z tagging.
   b. What is the MGF of the number of realizations of node *3*?
   c. What is the probability that node *3* will be visited 6 times before node *4* is reached? What is the MGF of the time to that event?
8. A camera shop handles a certain kind of professional camera. The

demand per week is as follows:

Quantity	0	1	2	3
Probability of demand	0.1	0.2	0.4	0.3

Customers usually do not wait; if a customer does not find a camera in the shop he usually goes to another. Therefore, demand that is not satisfied represents lost sales. The shop owner adopted the replenishment policy that if at the end of each week when he counts the cameras he has in stock, one or less are in hand, he orders three cameras; otherwise, he does not replenish his stock.

a. Assume that replenishment of his stock takes exactly one week (i.e., items ordered on Friday P.M. will arrive the following Friday P.M. for the sales starting the following Monday A.M.). Draw the SFG (or GERT) and determine: (i) the average time between customer arrivals, (ii) the average size of the stock, and (iii) the probability that the storekeeper will always satisfy his customers.

b. If the time of replenishment is itself a RV (integer) that varies uniformly between one and four weeks, rework points (i)–(iii) above [Whitehouse (1973b), p. 370].

9. a. Consider the following reliability problem. A device is being developed for a given application. When put into operation, the device will either function correctly or not at all. There can be only one thing wrong with the device that will cause it not to function, and it is the correction of this fault that requires developmental effort. The cycle of development is as follows: test the device, if it operates satisfactorily, then its design is effective and it is declared functional. However, if it fails it is declared defective and developmental work is done on it, whose probability of success is p on any trial, independent of the number of trials. Then it is tested, and so on. When tested, a device may exhibit both types of error; it may be "good" but fail the test, which occurs with a small probability a, or it may be "defective" and pass the test, which occurs with even smaller probability b. Hypothesize durations for the various development and test activities (constants or RVs). Construct the GERT network and then determine: (i) the MGF of the duration to the success state and (ii) the MGF of duation to success after having failed three times.

   b. Suppose that the device has *two* elements that can "go wrong" instead of only one. If it fails, developmental effort is devoted to either aspect of its design, or to both aspects, before the device is resubmitted to testing. The durations of development are in consonance with expectations; concentrating on either one aspect would take more than half the time and the cost of working on both aspects simultaneously (due to the possibility of running some activities in parallel). Redo (i) and (ii) of part a.

10. A trucking company, fearful of the collapse of its negotiations with the union over the number of days a driver is "away from his home city," wants to analyze the movement of the trucks over its network. If the cities are numbered $1, 2, \ldots, n$ in round-the-circle fashion (so that city n is adjacent to city *1*), then the probability that a truck is requested to deliver a load from city i to city $i+1$ is 0.5 and to city $i-1$, is 0.4, and to stay in city i for one day is 0.1. (Since city n is adjacent to city *1*, city n is city *0* when the truck is in city *1*. Similarly, city *1* is city $n+1$ when the truck is in city n.) The time of travel in days from city i to city j is a RV, denoted by Y_{ij}, assumed Poisson distributed with the following means (the table is constructed for six cities):

Source \ Destination (μ_{ij})	1,2, or 3	4,5, or 6
1,2, or 3	3	4
4,5, or 6	5	3

You are asked to construct a GERT model for a system with $n=5$ cities, then determine the MGF of the first return to the base city (say city *1*).

What is the probability of a driver being away from the base city more than one week ($=5$ days), and what is the probability that the driver will visit city *2* exactly twice before returning to his base city for the first time?

11. Reconstruct the GERT model of contract negotiations given in Example 5.4 in which you enrich the model with the following real-life considerations.
    a. The probability of request modification by the customer decreases monotonically with the number of requests made to date.
    b. A corporate-level planning conference (activity *21*) will take place only if a favorable engineering report *and* and a favorable financial report are received, irrespective of the nature of the reports in other areas.
    c. The cycle of activities represented by nodes *19, 20, 21, 18, 19* representing the implementation of modifications requested by the

customer is permitted to occur only three times, after which the bid is either accepted "as is" or rejected. [Coordinate this modification with (a) above.]

12. The central "dispatcher" of a trucking firm serving several cities is faced with the following problem on a daily basis.

 Demand for trucks initiates at any city for shipment of merchandise to any other city and is assumed to be measured in full truckloads. The demand from city i to city j is a RV denoted by X_{ij}; $i,j=1,2,\ldots,n$; $i \neq j$. If there is more demand than available trucks at a particular city all of the trucks are used and the excess demand represents "lost sales" (i.e., it is not carried over to the following day). However, if there are more trucks than the realized demand at any city, then some or all the excess trucks are sent as "empties" to some other city (or cities) in anticipation of future shortages at that city (or cities). The decision on shipment of "empties" is one that the dispatcher must make.

 The problem is compounded by the fact that the duration of the trip between any two cities i and j is a RV, since it depends on the weather and the traffic conditions throughout the trip. Denote this duration by Y_{ij}.

 The distribution of the fleet of trucks over the cities varies from day to day according to the movement of the trucks (both loaded and empty) as well as on the realization of demand at the various cities. It is easy to see that a city i may be "starving" for trucks while, on the same day, another city j may be experiencing "excess" trucks. This would happen if prior shipments into city i were sparse, while shipment out of it were abundant, and the converse is true for city j.

 The problem in its general context is of formidable magnitude. Here, you are asked to analyze a much simplified version.

 Assume there are only three cities; call them *1*, *2*, and *3*. Let the total fleet size be five trucks. The demand for trucks is assumed geometrically distributed, that is,

$$Pr[X_{ij}=n]=(1-p_{ij})p_{ij}^n$$

 between all cities, with the following parameters p_{ij}:

Source \ Destination	Parameter p_{ij} 1	2	3
1	—	0.5	0.75
2	0.75	—	0.80
3	0.80	0.75	—

The interpretation of this table is as follows. Consider city 2; $X_{2,3}$ has the probability mass function $0.2(.8)^n$. It is easy to verify that the expected demand from city i to city j is $p_{ij}/(1-p_{ij})$.

The duration, in days, of the trip from city i to city j is the RV Y_{ij}, which is assumed Poisson distributed with mean μ_{ij} given by the following table:

Source \ Destination	The mean μ_{ij} 1	2	3
1	—	3	1
2	2	—	1
3	3	2	—

Can this problem be modeled by GERT? If your answer is "yes," state your model (or a facsimile of it, if that is sufficient to explain its structure) and comment on its computational requirements. If your answer is "no," explain your reasons. Could you suggest another approach to the modeling and analysis of this problem?

13. Study the operations of a library (university or public) from the point of view of book lending, withholding by the borrower, return or loss, and the durations of each activity. Construct a GAN, GERT, or GERTS model for the process and comment briefly.

14. In a study of the population in the correctional institutions (i.e., prison inmates and parolees), the following simplified view of the system seemed to offer a good deal of insight into its inner workings and future trends.

Suppose that attention is focused only on the male population. This is divided between youth offenders and adults. Depending on the crime committed, there are two custody levels for youth offenders and four custody levels for adults: (a) minimal and medium for youth offenders and (b) minimal, medium, close, and maximal for adults. An inmate may be "promoted" from one level to a higher one, or "demoted" to a lower one, according to his behavior (which includes the *number* of infractions committed as well as their *kind*). Youth offenders are transferred laterally (i.e., at the same custody level) on reaching the age of 21 years.

An inmate is eligible for parole after serving 25% of his sentence. However, eligibility does not imply success in securing parole at the time of consideration. As a matter of fact, approximately only 20% of those eligible are successful in being released on parole. Those denied parole are reviewed at more-or-less regular intervals averaging 6 months apart, but varying between 5 and 10 months. A parolee may lose his parole status and return to custody if he commits certain infractions.

You are asked to hypothesize the various parameters needed to construct a GERTS model of the system for the purpose of studying the system's behavior under varying policy and environmental conditions (e.g., the "stiffening" of the length of sentences or the elimination of parole, both of which have been suggested).

The following four problems require that you derive some parameters of the PDFs of networks composed of arcs in parallel and of loops. This should enhance your analytical capability with GERT-type networks.

15. Consider a simple self-loop as shown in the figure, where p is the probability of exiting from the loop and $q=1-p$ is the probability of going one more time around the loop. Suppose the system starts in node *1*, and let N denote the number of times the loop is traversed. Show that N has the "geometric distribution," with mean $\mathcal{E}(N)=q/p$ and var $(N)=q(1+q^2)/p^3$.

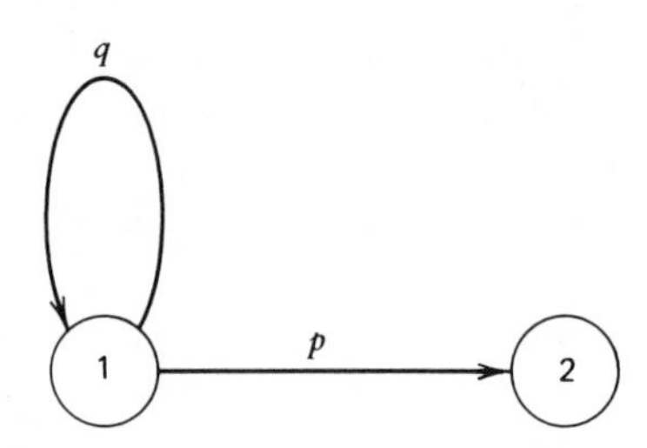

16. Consider the case of n arcs in parallel between two nodes, *1* and *2*, as shown in the figure. Suppose that the probability that arc i is traversed is $p_i>0$, and $\Sigma_{i=1}^{n} p_i=1$; and that the duration of activity i is a RV X_i of known discrete PDF. Let T denote the time of realization of node *2*, assuming that the system had started in node *1* at time 0. It is easy to see that the probability that $T=t$ is given by $Pr(T=t)=\Sigma_{i=1}^{n} p_i Pr\cdot(X_i=t)$. What is the expected value of T? What is the variance of T? Answer both questions for the case $X_i=c_i$, a constant.

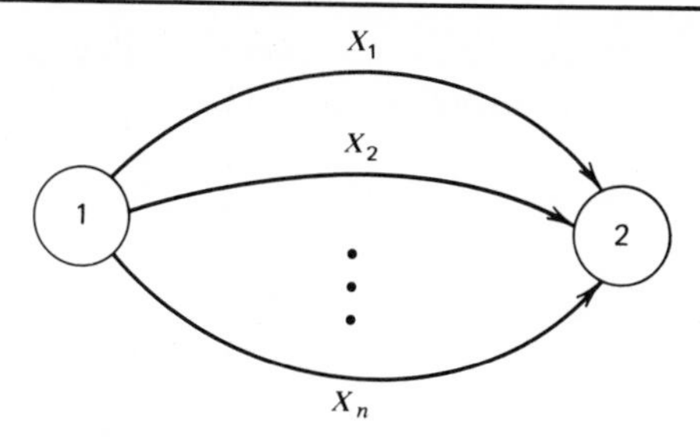

17. Consider a self-loop as shown, in which X designates the time to go around the loop once. X is a RV which is supposed to assume integer values, for simplicity. Instead of specifying the probability p of exiting to node *2*, as we did in Problem 15, we now wish to specify the PDF of N, the number of times the loop is traversed. Let T denote the time of realization of node *2*. Clearly, if the loop is traversed N times, then

$$T = X_1 + X_2 + \cdots + X_N$$

where the X_i's are identically distributed RVs. The total time T is a random sum of random variables and, therefore, has a compound probability distribution [see Feller (1968), p. 268 et. seq.]. Clearly,

$$Pr(T=t) = \sum_{n=1}^{\infty} Pr(T=t|N=n)Pr(N=n)$$

Show that, if $y(z)$ is the probability generating function of the RV Y, then $t(z) = n(x(z))$. Furthermore, show that if N is constant, then $t(z) = [x(z)]^n$.

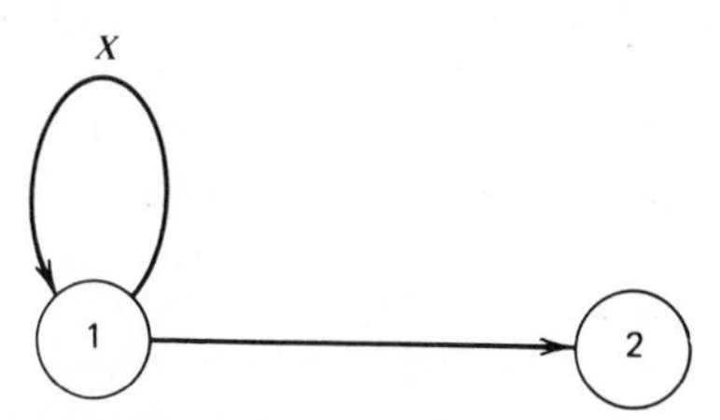

18. Consider the following simple GERT network in which the duration $t_a = 5$ and $t_b = 10$. Deduce the mean and variance of exiting to node *5* under the following conditions (N represents the number of times the cycle *1,2,3,4,1* is repeated):
(a) N is constant $= 5$
(b) N is geometrically distributed, $Pr(N=n) = \frac{1}{6}\left(\frac{5}{6}\right)^n$
(c) N is Poisson distributed with $\lambda = 4$ as follows: $Pr(N=n) = e^{-\lambda}\lambda^{n-1}/(n-1)!$ $n = 1, 2, 3, \ldots$.

Notice that in all three cases, the average value of N is equal to 5. [Hint: use the results of Problems 15–17.]

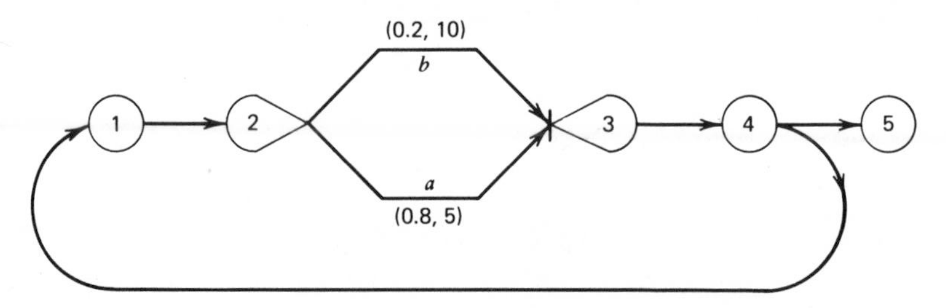

19. Consider the following research and development planning problem, which is a miniature version of a much larger problem. A research organization is concerned about bidding for three R&D projects. The company has two research teams of approximately equal capability. Each team can work independently on a research project in parallel with the other; the one that finishes first assumes responsibility for the third project.

    Grossly speaking, each project has five phases, which are (a) problem definition, (b) research activity, (c) evaluation of solution, (d) development of prototype, and (e) implementation of solution. Progress in any project is not necessarily in such linear sequence because of reversion to an earlier step due to dissatisfaction with the results obtained. In fact, is is possible that a whole project is abandoned if the "evaluation" phase results in a negative recommendation. On the other hand, if all five phases are successfully completed, the project is declared a success. Whether a success or a failure, a completed R&D project releases the research team to work on the remaining project if it is still available. If the third project has already been accepted by the other research team because of their earlier availability, the research team is idle.

    The GERT diagram of operation on the three research projects is depicted in the figure, together with the relevant parameters. (Note that all durations which are random variables are assumed to be uniformly distributed between two limits a and b.)

    You are asked to analyze this problem relevant to probabilities, durations, and costs. (For example, determine the probabilities of no success, one, two, or three successful completions of projects; the durations to landmark events such as project completion and project abandonment; and the costs incurred to reach such landmark events.)

    Attempt to answer as many questions as possible analytically before resorting to Monte Carlo sampling (simulation):

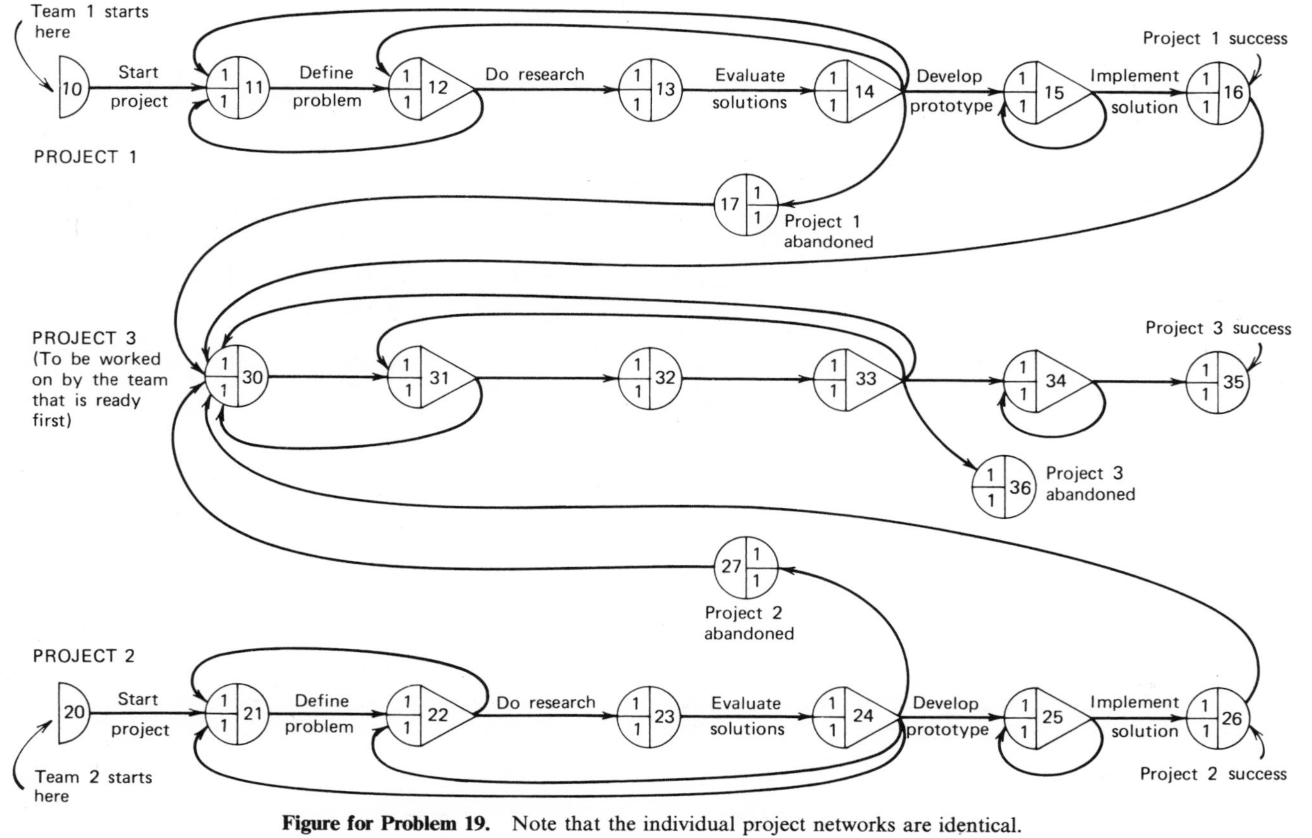

Figure for Problem 19. Note that the individual project networks are identical.

Activity	Transition Probabilities[†]	Durations Constant	Probabilistic a	b	Cost
10–11		0			10,000
11–12			20	50	450
12–13	0.80		60	120	2,300
12–11	0.20	0			0
13–14			8	20	500
14–11	0.10	3			0
14–12	0.10	0			0
14–15	0.60		60	100	7,600
14–17	0.20	0			1,000
15–15	0.40		60	100	1,600
15–16	0.60		75	130	3500
16–30		10			2,000
17–30		10			2,000
20–21		0			13,500
21–22			20	60	475
22–21	0.10	0			0
22–23	0.90		21	55	4,000
23–24			14	45	625
24–21	0.10	0			0
24–22	0.18	0			0
24–25	0.70		50	150	4,500
24–27	0.02	0			1,000
25–25	0.40		50	150	1,000
25–26	0.60		35	100	3,250
26–30		10			2,000
30–31			25	120	18,525
31–30	0.50	0			0
31–32	0.50		80	200	4,925
32–33			20	50	625
33–30	0.05	3			0
33–31	0.10	0			0
33–34	0.55		100	210	7,100
33–36	0.30	0			2,000
34–34	0.25		105	210	1,600
34–35	0.75		60	200	6,000

[†]Source: Moore and Taylor (1976). A no-entry designates probability 1.

REFERENCES

Arisawa and Elmaghraby, 1970.
Bellas and Samli, 1973.
Bird, Clayton, and Moore, 1973.
Crowston, 1970.
Crowston and Thompson, 1967.
Dean, 1966.
Drezner and Gatto, 1965.
——— and Pritsker, 1965.
——— and ———, 1966.
Eisner, 1962.
Elmaghraby, 1964.
———, 1966.
Epstein, 1948.
Feller, 1968.
Frosch, 1969.
Furniss, 1973.
Halpin and Happ, 1971.
Happ, 1964.
Howard, 1964.
———, 1971.
Kelly, 1976.
Lorens, 1966.
Mason, 1953.
Moder, Clark, and Gomez, 1971.
Moore and Clayton, 1976.
——— and Taylor, 1976.
Neumann, 1975.
Pritsker and Happ, 1966.
——— and Whitehouse, 1966.
——— and Kiviat, 1969.
——— and Sigal, 1974.
Raju, 1970.
———, 1971.
Rektorys, 1969.
Samli and Bellas, 1971.
Sneddon, 1972.
Thompson, 1968.
Warfield, 1971.
——— and Hill, 1976.
White, 1970.
Whitehouse, 1973.
——— and Pritsker, 1969.

BIBLIOGRAPHY

BOOKS ON ACTIVITY NETWORKS

Archibald and Villoria, 1967.
Barnetson, 1969.
Barr and Howard, 1961.
Battersby, 1967.
Battersby, 1970.
Benjamin, 1968.
Berge, 1958.
Eris and Baker, 1964.
Evarts, 1964.
Horowitz, 1967.
Kaufmann and Desbazeille, 1969.
Levin and Kirkpatrick, 1966.
Lockyer, 1970.
Lombaers, 1969.
Miller, 1963.
Moder and Phillips, 1964.
Morris, 1967.
O'Brien, 1971.
Roy, 1964.
Shaffer, Ritter, and Meyer, 1965.
Sicard, 1970.
Thornley, 1968.
Villemain and Le Borgne, 1966.
Wiest and Levy, 1977.
Woodgate, 1964.
———, 1970.

GENERAL REFERENCES

Agin, 1966.
Beightler, Mitten, and Nemhauser, 1961.
Bigelow, 1962.
Bonnesen and Fenchel, 1948.
Brand, Meyer, and Shaffer, 1964.
Busacker and Gowen, 1961.
Carruthers and Battersby, 1966.
Cramer, 1957.
Dantzig, Ford, and Fulkerson, 1956.
Dantzig and Wolfe, 1960.
Davies, 1973.
Davis, 1976.
Dijkstra, 1959.
Elmaghraby, 1966.
———, 1970.
Falk and Soland, 1969.
Feller, 1968.
———, 1971.
Fenves, 1967.
Ford and Fulkerson, 1962.
——— and ———, 1958.
Fry, 1963a.
———, 1963b.
Hadley, 1964.
Hald, 1952.
Hammersley and Handscomb, 1967.
Hartley and Hocking, 1963.
Helm, 1959.
Hoel, 1954.
IBM Applications Program (1975) Mathematic Programming System/360 Linear and Separable Programming—User's Manual. (H20-0476-0-3); Ch. 5.
Krishnamoorthy, 1968.
Kuhn and Tucker, 1951.
Lerda-Olberg, 1966.
Levy, Thompson, and Wiest, 1963.
McLaren and Buesnel, 1969.
Norton, 1960.
Pearce, 1968.

Polya, 1957.
Seshu and Reed, 1961.
Shrieder, 1966.
Staffurth and Walton, 1974.
Van Der Pol and Bremmer, 1955.
Vazsonyi, 1970.
Walton, 1964.
Whitehouse, 1973.
Wilks, 1962.

REFERENCES

Abernathy, W. J. and J. S. Demski (1973), Simplification of activities in a network scheduling context, *Management Sci.*, **19** (9), 1052–1062.

Agard, J. and G. Gamot (1966), Ordonnancement du grand entretien d'un avion avec egalisation des effectifs d'entretien, *Revue Francaise de Recherche Operationnelle*, (38), 39–54.

Agin, Norman (1966), Optimum seeking with branch-and-bound, *Management Sci.*, **13** (4), B176–B185.

Alpert, L. and D. S. Orkand (1962), A time-resource trade-off model for aiding management decisions, Operations Research Inc., Report No. 12, Silver Springs, Maryland.

Alsaker, E. T. (1962), The Basic Technique: Network Analysis, in J. W. Blood (ed.), *PERT, A New Management Planning and Control Technique*, American Management Association, pp. 37–60.

Antill, James M. and Ronald W. Woodhead (1965), *Critical Path Methods in Construction Practice*, Wiley, New York.

Archibald, R. D. and R. L. Villoria (1967), *Network-based Management Systems*, Wiley, New York.

Arisawa, S. and S. E. Elmaghraby (1970), On hyperbolic programming with a single constraint and upper-bounded variables. *Management Sci.*, **19** (1), 42–45.

Arthanari, T. S. and K. G. Ramamurthy (1970), A branch and bound algorithm for sequencing n jobs on m-parallel processors, *Opsearch*, **7** (3), 147–156.

Avots, I. (1962), The management side of PERT, *Calif. Management Rev.*, **4** (2), 16–27.

Balas, E. (1970), Project scheduling with resource constraints, Section 5 in E. M. L. Beale (ed.), *Applications of Math. Programming Techniques*, pp. 187–200.

Barnes, N. M. L. and J. S. Gillespie (1972), *Computer-based Cost Model for Project Management*, Book III, INTERNET 72, Third International Congress on Project Planning by Network Techniques, Stockholm, Sweden, pp. 37–58.

Barnetson, P. (1969), *Critical Path Planning*, Newnes-Butterworths, London.

Barr, J. and W. E. Howard (1961), *Polaris*. Harcourt, Brace, Jovanovich, New York.

Battersby, A. (1967), *Network Analysis*, Macmillan, New York.

——— (1970), *Network Analysis for Planning and Scheduling*, St. Martin's Press, New York, third edition.

Becker, A. M. (1969), DYNET—A dynamic network planning technique, *Datamation*, **15** (4), 113–114 and 119–122.

Beckwith, R. E. (1962), A cost control extension of the PERT system, *IRE Trans. Eng. Management*, 147–149.

Beightler, C. S., L. G. Mitten, and G. L. Nemhauser (1961), A short table of z-transforms and generating functions, *Oper. Res.*, **9** (4), 574–578.

Bellas, C. J. and A. C. Samli (1973), Improving new product planning with GERT simulation, *Calif. Management Rev.*, **15** (4), 14–21.

Benjamin, A. (1968), Critical path methods used by a government works department and its contractors, *Applications of Critical Path Techniques*, The English Universities Press Ltd., London, England.

Bennett, F. L. (1966), Some approaches to the critical path scheduling resource allocation problem, unpublished Ph.D. thesis, Cornell University, Ithaca, New York.

——— (1968), Critical path resource scheduling algorithm. *J. Const. Div., ASCE Proc.*, **94** (CO2), 161–180.

Bennington, G. E. and L. F. McGinnis (1973), A critique of project planning with constrained resources, in S. E. Elmaghraby, (ed.), *Symposium on Theory of Scheduling and Its Applications*, Springer-Verlag, New York, 1–28.

Berge, C. (1958), *Theorie des Graphes et ses Applications*, Dunod, Paris.

Berman, E. B. (1964), Resource allocation in a PERT network under continuous time–cost functions, *Management Sci.*, **10** (4), 734–745.

Berman, H. (1961), The critical path method for project planning and control, *The Constructor*, **43** (9), 24–27.

Bigelow, O. J. (1962), Bibliography on project planning and control by network analysis: 1959–1961, *Oper. Res.*, **10** (5), 728–731.

Bildson, R. A. and J. R. Gillespie (1962), Critical path planning-PERT integration, Letter to the Editor, *Oper. Res.*, **10** (6), 909–912.

Bird, M. M., E. R. Clayton, and L. J. Moore (1973), Sales negotiation cost planning for corporate level sales, *J. Market.*, **37** (2), 7–13.

Black, Owen J. (1965), An Algorithm for Resource Leveling in Project Networks, unpublished report, Department of Industrial Administration, Yale University, New Haven, Connecticut.

Blair, R. J. (1963), Critical path resources simulation and scheduling, *IEEE Trans. Eng. Management*, **EM-10**, 100–103.

Blanning, R. W. and A. G. Rao (1965), A note on "decomposition of project networks," *Management Sci.*, **12** (1), 145–148.

Boehm, G. A. W. (1962), Helping the executive to make up his mind, *Fortune*, **65** (4), 128–131.

Bonnesen, T. and W. Fenchel (1948), *Theory of Convex Sets*. Chelsea, New York.

Boverie, R. T. (1963), The practicalities of PERT, *IEEE Trans. Eng. Management*, **EM-10** (1), 3–5.

Brand, J. D., W. L. Meyer and L. R. Shaffer (1964), The resource scheduling problem in construction, *Civil Engineering Studies*, Report No. 5, Dept. of Civil Engineering, University of Illinois, Urbana.

Burgess, A. R. and J. B. Killebrew (1962), Variation in activity level on a cyclical arrow diagram, *J. Ind. Eng.*, **13** (2), 76–83.

Burt J. M., Jr. and M. B. Garman (1971). Conditional Monte Carlo: a simulation technique for stochastic network analysis, *Management Sci.*, **18** (3), 207–217.

——— and ——— (1971), Monte Carlo techniques for stochastic PERT network analysis, *INFOR*, **9** (3), 248–262.

———, D. P. Gaver, and M. Perlas (1970), Simple stochastic networks: some problems and procedures, *Nav. Res. Log. Quart.*, **17**, (4), 439–460.

Busacker, R. G. and P. J. Gowen (1961), A Procedure for determining a family of minimal-cost network flow patterns, Tech. Report No. 15, Operations Research Office, Johns Hopkins Univ., Baltimore, Maryland.

Butcher, William S (1967), Dynamic programming for project cost–time curves, *J. Const. Div., ASCE Proc.*, **93** (CO1), 59–73.

Calica, A. B. (1965), Project network algorithms for use in shop loading and sequencing, *IBM Systems J.*, **4** (3), 225–240.

Carré, D. (1964), Ordonnancement des travaux d'entretien dans une raffinerie, in B. Roy (ed.), *Les Problemes d'Ordonnancement, Applications et Methodes*, Dunod, Paris, pp. 67–84.

Carruthers, J. A. (1968), Probabilistic times, Chapter 8 in Gail Thronley (ed.), *Critical Path Analysis in Practice,* Tavistock, London.

——— and Albert Battersby (1966), Advances in critical path methods. *Oper. Res. Quart.* **17** (4), 359–380.

Chapman, C. B. (1970), The optimal allocation of resources to a variable timetable. *Oper. Res. Quart.*, **21** (1), 81–90.

Charnes, A. and W. W. Cooper (1962), A network interpretation and a directed sub-dual algorithm for critical path scheduling, *J. Ind. Eng.*, **13** (4), 213–219.

———, ———, and G. L. Thompson (1964), Critical path analysis via chance constrained and stochastic programming, *Oper. Res.*, **12** (3), 460–470.

Chaudhuri, A. R. (1968), On the problem of critical path method, *Opsearch*, **5** (4), 207–213.

Clark, Charles E. (1961), The optimum allocation of resources among the activities of a network. *J. Ind. Eng.*, **12** (1), 11–17.

——— (1961), The greatest of a finite set of random variables, *Oper. Res.*, **9** (2), 146–162.

——— (1962), The PERT model for the distribution of an activity time. *Oper. Res.*, **10** (3), 405–406.

Clark, Wallace (1954), *The Gantt Chart*, Pitman, New York.

Clingen, C. T. (1964), A modification of Fulkerson's PERT algorithm, Letter to the Editor, *Oper. Res.*, **12** (4), 629–632.

Codier, E. O. (1962), Fundamental principles and applications of PERT, in J. W. Blood (ed.), *PERT a New Management Planning and Control Technique*, American Management Association, pp. 61–71.

Coon, Helen (1965), Note on William A. Donaldson's paper, the estimation of the mean and variance of a PERT activity time, *Oper. Res.*, **13** (3), 386–387.

Cooper, Dale F. (1976), Heuristics for scheduling resource-constrained projects: An experimental Investigation, *Management Sci.*, **22** (11), 1186–1194.

Cramer, Harold (1957), *Mathematical Methods of Statistics*, Princeton U. P., Princeton, New Jersey.

Croft, F. Max (1970), Putting a price tag on PERT activities, *Ind. Eng.*, 19–21.

Crowston, W. (1970), Decision CPM: Network reduction and solution, *Oper. Res. Quart.*, **21** (4), 435–452.

——— (1971), Models for project management, *Sloan Management Rev.*, **12** (3), 25–42.

——— and G. L. Thompson (1967), Decision CPM: a method for simultaneous planning, scheduling, and control of projects, *Oper. Res.*, **15** (3), 407–426.

——— and M. H. Wagner (1970), A comparison of tree search schemes for decision networks, Research Report, Alfred P. Sloan School of Management, Massachusetts Inst. of Tech., Cambridge, Mass.

Dantzig, G. B., L. R. Ford, Jr. and D. R. Fulkerson (1956), A primal–dual algorithm for linear programs, in H. W. Kuhn and A. W. Tucker (eds.), *Linear Inequalities and Related Systems, Annals of Mathematics Study No. 38*, Princeton U. P., Princeton, New Jersey, pp. 171–181.

——— and P. Wolfe (1960), Decomposition principle for linear programs. *Oper. Res.*, **8** (1), 101–111.

Davies, R. B. (1973), Numerical inversion of a characteristic function. *Biometrika*, **60** (2), 415–417.

Davis, Edward W. (1966), Resource allocation in project network models—a survey, *J. Ind. Eng.*, **17** (4), 177–188.

——— (1968), An exact algorithm for the multiple constrained-resource project scheduling problem, Ph.D. dissertation, Department of Administrative Science, Yale Univ., New Haven, Connecticut.

——— (1972), How to describe a project network, paper presented to the ORSA National Meeting, New Orleans.

——— (1973), Project scheduling under resource constraints—historical review and categorization of procedures, *IE Trans.*, **5** (4), 297–313.

——— (1974), Network resource allocation, *IE Trans.*, **6** (4), 22–32.

——— (1976), Project management: techniques, applications, and managerial issues, *Publication No. 3 of Production Planning and Control Division*, AIIE.

——— and G. E. Heidorn (1971), An algorithm for optimal project scheduling under multiple resource constraints. *Management Sci.*, **17** (12), B803–B816.

——— and J. H. Patterson (1975), A comparison of heuristic and optimum solutions in resource-constrained project scheduling, *Management Sci.*, **21** (8), 944–955.

Davis, G. B. (1963), Network techniques and accounting with an illustration, *N.A.A. Bulletin.*

Dean, Burton V. (1966), Stochastic networks and research planning, *Proceedings of the Fourth International Conference on Operations Research*, Wiley, New York, pp. 215–234.

De Witte, L. (1964), Manpower leveling of PERT networks, *Data Process. Sci./Eng.*, **2** (2), 29–38.

Digman, Lester A. (1967), PERT/LOB: life cycle technique, *J. Ind. Eng.*, **18** (2), 154–158.

Dijkstra, E. W. (1959), A note on two problems in connection with graphs. *Numerische Mathematik*, **1** (5), 269–271.

Dike, S. H. (1964), Project scheduling with resource constraints, *IEEE Trans. Eng. Management*, **EM-11** (4), 155–157.

Dimsdale, B. (1963), Computer construction of minimal project networks, *IBM Systems J.*, **2**, 24–36.

DOD and NASA Guide (1962), PERT-Cost System Design. Department of Defense and National Aeronautics and Space Administration.

Doersch, R. H. and J. H. Patterson (1977), Scheduling a project to maximize its present value: A zero-one programming approach, *Management Sci.*, **23** (8), 882–889.

Dogrusoz, Halim (1961), Development programming, *Oper. Res.*, **9** (5), 753–755.

Donaldson, William A. (1965), The estimation of the mean and variance of a PERT activity time, *Oper. Res.*, **13** (3), 382–385.

Dooley, A. R. (1964), Interpretations of PERT, *Harv. Bus. Rev.*, **42** (2), 160–172.

Drezner, Steven M. and O. T. Gatto (1965), Computer-assisted countdown: preliminary report on a test of early capability, RAND Memo., RM-4565-NASA, The RAND Corp., Santa Monica, California.

——— and A. A. B. Pritsker (1965), Use of generalized activity networks in scheduling, RAND Paper P-3040, the RAND Corp., Santa Monica, California.

——— and ——— (1966), Network analysis of countdown, RAND Memo., RM-4976-NASA, The RAND Corp., Santa Monica, California.

Eisner, Howard (1962), A generalized network approach to the planning and scheduling of a research program, *Oper Res.*, **10** (1), 115–125.

Elmaghraby, S. E. (1964), An algebra for the analysis of generalized activity networks, *Management Sci.*, **10** (3), 494–514.

——— (1966), On generalized activity networks. *J. Ind. Eng.*, **17** (11), 621–631.

——— (1966), *The Design of Production Systems*. Reinhold, New York.

——— (1967), On the expected duration of PERT type network. *Management Sci.*, **13** (5), 299–306.

——— (1968), The determination of optimal activity duration in project scheduling. *J. Ind. Eng.*, **19**, (1), 48–51.

——— (1968), The machine sequencing problem: review and extensions. *Nav. Res. Log. Quart.*, **15** (2), 205–232.

——— (1969), The sequencing of *n* jobs on *m* parallel processors with extensions to the scarce resource problem of activity networks, *Proceedings Inaugural Conference*, The Scientific Computation Center, Cairo, Egypt, Dec. 17–20.

——— (1970), *Some Network Models in Management Science*, Springer-Verlag, New York. Lecture Notes in Operations Research and Mathematical Systems, Vol. 29.

——— and A. N. Elshafei (1976), Branch and bound revisited: a survey of basic concepts and their applications in scheduling, Chapter 8 in W. H. Marlow (Ed.), *Modern Trends in Logistics Research*, the MIT Press, Cambridge, Mass.

Emmons, Hamilton (1968), One machine sequencing to minimize total tardiness, Report No. 52, Department of Operations Research, Cornell Univ., Ithaca, New York.

Epstein, B. (1948), Uses of the mellin transform in statistics, *Ann. Math. Stat.*, **19**, 370–379.

Eris, Rene L. and Bruce N. Baker (1964), *Introduction to PERT-CPM*, Irwin.

Evarts, Harry F. (1964), Introduction to PERT, Allyn and Bacon, Boston, Mass.

Falk, J. E. and J. L. Horowitz (1972), Critical path problems with concave cost–time curves, *Management Sci.*, **19** (4), 446–455.

——— and R. M. Soland (1969), An algorithm for separable nonconvex programming problems, *Management Sci.*, **15** (9), 550–569.

Feiler, A. M. (1974), PERA (CV) Project Risk Management, UCLA-ENG-7445, School of Engineering and Applied Sciences, UCLA, California.

——— (1976), Project management through simulation, Report No. UCLA-ENGR-76-119, University of California at Los Angeles.

Feller, W. (1968), *An Introduction to Probability Theory and Its Applications*, Vol. 1, 3rd ed., Wiley, New York.

——— (1971), *An Introduction to Probability Theory and Its Applications*, Vol. 2, 2nd ed., Wiley, New York.

Fendley, L. G. (1968), Toward the development of a complete multiproject scheduling system, *J. Ind. Eng.*, **19** (10), 505–515.

Fenves, Steven J. (1967), *Computer Methods in Civil Engineering*, Prentice-Hall, Englewood Cliffs, New Jersey.

Fisher, Carolyn and George L. Nemhauser (1967), Multicycle project planning, *J. Ind. Eng.*, **18** (4), 113–114, 119–122.

Fisher, Marshall L. (1973), Optimal solution of scheduling problems using Lagrange multipliers, Part I, *Oper. Res.*, **21** (5), 1114–1127.

——— (1973), Optimal solution of scheduling problems using Lagrange multipliers, Part II, in S. E. Elmaghraby (ed), *Proceedings of Symposium on Scheduling Theory and its Application*, Springer-Verlag, New York, pp. 294–318.

Ford, L. R., Jr. and D. R. Fulkerson (1962), *Flows in Networks*, Princeton U. P., Princeton, New Jersey.

——— and ——— (1958), A suggested algorithm for maximal multicommodity network flows, *Management Sci.*, **5** (1), 97–101.

Francis, H. G. (1962), Practical Advice for the Use of PERT, in J. W. Blood (Ed.), *PERT a New Management Planning and Control Technique*, American Management Association, pp. 125–131.

Frank, H. (1969), Shortest paths in probabilistic graphs, *Oper. Res.*, **17** (3), 583–599.

Freeman, D. R. and J. V. Tucker (1967), The line balancing problem, *J. Ind. Eng.*, **18** (6), 361–364.

Freeman, Raoul J. (1960a), A generalized PERT. *Oper. Res.*, **8** (2), 281.

——— (1960b), A generalized network approach to project activity sequencing. *IRE Trans. Eng. Management*, **EM-7** (3), 103–107.

Frosch, R. A. (1969), A new look at systems engineering, *IEEE Spectrum*, **6** (9), 24–28.

Fry, B. L. (1963a), Network-type management control systems bibliography, RAND Memo —RM 3074-PR, the RAND Corp., Santa Monica, California.

——— (1963b), Selected references on PERT and related techniques, *IEEE Trans. Eng. Management*, **EM-10** (3), 150–151.

Fulkerson, D. R. (1961), A network flow computation for project cost curve, *Management Sci.*, **7** (2), 167–178.

——— (1962), Expected critical path lengths in PERT networks, *Oper. Res.*, **10** (6), 808–817.

——— (1964), Scheduling in project networks, RAND Memo., RM-4137-PR, RAND Corp. Santa Monica, California.

Galbreath, Robert V. (1956), Computer program for leveling resource usage. *J. Const. Div., ASCE Proc.*, **91** (CO14319), 107–124.

Garman, M. B. (1972), More on conditional sampling in the simulation of stochastic networks, *Management Sci.*, **19** (1), 90–95.

Geller, Dennis P. and Frank Harary (1968), Arrow diagrams and line digraphs, *SIAM J.*, **16** (6), 1141–1145.

Gessford, P. M. (1966), An investigation of the utility of critical path time and cost analysis as an aid to budgeting and administrative control by construction firms, Unpublished Ph.D. thesis, Univ. of California at Los Angeles, California.

Ghare, P. M. (1965), Optimal resource allocation in activity networks, Ph.D. thesis, Oklahoma State Univ., Stillwater, Oklahoma.

——— (1965), Optimal resource allocation in activity networks, *Proc. ORSA 28th Nat. Mtg.*, Vol. 13, Supplement No. 2, Event No. D-2.8, Fall.

Goldberg, C. R. (1964), An algorithm for the sequential solution of schedule networks. *Oper. Res.*, **12** (3), 499–502.

Gonguet, L. (1969), Comparison of three heuristic procedures for allocating resources and producing schedule, *Project Planning by Network Analysis*, North-Holland, Amsterdam.

Gorenstein, S. (1972), An algorithm for project (job) sequencing with resource constraints, *Oper. Res.*, **20** (4), 835–850.

Goyal, S. K. (1975), A note on a simple CPM time–cost tradeoff algorithm, *Management Sci.*, **21** (6), 718–722.

Graham, Pearson (1965), Profit probability analysis of research and development expenditures, *J. Ind. Eng.*, **16** (3), 186–191.

Grubbs, F. E. (1962), Attempts to validate PERT statistics or "picking on PERT," Letter to the Editor, *Oper. Res.*, **10** (6), 912–915.

Gutjahr, A. L. (1963), An algorithm for the assembly line balancing, Master's thesis, The Johns Hopkins University, Baltimore, Maryland.

——— and G. L. Nemhauser (1964), An algorithm for the line balancing problem, *Management Sci.*, **11** (2), 308–315.

Gutsch, Roland W. (1969), Project management by means of a network systems of the third generation, *Project Planning by Network Analysis*, Lombaers (ed.), North-Holland, Amsterdam, pp. 369–380.

Hadley, G. (1964), *Nonlinear and Dynamic Programming*, Addison-Wesley, Reading, Massachusetts, Section 8.8, pp. 263–267.

Hald, A. (1952), *Statistical Theory with Engineering Applications*, Wiley, New York.

Halpin, D. W. and W. W. Happ (1971), Digital simulation of equipment allocation for Corps of Engineers construction planning, *Tech. Paper, CERL*, Champaign, Ill.

Hammersley, J. M. and D. C. Handscomb (1967), Monte Carlo Methods, Methuen, London, England.

——— and J. G. Mauldon (1956), General Principles of antithetic variates, *Proc. Cambridge Phil. Soc.*, **52**, 476–481.

Happ, W. W. (1964), Applications of flowgraph technique to the solution of reliability problems, in M. F. Goldberg and J. Vaccaro (eds.), *Physics of Failure in Electronics*, U.S. Dept. of Commerce, Office of Technical Services, AD-434/329; pp. 375–423.

Harary, Frank and R. Z. Norman (1961), Some properties of line digraphs, *Rend. Circ. Mat. Palermo*, **9**, 161–168.

Hartley, H. O. and R. R. Hocking (1963), Convex programming by tangential approximation, *Management Sci.*, **9** (4), 600–612.

——— and A. W. Wortham (1966), A statistical theory for PERT critical path analysis, *Management Sci.*, **12** (10), B469–B481.

Hastings, N. A. J. (1972), On resource allocation in project networks, *Oper. Res. Quart.*, **23** (2), 217–221.

Hayes, R. D., C. A. Komar, and Jack Byrd, Jr. (1973), The effectiveness of three heuristic rules for job sequencing in a single production facility, *Management Sci.*, **19** (5), 575–580.

Healy, Thomas (1961), Activity subdivision and PERT probability statements, *Oper. Res.*, **9** (3), 341–348.

Hegelson, W. B. and D. P. Birnie (1961), Assembly line balancing using the ranked positional weight technique, *J. Ind. Eng.*, **13** (6), 394–398.

Helm, H. A. (1959), The Z-transformation, *Bell Syst. Tech. J.*, **38** (1), 177–196.

Hill, Larry S. (1966), Some cost accounting problems in PERT/cost, *J. Ind. Eng.*, **17** (2), 87–91.

Hoel, Paul G. (1954), *Introduction to Mathematical Statistics*, 3rd ed., Wiley, New York.

Hooper, P. C. (1965), Resource allocation and levelling, *Proc. Third CPA Symposium, Oper. Res. Soc.*, London.

Horowitz, J. (1967), *Critical Path Scheduling*, Ronald Press, New York.

Howard, Ronald A. (1964), System analysis of semi-Markov processes, *IEEE Trans., Professional Group Mil. Electron.*, **MIL-8** (2), 114–124.

——— (1971), Dynamic Probabilistic Systems, Vols. 1 and 2, Wiley, New York.

Howard, S. A. (1971), The application of PERT to a multidiscipline project, *Telecommun. J.* (Austr.), pp. 131–136.

Hsu, J. P. (1969), Project resource balancing by project progress curve analysis, Unpublished M.Sc. thesis, Univ. of Houston, Houston, Texas.

IBM Applications Program (1975), *Mathematical Programming System/360 Linear and Separable Programming—User's Manual*, H20-0476-0-3 Chapter 5.

Jenett, Eric (1969), Experience with and evaluation of critical path methods, *Chem. Eng.*, **76** (3), 96–106.

Jewell, W. (1965), Risk taking in critical path analysis. *Management Sci.*, **11** (3), 438–443.

——— (1965), Divisible activities in critical path analysis, *Oper Res.*, **13** (5), 747–760.

——— (1971), Divisible and moveable activities in critical path analysis, *Oper. Res.*, **19** (2), 323–348.

Johnson, T. J. R. (1967), An algorithm for the resource-constrained project scheduling problem, Unpublished Ph.D. thesis, Sloan School of Management, MIT, Cambridge, Massachusetts.

Joyce, P. J. (1974), Management controls for constructing new facilities, *IFAC Workshop on Corporate Control Systems*, Enschede, The Netherlands, April 2–4.

Kapper, F. B. (1966), A feasible solution to the resource allocation problem of network scheduling, Unpublished Ph.D. thesis, St. Louis Univ., Missouri.

Kapur, K. C. (1973), An algorithm for project cost-duration analysis problem with quadratic and convex cost functions, *IE Trans.*, **5** (4), 314–322.

Karush, W. (1964), On scheduling a network of activities under resource constraints over time, SP-2654, Systems Development Corporation, Santa Monica, California.

——— (1967), On scheduling a network of activities under resource restraints over time, USGRDR, AD-648749, Vol. 67, p. 101, May 25.

Kaufmann, A. and G. Desbazeille (1969), *La Methode du Chemin Critique*, 2nd ed., Dunod, Paris.

Kedia, S. K. and Gui Ponce-Campos (1976), Looping relationships in precedence networks, Paper presented at the ORSA/TIMS Special Interest Conference on the Theory and Applications of Scheduling, Orlando, Florida, February 4–6.

Kelley, J. E. Jr. (1961), Critical path planning and scheduling, mathematical basis, *Oper. Res.*, **9** (3), 296–320.

——— (1963), The critical path method: resources planning and scheduling, Chapter 21 in Muth and Thompson (eds.), *Ind. Scheduling*, Prentice-Hall, Englewood Cliffs, New Jersey.

——— and M. R. Walker (1959), Critical path planning and scheduling, *Proc. Eastern Joint Comp. Conf.*, **16**, 160–172.

Kelly, John C. (1976), Repetition networks and their application to computer programs, Fed. Comp. Performance Evaluation & Simulation Center, Washington, D.C.

Kilbridge, M. D. and L. Wester (1962), A review of analytical system of line balancing, *Oper. Res.*, **10**, (5), 626–638.

King, W. R. (1971), Network simulation using historical estimating behavior, *IE Trans.*, **3** (2), 150–155.

——— and Paul A. Lukas (1973), An experimental analysis of network planning, *Management. Sci.*, **19** (12), 1423–1433.

——— and T. Wilson (1967), Subjective time estimates in critical path planning; A preliminary analysis, *Management Sci.*, **13** (5), 307–320.

———, Donald M. Wittevrongel, and Karl D. Hezel (1967), On the analysis of critical path time estimating behavior, *Management Sci.*, **14** (1), 79–84.

Klass, Ph. J. (1960), PERT/PEP management tool use grows, *Aviation Week*, November, 85–91.

Klein, M. M. (1967), Scheduling project networks, *Comm. of the ACM*, **10** (4), April, 225–231.

Klingel, A. R., Jr. (1966), Bias in PERT project completion time calculations for a real network, *Management Sci.*, **13** (4), 194–201.

Kotiah, T. C. T. and N. D. Wallace (1973), Another look at the PERT assumptions, *Management Sci.*, **20** (1), 44–49.

Krishnamoorthy, M. (1968), Critical path method: A review, Tech. Report No. 1968-4, Department of Industrial Engineering, Univ. of Michigan, Ann Arbor, Michigan.

Kuhn, Harold W. and A. W Tucker (1951), Non-linear programming, *Proceedings of Second Berkeley Symposium on Mathematical Statistics and Probability*, California U. P., Berkeley, California.

Lamberson, L. R. and R. R. Hocking (1970), Optimum time compression in project scheduling, *Management Sci.*, **16** (10), B597–B606.

Lambourn, S. (1963), RAMPS—a new tool in planning and control, *Computer J.* **5** (4), 300–304.

Lave, H. J. (1968), Efficient methods for the allocation of resources in project networks, *Unternehmensforschung*, **12**, 133–143.

Lawler, E. L. and J. M. Moore (1969), A functional equation and its applications to resource allocation and sequencing problems, *Management Sci.*, **16** (1), 77–84.

Lenstra, Jan Karel (1976), Sequencing by enumerative methods, Unpublished Ph.D. dissertation, University of Amsterdam, Holland.

Lerda-Olberg, Sergio (1966), Bibliography on network-based project planning and control techniques: 1962–1965, *Oper. Res.*, **14** (5), 925–931.

Levin, R. I. and C. A. Kirkpatrick (1966), *Planning and Control with PERT/CPM*, McGraw-Hill, New York.

Levitt, Harry P. (1968), Computerized line of balance technique, *J. Ind. Eng.*, **14** (2), 61–67.

Levy, F. K., G. L. Thompson, and J. D. Wiest (1962), Multi-ship, multi-shop, workload smoothing program, *Nav. Res. Log. Quart.*, **9** (1), 37–44.

———, ———, and —— (1963), Mathematical basis of the critical path method, Chapter 22 in J. F. Muth and G. L. Thompson (eds.), *Industrial Scheduling*, Prentice-Hall, Englewood Cliffs, New Jersey.

———, ———, and ——— (1963), The ABCs of the critical path method, *Harvard Bus. Rev.*, **41** (5), September–October, 98–108.

Lindsey, J. H., II (1972), An estimate of expected critical-path length in PERT networks, *Oper. Res.*, **20** (4), 800–812.

Litsois, Socrates (1965), A resource allocation problem, *Oper. Res.*, **13** (6), 960–988.

Lockyer, K. G. (1970), *An Introduction to Critical Path Analysis*, Pitman, 3rd edition, London, England.

Lofts, N. R. (1974), Multiple allocation of resources in a network—an optimal scheduling algorithm, *INFOR*, **12** (1), 25–38.

Lombaers, H. J. M. (ed.) (1969), *Project Planning by Network Analysis*, North-Holland, Amsterdam, The Netherlands.

Lorens, C. S. (1966), Flowgraphs applied to continuous generating functions, *Proceedings of the Fourth International Conference on OR*, Wiley, New York, 205–214.

Lukaszewicz, Josef (1965), On the estimation of errors introduced by standard assumptions concerning the distribution of activity duration in PERT calculations, *Oper. Res.*, **13** (2), 326–327.

Lyden, Fremont and Ernest G. Miller (1967), *Planning, Programming, Budgeting: A Systems Approach to Management*, Markham, Chicago.

McBride, W. and C. McClelland (1967), PERT and the beta distribution, *IEEE Trans. Eng. Management*, **EM-14** (4), 166–169.

McClellan, H. S., Jr. (1969), Bounds for use in stochastic network analysis, Report No. GSA/SM/69-13, Air Force Institute of Technology, Wright Patterson AFB, Ohio.

McCoy, Paul F. (1972), Project scheduling with limited resources by extreme point enumeration, Paper presented at the joint ORSA–TIMS meeting, Atlantic City, New Jersey.

McGee, A. A. and M. D. Markarian (1962), Optimum allocation of research/engineering manpower within a multi-project organizational structure, *IRE Trans. Eng. Management*, **EM9**, 104–108.

McGinnis, Leon F. (1973), A schedule modifying algorithm for project planning with resource constraints, Unpublished Masters thesis, North Carolina State University, Raleigh, North Carolina.

McLaren, K. G. and E. L. Buesnel (1969), *Network Analysis in Project Management*, Cassell Management Studies, Cassell & Company Ltd., London, England.

McNeil, J. F. (1964), Program cost control system, *N.A.A. Bull.*, **20**, January, 11–20.

MacCrimmon, K. R. and C. A. Ryavec (1964), An analytical study of PERT assumptions, *Oper. Res.*, **12** (1), 16–37.

Madansky, A. (1960), Inequalities for stochastic linear programming problems, *Management Sci.*, **6** (2), 197–204.

Malcolm, D. G. (1963), Reliability maturity index (RMI)—an extension of PERT into reliability management, *J. Ind. Eng.*, **14** (1), 3–12.

———, J. H. Roseboom, C. E. Clark, and W. Fazar (1959), Applications of a technique for research and development program evaluation, *Oper. Res.* **7** (5), 646–669.

Mansoor, E. M. (1964), Assembly line balancing—an improvement on the ranked positional weight technique, *J. Ind. Eng.*, **15** (2), 73–77.

Marimont, R. B. (1959), A new method of checking the consistency of precedence matrices, *J. ACM*, **6** (2), 164–171.

Martin, J. J. (1965), Distribution of the time through a directed acyclic network, *Oper. Res.*, **13** (1), 46–66.

Martino, R. L. (1963), Applying critical path method to system design and installation, *Control Eng.*, **10** (2), 93–98.

——— (1965), *Project management and control, Vol. II: applied operational planning*, Amer. Management Association, New York.

——— (1965), *Project Management and Control, Vol.3, Resource Allocation and Scheduling*, American Management Association, New York.

Mason, S. J. (1953), Feedback theory: some properties of signal flow graphs, *Proc. Inst. Radio Eng.*, **41** (9), 1144–1156.

Mason, Thomas A. and Colin L. Moodie (1971), A branch and bound algorithm for minimizing cost in project scheduling, *Management Sci.*, **18** (4), Part 1, B158–B173.

Mauchly, J. W. (1962), Critical path scheduling, *Chem. Eng.*, **69** (8), 139–154.

Meyer, Karl Heimy F. (1972), *Interdependent Activities in Network Planning*, Book III, pp. 256–264, INTERNET 72 Third International Congress of Project Planning by Network Techniques, Stockholm.

Meyer, W. L. and L. R. Shaffer (1963), Extensions of the critical path method through the application of integer programming, Department of Civil Engineering, Univ. of Illinois, Chicago, Ill.

Miller, Robert W. (1962), How to plan and control with PERT, *Harv. Bus. Rev.*, **40** (2), 93–104.

——— (1963), *Schedule Cost and Profit Control with PERT*, McGraw-Hill, New York.

Minieka, Edward (1975), The double sweep algorithm for finding all *k*th shortest paths. Department of Quantitative Methods, Univ. of Illinois, Chicago, Ill.

Minty, G. J. (1957), A comment on the shortest route problem, *Oper. Res.*, **5** (5), 724.

Mitten, L. G. and A. R. Warburton (1973), Implicit enumeration procedures, Research Memorandum, University of British Columbia, Vancouver, B.C., Canada.

Mize, J. H. (1964), A heuristic scheduling model for multi-project organizations, Unpublished Ph.D. thesis, Purdue Univ., Lafayette, Indiana.

Moder, J. J. (1975), Private communication.

——— and C. R. Phillips (1964), *Project Management with CPM and PERT*, Reinhold, New York.

———, R. A. Clark, and R. S. Gomez (1971), Application of a GERT simulator to a repetitive hardward development type project, *IE Trans.*, **3** (4), 271–280.

——— and E. G. Rogers (1968), Judgment estimates of the moments of PERT type distributions, *Management Sci.*, **15** (2), B76-B83.

Moodie, C. L. and D. E. Mandeville (1966), Project resource balancing by assembly line balancing techniques, *J. Ind. Eng.*, **17** (7), 377–383.

Moore, J. Michael (1968), An *n*-job, one machine sequencing algorithm for minimizing the number of late jobs, *Management Sci.*, **15** (1), 102–109.

Moore, L. J. and E. R. Clayton (1976), *GERT Modeling and Simulation: Fundamentals and Applications*, Petrocelli-Charter, New York.

——— and B. W. Taylor, III (1976), Multiteam, multiproject research and development planning with GERT, Department of Business Administration, Virginia Polytechnic Institute and State University, Blacksburg, Va.

Morris, L. N. (1967), *Critical Path; Construction and Analysis*, Pergamon, London.

Morris, R. G. and R. G. Denison (1967), Simulation of critical path networks using historical time estimate data, Student thesis, Air Force Institute of Technology, Wright Patterson AFB, Ohio.

Moshman, J., J. Johnson, and M. Larsen (1963), RAMPS—A technique for resource allocation and multi-project scheduling, *Proc. Spring Joint Comp. Conf.*, Spartan Books, Baltimore, Maryland, pp. 17–28.

Murray, J. E. (1963), Considerations of PERT assumptions, *IEEE Trans. Eng. Management*, **EM-10** (3), September, 94–99.

Muth, J. F. and G. L. Thompson (1963), *Industrial Scheduling*, Prentice-Hall, Englewood Cliffs, N. J.

Nepomiastchy, Pierre (1974), Un probleme d'ordonnancement a rescources variables resolu par une méthode non combinatoire, Working Paper 74-21, The European Institute for Advanced Studies in Management, Brussels, Belgium.

Neumann, Klaus (1975), Structure, evaluation, and control of decision activity networks, Report ORC 75-11, The Operations Research Center, Univ. of California at Berkeley, California.

Nevill, G. and D. Falconer (1962), Network models for planning & scheduling can control small projects as well as large ones, *Int. Sci. Technol.*, **10**, October, 43–49.

Noettl, J. N. and P. Brumbaugh (1967), Information concepts in network planning, *J. Ind. Eng.*, **18** (7), 428–435.

Norden, P. V. (1963), Resource usage and network planning, in E. V. Dean (Ed.), *Operations Research in Research and Development*, Wiley, New York, 149–169.

Norton, P. V. (1960), On the anatomy of development projects, *IRE Trans. on Eng. Mgt.*, **EM-7** (1), March, 34–42.

Novick, David (1965), Program Budgetting, U.S. Government Printing Office, 65–18965, Washington, D.C.

O'Brien, James J. (1971), *CPM in Construction Management*, McGraw-Hill, New York.

Page, E. S. (1965), On Monte Carlo methods in congestion problems: II, simulation of queuing systems, *Oper. Res.*, **13** (2), 300–305.

Paige, Hilliard W. (1963), How PERT-cost helps the general manager, *Harvard Bus. Rev.*, **41** (6), 87–95.

Parikh, Shailendra C. and William S. Jewell (1965), Decomposition of project networks, *Management Sci.*, **11** (3), 444–459.

Parzen, Emanuel (1960), *Modern Probability Theory and Its Applications*, Wiley, New York.

Pascoe, T. L. (1965), An experimental comparison of heuristic methods for allocating resources, Ph.D. thesis, Cambridge Univ., Engineering Department, Cambridge, England.

——— (1966), Allocation of resources CPM, *Revue Française de Recherche Operationnelle*, (38), 31–38.

Patterson, James H. (1973), Alternate methods of project scheduling with limited resources, *Nav. Res. Log. Quart.*, **20** (4), 767–784.

Patterson, J. H. (1976), Project Scheduling: The effects of problem structure on heuristic performance, *Nav. Res. Log. Quart.*, **23** (1), 95–124.

——— and W. D. Huber (1974), A horizon-varying, zero–one approach to project scheduling, *Management Sci.*, **20** (6), 990–998.

——— and G. W. Roth (1976), Scheduling a project under multiple resource constraints: A zero-one programming approach, *IEEE Trans.*, **8** (4), 449–455.

——— and J. A. Werne (1972), A heuristic model for multi-project scheduling with limited resources, Paper presented to 41st National Meeting, ORSA, New Orleans.

Patton, G. T. (1968), Optimal scheduling of resource constrained projects, Ph.D. Dissertation, Stanford Univ., California.

Pearce, W. G. (1968), Canadian national railways' experience with critical path methods, *Applications of Critical Path Techniques*, The English Universities Press Ltd., London.

Petrovic, Radivoj (1968), Optimization of resource allocation in project planning, *Oper. Res.*, **16** (3), 559–568.

Pocock, J. W. (1962), PERT as an analytical aid for program planning—its payoffs and problems, *Oper. Res.*, **10** (6), 893–903.

Polya, Gyorgy (1957), *How to Solve It*, Doubleday-Anchor Books, New York, p. 113.

Pospelov, G. S. and V. A. Barishpolets (1966), A stochastic PERT systems, I, *Eng. Cybern.* **6**, 14–24.

Prager, William (1963), A structural method of computing project cost polygons, *Management Sci.*, **9** (3), 394–404.

Pritsker, A. A. B. and W. W. Happ (1966), GERT: graphical evaluation and review technique —Part I, fundamentals, *J. Ind. Eng.*, **17** (5), 267–274.

——— and Philip J. Kiviat (1969), *Simulation with GASP II*, Prentice-Hall, Inc., Englewood Cliffs, N.J.

——— and C. E. Sigal (1974), The GERT III-Z User's Manual, Pritsker and Assoc., Inc., 1201 Wiley Drive, West Lafayette, Indiana.

———, L. J. Watters, and P. M. Wolfe (1969), Multiproject scheduling with limited resources: a zero–one programming approach, *Management Sci.*, **16**, (1),93–108.

———and G. E. Whitehouse (1966), GERT: graphical evaluation and review technique, Part II, probabilistic and industrial engineering applications, *J. Ind. Eng.*, **17** (6), 293–301.

Raju, G. V. S. (1971), Sensitivity analysis of GERT networks, *IEEE Trans.*, **3** (2), 133–141.

Rao, C. Radhakrishna (1973), *Linear Statistical Inference and Its Applications*, second edition, Wiley, New York.

Rech, P. and L. G. Barton (1970), A non-convex transportation algorithm, in E. M. L. Beale, (ed.), *Applications of Mathematical Programming Techniques*, Amer. Elsevier Co., New York, 250–260.

Rektorys, Karel (ed.) (1969), *Survey of Applicable Mathematics*, MIT Press, Cambridge, Massachusetts, pp. 1125–1135.

Richards, John L. (1973), An integrated approach to construction management, Tech. MS. P-19, Construction Engineering Research Laboratory, Champaign, Ill.

Richards, Paul I. (1967), Precedence constraints and arrow diagrams, *SIAM Review*, **9** (3), 548–553.

Richardson, G. L., C. W. Butler, and P. B. Coup (1972), LOB/budget: An algorithm for product analysis, Paper presented at ORSA 41st National Meeting, New Orleans.

Ringer, L. J. (1966), Analytic and Monte Carlo Distribution Theory for PERT, Unpublished Ph.D. thesis, Texas A & M Univ., College Station, Texas.

——— (1969), Numerical operators for statistical PERT critical path analysis, *Management Sci.*, **16** (2), B136–B143.

——— (1971), A statistical theory for PERT in which completion times of activities are interdependent, *Management Sci.*, **17** (11), 717–723.

Robillard, Pierre and Michel Trahan (1976), Expected completion time in PERT networks, *Oper. Res.*, **24** (1), 177–182.

——— and ——— (1977), The completion time of PERT networks, *Oper. Res.*, **25** (1), 15–29.

Robinson, Don R. (1975), A dynamic programming solution to cost–time trade-off for CPM, *Management Sci.* **22** (2), 158–166.

Robinson, I. G. (1965), The PERT approach to plant layout, *Factory Management*, (123), September, 104–105.

Roper, D. E. (1964), Critical path scheduling, *J. Ind. Eng.*, **15** (2), 51–59.

Rosenbloom, R. S. (1964), Notes on the development of network models for resource allocation in R & D projects, *IEEE Trans. Eng. Management*, 58–62.

Roy, B. and B. Sussmann (1964), Les problemes d'ordonnancement avec constraintes disjunctives, Research Report No. 9, Direction Scientifique SEMA, Paris, France.

——— (Ed.) (1964a), *Les Problemes d'Ordonnancement, Applications et Methodes*, Dunod, Paris, France.

——— (1964b), Contribution de la theorie des graphes a l'etude des problems d'ordonnancement, in *Les Problemes d'Ordonnancement-Applications et Methodes*, Dunod, Paris, France, pp. 109–125.

Russell, A. H. (1970), Cash flows in networks, *Management Sci.*, **16** (5), 357–373.

Samli, A. C. and C. Bellas (1971), The use of GERT in the planning and control of marketing research, *J. Mark. Res.*, **8**, 335–339.

Scherer, F. M. (1966), Time–cost trade-offs in uncertain empirical research projects, *Nav. Res. Log. Quart.*, **13** (1), 71–82.

——— (1966), A note on time–cost trade-offs in uncertain empirical research projects, *Nav. Res. Log. Quart.*, **13** (3), 349–350.

Schick, G. I. and J. L. Maybell (1968), When budget variances warrant investigation by engineering project management, *Soc. Automot. Eng.*, Paper No. 680683.

Schoderbek, P. P. (1966), Overcoming resistance to PERT. Group involvement is effective in furthering acceptance of a valuable new technique, *Bus. Topics*, **14** (2), 50–56.

——— and L. A. Digman (1970), Third generation PERT/LOB, *Harvard Bus. Rev..*, **48** (5), 100–110.

Schrage, Linus (1970), Solving resource-constrained network problems by implicit enumeration—nonpreemptive case, *Oper. Res.*, **18** (2), 263–278.

——— (1972), Solving resource-constrained network problems by implicit enumeration, preemptive case, *Oper. Res.*, **20** (3), 668–677.

Schultis, R. L. (1962), Applying PERT to standard cost revisions, *N. A. A. Bull.*, September.

Seshu, Sundaram and Myril B. Reed (1961), *Linear Graphs and Electrical Networks*, Addison-Wesley, Reading, Mass.

Shackelton, N. J., Jr. (1973), Minimization of project duration when resources are limited, Paper presented to 43rd National Meeting of ORSA, Milwaukee, Wisconsin, May 9–11.

Shaffer, L. R., J. B. Ritter, W. L. Meyer, (1965), *The Critical Path Method*, McGraw-Hill, New York.

Sherrard, W. R. and Fred Mehlick, (1972) PERT: A Dynamic Approach, *Dec. Sc.*, **3** (2), 14–26.

Shier, Douglas R. (1974), *Computational Experience with an Algorithm for Finding the k Shortest Paths in a Network,* Institute for Basic Standards, National Bureau of Standards, Rockville, Md.

——— (1974), Iterative Methods for determining the *k* shortest paths in a network, Applied Mathematics Division, National Bureau of Standards, Rockville, Md.

Shrieder, Y. A. (ed.) (1966), *The Monte Carlo Method*, Pergamon, Oxford, England.

Shwimer, Joel (1972), On the *N*-job, one machine sequence-independent scheduling problem with tardiness penalties: a branch–bound solution, *Management Sci.*, **18** (6), B301–B313.

Sicard, P. (1970), *Pratique due PERT* (*Methode de Controle des Delais et des Couts*), Dunod, Paris.

Sielken, R. L., H. O. Hartley, and E. Arseven (1975), Critical path analysis in stochastic networks, Research Memorandum, Texas A&M Univ., College Station, Texas.

Siemens, Nicolai (1971), A simple CPM time–cost tradeoff algorithm, *Management Sci.*, **17** (6), B354–B363.

Simon, Herbert A. and A. Newell (1958), Heuristic problem solving: the next advance in operations research, *Oper. Res.*, **6** (1), 1–10.

——— and A. Newell (1964), Information processing in computer and man, *Amer. Sci.*, **52** (3), 281–300.

Simonnard, Michel (1966), *Linear Programming*, translated by W. S. Jewell, Prentice-Hall, Englewood Cliffs, N.J.

Sneddon, Ian H. (1972), *The Use of Integral Transforms*, McGraw Hill, New York.

Sobczak, Th. V. (1962), A look at network planning, *IRE Trans. Eng. Management*, 113–116, September.

——— (1963), A statistical analysis of the general characteristics of a PERT technician, *IEE Trans. on Eng. Management*, March, 25–28.

Soistman, E. C. (1962), Input-output; management control at Martin, *Aerospace Management*, **5** (4), April, 34–37.

——— (1965), How Martin Company controls a single research project, *Soc. Automotive Eng. J.*, **73** (5), 63–65.

——— (1966), Research and development can be controlled, *Research Management*, **9** (1), 15–27.

Special Projects Office, (1958); PERT, Phase I Summary Report, Bureau of Ordnance, Department of the Navy, Washington, D. C.

Staffurth, C. and H. Walton, (ed.) (1974), *Programs for Network Analysis*, Joint Report by INTERNET (UK) and Loughborough University of Technology, The National Computing Center, Quay House, Quay St., Manchester M3-3HU, U. K.

Steinberg, R. (1963), Some notes on the similarity of three management science models and their analysis by connectivity matrix techniques, *Management Sci.*, **9** (2), 341–343.

Sumray, E. J. (1965), Planning with critical path; an introduction, *Process Control Autom.*, **12** (4), 149–152.

Sunaga, Teruo (1970), A method of the optimal scheduling for a project with resource restrictions, *J. Oper. Res. Soc. Jap.*, **12** (2), 52–64.

Swanson, L. A. and H. L. Pazer (1971), Implications of the underlying assumptions of PERT, *Dec. Sci.*, **2** (4), October, 461–480.

Thomas, Warren (1969), Four float measures for critical path scheduling, *J. Ind. Eng.*, **1** (10), 19–23.

Thompson, G. L. (1968), CPM and DCPM under risk, *Nav. Res. Log. Quart.*, **15** (2), 233–240.

Thompson, Robert E. (1966), Adjusting network plans with "PERT" slack bonus, *J. Ind. Eng.*, **17** (3), 145–149.

Thornley, Gail (ed.) (1968), Critical path analysis in practice, Tavistack Publications, London, England.

Tippett, L. H. C. (1925), On the extreme individuals and the range of samples taken from a normal population, *Biometrika*, **17**, 364–377.

Trilling, D. R. (1966), Job shop simulation of orders that are networks, *J. Ind. Eng.*, **17** (2), 59–71.

Turban, Efraim (1968), The line of balance—a management by exception tool, *J. Ind. Eng.*, **19** (9), 440–448.

Van Der Pol, Balth and H. Bremmer (1955), *Operational Calculus*, Cambridge U. P. Cambridge, England.

Van Slyke, Richard M. (1963), Monte Carlo methods and the PERT problem, *Oper. Res.*, **11** (5), 839–860.

Vazsonyi, Andrew (1970), L'histoire de grandeur et de la decadence de la méthode PERT, *Management. Sci.*, **16** (8), B449–455.

Ventura, E. (1965), L'introduction de l'aleatoire dans les reseaux PERT, *Gestion*, June, 326–333.

Verhines, D. R. (1963), Optimum scheduling of limited resources, *Chem. Eng. Prog.*, **59** (3), 65–67.

Villemain, Ch. A. and N. LeBorgne (1966), *Methode de Planification d'Ensembles par Reseaux Lineaires*, Dunod, Paris, France.

Voraz, Charles (1966), *Le PERT COST* (*Elaboration des Budgets et Controle des Depenses*), Entreprise Moderne d'Edition, 4 rue Cambon, Paris.

Voronov, A. A. and E. P. Petrushinin (1966), Solving optimal resource allocation by quadratic programming, *Autom. Remote Control*, **27** (11), 1921–1934.

Walton, H. (1964), Experience of the application of critical path method to plant construction, *Oper. Res. Quart*, **15** (1), 9–16.

Warfield, John N. (1976), *Societal Systems: Planning, Policy and Complexity*, Wiley, New York, Chapter 16.

——— and J. D. Hill (1971), The DELTA chart: a method for R&D Project portrayal, *IEEE Trans. Eng. Management*, **EM-18**, (4) 131–139.

Welsh, D. J. A. (1965), Errors introduced by a PERT assumption, *Oper. Res.*, **13** (1), 141–143.

——— (1966), Super-critical arcs of a PERT network, *Oper. Res.*, **14**, (1) 173–174.

White, John A. (1970), On absorbing Markov chains and optimum batch production quantities, *I.E. Trans.*, **2** (1), March, 82–88.

Whitehouse, G. E. (1973a), Project management techniques, *Ind. Eng.*, **5** (5), 24–29.

——— (1973b), *Systems Analysis and Design Using Network Techniques*, Prentice-Hall, Englewood Cliffs, New Jersey.

——— and A. A. B. Pritsker (1969), GERT: Part III—further statistical results; counters, renewal times and correlations, *IE Trans.*, **1**, 45–50.

Wiest, Jerome D. (1964), Some properties of schedules for large projects with limited resources, *Oper. Res.*, **12** (3), 395–418.

——— (1966), Heuristic programs for decision making, *Harvard Bus. Rev.*, **44** (5), Sept–Oct., 129–143.

——— (1967), A heuristic model for scheduling large projects with limited resources, *Management Sci.*, **13** (6), B359–B377.

——— and F. K. Levy (1977), *A Management Guide to PERT/CPM*, 2nd edition, Prentice-Hall, Englewood Cliffs, New Jersey.

Wilks, Samuel W. (1962), *Mathematical Statistics*, Wiley, New York.

Wilson, R. C. (1964), Assembly line balancing and resource leveling, Univ. Michigan Summer Conference, Production and Inventory Control, Ann Arbor, Michigan.

Woodgate, H. S. (1964), *Planning by Network*, Business Publications, Ltd., London, England.

——— (1968), Planning networks and resource allocation, *Datamation*, **14** (1), 36–43.

——— (1970), (J. E. Leymarie, trans.), *Comment Utiliser les Plannings par Reseaux*, Les Editions D'Organisation—5, rue Rousselet, Paris.

Woodworth, B. M. and C. W. Dane (1974), Multi-project network analysis with resource leveling: state-of-the-art and a governmental application, Department of Business Administration, Oregon State Univ., Corvallis; also presented at the joint National Meeting of ORSA/TIMS, April 22–24.

Zaloom, V. (1971), On the resource-constrained project scheduling problem, *IE Trans.*, **3** (4), 302–305.

AUTHOR INDEX

SUBJECT INDEX